How to Critically Read the Scientific Research Literature

Unlock the secrets of scientific articles with Claim, Evidence, Reasoning, Implications, and Context (CERIC). This approachable guide helps readers break down dense articles into their core arguments using a focused hunt-and-seek approach, enabling deeper insight and engagement with the research literature. Each chapter features worked examples drawn from multiple scientific disciplines, preempts common misunderstandings, and provides knowledge checks to reinforce learning. Readers emerge able to identify and evaluate claims and evidence, spot gaps in reasoning, and articulate their findings through presentations and literature critiques — skills essential for success in higher education, industry, and informed citizenship. Whether you are an undergraduate tackling your first research article, a graduate student preparing a literature review, or an instructor teaching scientific literacy, the evidence-based CERIC method transforms reading apprehension into confidence. Accompanying student and instructor supplements can be found online, with further discipline-specific examples and guidance on course preparation and professional development.

Genevive Bjorn, EdD, directs the nonprofit Higher Learning Lab Co. and is a former workforce researcher at Rutgers University. Holding a doctorate from the Johns Hopkins School of Education, her scholarship focuses on critical thinking, online learning, and career development. Author of more than 50 articles appearing in publications such as *Nature*, *Science*, and *The New York Times*, and a recipient of a 2017 NSTA Teacher Award and a 2012 Knight Digital Media Fellowship, she champions innovative and transformative pedagogy that exposes hidden curricula and centers on critical reading.

Adam J. Burgasser, PhD, is a professor of astronomy and astrophysics and director of the Cool Star Lab at the University of California San Diego. Holding a doctorate from Caltech, his research focuses on the coldest stars and exoplanets, and he has authored over 350 peer-reviewed publications on scientific and educational research. He is a recipient of multiple teaching awards, a Fulbright Scholar, and an AAS Fellow. Dr. Burgasser innovates in teaching and mentorship by blending astrophysics research with critical pedagogy.

"An essential guide that finally codifies what expert researchers do instinctively when reading scientific papers. The CERIC method transforms reading comprehension into critical analysis, offering students and early-career researchers a systematic approach to understanding and evaluating scientific arguments across disciplines. Particularly timely is the book's thoughtful integration of generative AI tools, demonstrating how to leverage these technologies while maintaining rigorous critical thinking. This book will strengthen scientific literacy in classrooms and laboratories alike."

Keivan G. Stassun, Vanderbilt University

"Community college students, particularly those pursuing STEM majors, are at a severe competitive disadvantage when they transfer to 4-year+ institutions. As a chemistry instructor at a community college teaching primarily economically disadvantaged, often English-learning, first-generation students, I believe that the CERIC approach to engaging primary literature presented in great detail in *How to Critically Read the Scientific Research Literature: Introducing the CERIC Method* provides instructors a systematic roadmap of transformative tools and strategies that, when integrated into the curriculum, not only will help level the playing field for our transfer students, but will keep them excited and engaged in their studies and research with the ultimate goal of supporting and reinforcing their desire to continue to pursue STEM careers."

David Hecht, Southwestern College

"The CERIC method distills the invisible architecture of scientific argument into five manageable elements, representing what experts do intuitively and empowering novices to practice those moves deliberately. What makes this book special for me is the intellectual authority of its authors, who draw on decades of teaching, mentoring, and research to offer an approach that is at once rigorous and welcoming. They remind us that systemic barriers may affect access to higher education, yet instruction in essential scholarly skills should remain available to every learner. By demystifying research articles, this text widens the doorway to scientific culture and invites a more diverse chorus of voices to enter. Having incorporated the CERIC method into my own teaching, I wholeheartedly recommend this book to anyone who wishes to learn how to read research papers both competently and critically."

Kevin Heng, Ludwig Maximilian University

How to Critically Read the Scientific Research Literature

Introducing the CERIC Method

GENEVIVE BJORN
Higher Learning Lab Co.

ADAM J. BURGASSER
University of California, San Diego

Shaftesbury Road, Cambridge CB2 8EA, United Kingdom

One Liberty Plaza, 20th Floor, New York, NY 10006, USA

477 Williamstown Road, Port Melbourne, VIC 3207, Australia

314–321, 3rd Floor, Plot 3, Splendor Forum, Jasola District Centre, New Delhi – 110025, India

Cambridge University Press is part of Cambridge University Press & Assessment,
a department of the University of Cambridge.

We share the University's mission to contribute to society through the pursuit of
education, learning and research at the highest international levels of excellence.

www.cambridge.org
Information on this title: www.cambridge.org/9781009631273
DOI: 10.1017/9781009631266

When citing this work, please include a reference to the DOI 10.1017/9781009631266

First published 2026

Cover image: Five watercolour brushstrokes © Yifei Fang / Moment / Getty Images

A catalogue record for this publication is available from the British Library

Library of Congress Cataloging-in-Publication Data
Names: Bjorn, Genevive author | Burgasser, Adam J. author
Title: How to critically read the scientific research literature : introducing the CERIC
method / Genevive Bjorn, Higher Learning Lab Co., Adam J. Burgasser, University
of California, San Diego.
Description: Cambridge, United Kingdom ; New York, NY : Cambridge University Press,
2026. | Includes bibliographical references and index.
Identifiers: LCCN 2025045397 (print) | LCCN 2025045398 (ebook) |
ISBN 9781009631273 paperback | ISBN 9781009631266 ebook
Subjects: LCSH: Research – Evaluation | Scientific literature – Evaluation |
Science – Study and teaching | Peer review
Classification: LCC Q180.55.E9 B56 2026 (print) | LCC Q180.55.E9 (ebook)
LC record available at https://lccn.loc.gov/2025045397
LC ebook record available at https://lccn.loc.gov/2025045398

ISBN 978-1-009-63127-3 Paperback

Additional resources for this publication at www.cambridge.org/bjornburgasser

For Laura Jeanne Quaynor (1982–2023), our coauthor, colleague, and friend, whose early passion for CERIC's promise motivated us to expose this hidden curriculum in higher education.

I guess the underlying assumption is that here's a paper, go read it, and if you recognize words, you can read it well enough. I think that has always been the assumption in undergrad training and graduate training, which is so sad, because that's so not true. I wish I would have learned this method when I first had to read primary lit as an undergrad. I would have struggled much less and probably felt like less of a fraud in grad school.

—Doctoral Participant 23 (Bjorn, 2024, p. 16)

Contents

Foreword

Learning how to read, dissect, and interpret peer-reviewed scientific literature is an indispensable skill for any academic researcher and scholar. One cannot possibly judge if a research idea is exciting, timely, and impactful unless one knows the "lay of the land." Yet, we are living in an age where the daily flow of information overwhelms the actual amount of nontrivial knowledge it contains. In astronomy and astrophysics alone, the online electronic archive (www.arxiv.org) averages about 30–50 preprints *daily*. It is abundantly clear that attempting to read all of these preprints in the usual way is unsustainable for any human being.

Typically, reading peer-reviewed literature is a skill that is not formally taught until one advances to a graduate program (master's degree and higher). In some graduate programs, this skill is not formally taught at all. The pervasive assumption is that one picks it up by doing research and by watching one's mentors perform a literature review. The present textbook addresses this need. It is a rare treasure trove of instructions that offers an explicit, well-marked trail for how to become an accomplished reader of research papers.

What makes the current textbook special for me is the intellectual authority of its authors. Professor Adam J. Burgasser is a renowned pioneer in the study of brown dwarfs, which are entities too massive to be planets and too diminutive to be stars. In fact, he is known to me and widely among other astrophysicists as "Mr. Brown Dwarf." He has pioneered methods for their classification, reported some of the first discoveries of low-mass brown dwarfs, and made fundamental contributions to understanding their physics and chemistry. Above all, he reports his findings in research papers of immense pedagogical value. Professor Burgasser is one of the most lucid and graceful speakers whom I have had the pleasure to listen to, and his voice is clearly heard in the pages of this textbook.

Dr. Genevive Bjorn pioneered the CERIC (Claim, Evidence, Reasoning, Implications, and Context) method for how to read, dissect, and interpret research papers as part of her doctoral dissertation. When she conducted a demonstration for my research center in Switzerland about a decade ago, I must admit that I was initially skeptical. How could any "canned" recipe possibly replace the actual experience of learning how to read papers the "hard way"? To my great surprise, the CERIC method essentially replicated what I had painstakingly acquired over a decade in academia from learning "at the feet" of several distinguished mentors. But it does so in a principled, systematic way that is certain to save any budding academic plenty of time and pain on the long road to mastery. I found the CERIC method so compelling that I incorporated it into one of the master's courses I am teaching at the University of Munich. It is now the standard procedure for teaching my master's students how to read research papers.

The authors draw on decades of teaching, mentoring, and research to offer an approach that is at once rigorous and welcoming. The CERIC method distills the invisible architecture of scientific argument into five manageable elements. By naming what experts do intuitively, they empower novices to practice those moves deliberately. The result is more than comprehension; it is participation. Readers who master CERIC can compare studies, craft incisive questions, and contribute thoughtfully to peer review and collaborative projects.

The design of the textbook strengthens its message. Structured sections invite quick reference. Worked examples model authentic practice across astronomy, biology, chemistry, earth science, and physics. Graphics translate abstract ideas into memorable images. Instructors will welcome the teaching callouts, knowledge checks, and sample rubrics that transform chapters into ready-to-use lessons. Librarians will value the alignment with information-literacy outcomes, especially the emphasis on ethical paraphrasing and transparent citation. Researchers will appreciate the guidance on turning CERIC notes into literature reviews that reveal new research opportunities.

The very writing of this textbook is a tribute to equity. The authors remind us that barriers such as health, finances, or caretaking responsibilities may affect access to higher education, yet instruction in essential scholarly skills, especially critical reading and critical thinking, should remain available to every learner. By demystifying research articles, this textbook widens the doorway to scientific culture and invites a more diverse chorus of voices to enter.

I wholeheartedly recommend this textbook to anyone who wishes to learn how to competently and critically read research papers. May it kindle curiosity, sharpen judgment, and nurture the next generation of careful, courageous scholars.

Kevin Heng
Chair Professor of Theoretical Astrophysics
Ludwig Maximilian University of Munich
Germany

Honorary Professor
University College London
United Kingdom

Honorary Professor
University of Warwick
United Kingdom

Preface

How to Critically Read the Scientific Research Literature: Introducing the CERIC Method grew from a single question that students kept asking in our classrooms and laboratories:

How do I actually understand this article?

There are many related questions, including:

How do I know this is a good article?
How do I know I'm not missing anything important?
What can I learn from this article?
How do I trust the results?
How do I incorporate this into my assignments and research?

Each time we handed out a research article – whether the discovery of brown dwarfs or a new development in *CRISPR* gene editing – learners anguished over columns of dense text and specialist jargon. Traditional instruction expected readers to absorb the essentials by assuming, incorrectly, that if readers understood the words on the page, they would also be able to analyze the argument from evidence. Many try, yet many more disengage, unsure of how to disentangle a claim from its evidence or trace a chain of reasoning across a single study, let alone compare multiple studies. Learning the structure of research articles is great for finding broad types of information, like the methods and discussion, but this traditional approach is nowhere near sufficient to understand an article's main argument.

We believed there had to be a smarter, better way. Drawing on Toulmin, Rieke, and Janik's (1984) model of scientific argument and decades of evidence-based teaching research, we identified and tested a categorical framework – CERIC – that names the five pillars of argumentation:

- Claim
- Evidence
- Reasoning
- Implications
- Context

These pillars act like cardinal directions on a compass. With them, any reader – undergraduate, graduate student, educator, or professional – can navigate the terrain of a research article with confidence, curiosity, and success.

Why This Book Now?

We see at least three significant reasons to share this guide now:

1. Exponential growth of literature: PubMed alone indexed more than one million new articles last year. Structured reading keeps overwhelm at bay.
2. Urgent need for information literacy: From climate forecasts to vaccine trials to record-levels of article retractions, public dialogue and policy hinge on careful evaluation of scientific claims.
3. Equity in education: Barriers such as health, finances, and caregiving may limit access to higher education, yet essential scholarly skills must remain within every learner's reach. These skills must be explicit. We want to live in a world where every scholar who wants to contribute new knowledge is empowered to make that contribution.

How the Book Is Organized

Part I: Critical Reading

1. Foundations (Chapters 1–2) – Introduce primary literature, the *IMRaD* blueprint, and the evolution from Toulmin to CERIC.
2. Anatomy of Argument (Chapters 3–7) – Devote a whole chapter to each CERIC element, with discipline-spanning worked examples and graphics.

Part II: Critical Thinking

3. Applied Practice (Chapters 8–10) – Guide readers through full CERIC reviews, comparative analyses, and peer critiques.
4. Communication & Synthesis (Chapters 11–12) – Translate CERIC insights into presentations and literature reviews.
5. Collaboration & Innovation (Chapters 13–14) – Embed CERIC in social annotation communities and explore its synergy with generative AI tools.

Every chapter features:

- Structured text and generous white space for quick scanning.
- Callout boxes that clarify common confusions.
- Knowledge checks that reinforce learning.
- Teaching notes that shorten preparation time for instructors.

Additional worked examples, resources, and instructional guidance are freely available on www.cambridge.org/bjornburgasser as:

- Instructor supplement
- Student supplement

We invite you to treat this book as a workshop rather than a monument. Write in the margins, test the templates, and adapt the examples to various fields, including those outside our expertise. We encourage you to be active, experimental, and share your experience with colleagues and peers. Our collective practice and insight are like a rising tide that lifts all boats.

We begin where many beginning researchers struggle: decoding dense and difficult text while hoping to understand its deeper meaning. CERIC treats every research article as an argument to be mapped so you, the reader, can ask better questions, decide if the argument holds up, and carry the sound arguments forward with confidence. CERIC's unique promise is an intelligent way into the conversation. You might not yet grasp every nuance of an article, but its main points will be revealed.

Part I, critical reading with CERIC, equips you to see how scientific arguments are built. First, you will practice identifying a paper's central claim. Next, you will distinguish evidence types and evaluate how well the article's evidence supports the claim through reasoning. You will then examine arguments for logic and coherence, probe implications for scope and limits, and situate each study within its context, including the research gap that motivated the work. By the end of Part I, you will read with a master's composure, able to reconstruct the article's argument with clarity, alert to its hidden holes.

Part II, critical thinking with applications of CERIC, prepares you to apply what you learned in Part I to typical research literature products. You will use the same argument elements to compare, critique, and present articles; develop proposals; and synthesize literature. You will also learn to skillfully use social collaborative annotation and generative AI to support your literature work. These practices will help you build a deep base of paraphrased, CERIC-based article notes that will serve you through high-stakes exams, like comprehensive and qualifying examinations. At the same time, techniques such as concept maps, comparison templates, and synthesis matrices make reasoning visible and your thinking defensible.

This book points you toward meeting the need shared by many of our students: An evidence-based method that will prepare you to critically read, think, and contribute to the scientific literature.

May CERIC help you read with rigor, teach with clarity, and contribute your unique voice to the ever-advancing conversation that is science.

Acknowledgments

Every book is a collective endeavor, and *How to Critically Read the Scientific Research Literature: Introducing the CERIC Method*, owes its existence to a generous community of mentors, colleagues, students, friends, and family who steadied our hands and sharpened our thinking at each stage.

We first thank Genevive's dissertation committee members, Karen Karp, Alan Reid, and Laura Quaynor, whose support and critical questioning helped to move this scholarship to a more rigorous stage. Kevin Heng kindly agreed to craft a foreword and, in doing so, framed our work within the broader tradition of scientific communication; his perspective elevates this volume.

Our gratitude extends to the students at University of California San Diego (UCSD), University of California Santa Cruz, Johns Hopkins University, and Southwestern College who field-tested draft concepts and chapters, asked disarmingly clear questions, and refined every example through their curiosity and practice. Members of the Cool Star Lab and the Higher Learning Lab communities offered practical feedback that enriched the discipline-spanning case studies. Our "red team" reviewers gave us thoughtful and valuable feedback that helped us refine many of the pages. Trisha Muro lent us a particularly valuable perspective as an author and physics teacher. We are equally indebted to colleagues who piloted CERIC in classrooms and journal clubs across astronomy, biology, chemistry, earth science, and physics; their classroom notes and feedback shaped the teaching callouts and knowledge checks you will find in these pages.

Cambridge University Press provided an editorial home that balances rigorous review with creative freedom. We thank our commissioning editor and the production team for shepherding the manuscript through peer review, copyediting, and design. Their expertise ensured that the book's "chunked" layout, graphics, and pedagogical features serve readers as intended.

Several organizations supported this project materially. The American Astronomical Society's education grant funded a workshop, where the earliest CERIC examples and templates were piloted and improved; special thanks to Tom Rice for seeing the value in this approach. We also acknowledge other professional societies – the American Chemical Society and National Science Teachers Association – for creating inclusive forums where interdisciplinary ideas flourish. Library professionals at UCSD's Geisel Library and the Sheridan Libraries at Johns Hopkins guided us through open-access considerations and citation management strategies that now appear in Chapter 12.

Writing a book is equal parts marathon and high-wire act, and countless people supplied us with emotional nourishment and inspiration. To Adam's and Genevive's friends and families near and far – especially Aria, who reminds us daily why equitable education matters – thank you for the patience, understanding, and laughter that sustained us. We also remember teachers and mentors who are no longer with us; their insights echo throughout these pages.

Finally, we thank every reader who picks up this book, hoping to understand and critique scientific research more deeply. This is slow and challenging work that we believe is worth doing, now more than ever. Your commitment to critical reading and thinking keeps the self-correcting engine of science turning.

We look forward to critically reading your research one day soon.

Any remaining oversights or errors are ours alone, and we welcome your insights as CERIC practice continues to grow. Any opinions, findings, or conclusions expressed herein are solely our own.

Authors' Note

The pages you are about to explore grew from a simple classroom puzzle: How can we help readers see the invisible logic that threads data, inference, and discovery together? Over many years of journal clubs, seminars, and our own research, we discovered that readers make measurable gains when they read with a categorical lens rooted in the way research is practiced. Those insights inspired the CERIC method – Claim, Evidence, Reasoning, Implications, and Context – which serves as the spine of this book.

This first edition reflects a dialogue among scientists, logicians, and educators who share a commitment to transparent, reproducible scholarship. We adopted Toulmin's foundational model of argument (Toulmin, Rieke, & Janik, 1984), modernized it, and extended it with applied examples in the natural sciences from astronomy to physics. Each chapter pairs narrative explanation with a worked example and knowledge check questions.

Although we have worked diligently to cross-check facts and align the manuscript with the Cambridge University Press Style Guide, science advances quickly. To keep the material current, we maintain a companion website that houses:

- updates on key disciplinary case studies,
- downloadable instructor and student supplements, and
- a feedback portal for corrections or pedagogical success stories.

Your comments will guide future printings and shape the second edition.

Every book is the culmination of a journey, and our gratitude runs deep for the support we've received along the way. We thank our anonymous peer reviewers whose careful critiques strengthened every chapter. We also celebrate the "red team" readers who critiqued early drafts of this book and challenged us to make the book clearer, leaner, and more reader-friendly. Finally, we appreciate the editorial and production teams at Cambridge University Press, whose expertise has guided this project from manuscript to publication.

May the CERIC method empower you to read primary literature critically, think more critically, teach with confidence, and contribute to a culture of arguing from evidence.

Let's go.

Part I

Critical Reading

1 The Role of Scientific Research Articles in the Natural Sciences

Reading and writing do not stand in a separate functional relationship with inquiry [i.e., research], but are constitutive of it – essential elements of the whole.

—Norris & Phillips, 2003, p. 226

1.1 Overview

Chapter 1 introduces the concept of research articles as the foundation of scientific knowledge, detailing their role in documenting discoveries and fostering reproducibility. It explains the structure of research literature, particularly the *IMRaD* format (Introduction, Methods, Results, and Discussion), which organizes articles into a logical and consistent framework for publication. This chapter emphasizes the importance of critical reading of research articles in scientific education and highlights common challenges, such as dense jargon, assumed background knowledge, and the nonlinear structure of research. It addresses the gap in formal instruction for critical reading, which leaves many readers unprepared to analyze the essential elements of claims, evidence, and reasoning. This chapter outlines various reading strategies, including structured frameworks and collaborative annotation methods, to enhance critical reading skills. These approaches empower readers to engage deeply with published research, overcome comprehension barriers, and contribute to scientific discourse.

1.2 Understanding Scientific Research Literature

Consider the moment a new scientific result enters the public record: A research article appears in an online journal, showing how a team transformed marine bacteria to "eat" microplastic waste in the ocean. If this lab discovery proves to be valid and scalable, the implications for ocean health are enormous (not to mention the news headlines and memes). Overnight, researchers worldwide download the article, eager to read the report, evaluate its argument, and consider the potential impact on their work. This firsthand research report is what scientists call a research article, and a collection of articles together forms the primary scientific literature – a genre of original, peer-reviewed research studies published in scientific journals, allowing others to critique, extend, and replicate the findings.

Learning to read these articles critically is a foundational skill for anyone who hopes to become a researcher and make a contribution to a scientific field. In this book, we focus on the natural sciences (i.e., astronomy, biology, chemistry, earth science, and physics). In Part I of this book, we aim to prepare student researchers at the undergraduate and graduate levels to analyze, critique, and synthesize original research. In Part II, we help you transfer these critical thinking skills to other literature-based work necessary for academic and career progress, including presentations and literature reviews.

1.2.1 Why We Care Enough about Reading Research Articles to Write a Book

Graduate school marks the transition from receiving knowledge to creating it, and critical reading is the lever that moves scholars across this divide (Norris & Phillips, 2003). The papers we dissect today become the proposals we draft tomorrow. Professional societies list "deep engagement with primary literature" among core research competencies. Yet, most programs devote far more time to teaching writing than to teaching reading – if they address reading at all (Kwan, 2008, 2009). Consequently, many graduate students reach their second year only to discover that a credible literature review requires analyzing every cited article, a task for which skimming abstracts and figures is wholly inadequate.

The curricular gap is evident in the data and in daily lab life. Reading-completion rates in higher education have declined for decades (Burchfield & Sappington, 2000; McMinn et al., 2009), and skimming has become a survival tactic (Miller & Murillo, 2012). Weak reading strategies ripple outward, leading to poor evidence evaluation, unclear literature reviews, and stalled research designs. Indeed, "reading" is the second-most common but least acknowledged reason students quit graduate programs (Bjorn, 2023). Journal clubs attempt to fill the void, yet their social dynamics often silence newcomers and reinforce the voices of the dominant group. Lab culture can amplify the problem: Hands-on bench work is labelled "real science," while time spent grappling with papers is relegated to the margins as something we should already know how to do (Gorzycki et al., 2019).

At the same time, mastery of disciplinary discourse – defined as "the interplay of words, speech, text, and reasoning within an academic domain" (Sherboboevna, 2020) – is a cornerstone of professional growth in every research field. Discourse varies by field but always includes shared background knowledge, specialized vocabulary, and preferred methodologies (Borg, 2003; Carter-Thomas & Rowley-Jolivet, 2008). We acquire these norms through guided reading, structured writing tasks, and dialogue with experienced colleagues; as we do, we shift from consuming concepts to producing knowledge (Brown & Renshaw, 2000). Iterative feedback from mentors and writing groups further refines both oral and written skills (Aitchison & Guerin, 2014; Spaulding & Rockinson-Szapkiw, 2012). Within this conversational cycle, well-composed research arguments act as bridges between current findings and future inquiries, positioning emerging scholars who can meaningfully critique research literature to steer its next wave.

The central claim of this book is that expert-level critical reading and thinking skills are learned, not innate. We, therefore, offer a systematic framework for identifying Claims, Evidence, Reasoning, Implications, and Context (CERIC) as a practical method for developing the critical reading and thinking skills that graduate education often leaves to chance. Chapter 2 begins that journey.

1.2.2 Calling Out the Hidden Curriculum

A hidden curriculum occurs in the form of unspoken expectations. The hidden curriculum for critical reading is that everyone who can read the words on the page can critique research articles fluently. This situation can fuel feelings of imposter syndrome (Parkman, 2016). Many people new to a research field may think the problem lies with them rather than with a lack of instruction. Instructors play a crucial role in demystifying these skills by designing targeted reading assignments, structuring group discussions of research articles, and offering explicit feedback on critical reading notes. However, many instructors do not assess critical reading – and, in fairness, most learned to critique research articles through trial and error and were never formally taught – which is why we are here now with this book as your guide. We want to break this cycle and provide a new model of clear, positive instruction.

In our experience, learners benefit from explicit support to boost confidence and develop solid critical reading habits. Tailored interventions – like reading workshops and structured peer groups and presentations – can reduce anxiety, boost confidence, and clear the path to becoming independent scholars (Bjorn, 2024). Learners from under-resourced backgrounds and who are first-generation scholars may benefit most. By treating critical reading as a foundational skill, not a magical talent, we empower ourselves and others to succeed.

It is high time to move past assumptions that student researchers will figure out critical reading of research articles on their own. Integrating explicit instruction in critical reading across the spectrum of research and learning activities helps everyone build stronger arguments and enriches the entire scholarly conversation. By equipping you with the tools to analyze data, evaluate interpretations, and draw meaningful conclusions, we create a path that we hope will help you produce new knowledge with confidence and depth.

Box 1.1 Common Confusion: A Research Article versus Primary Literature

- **A research article** refers to an individual study that presents new knowledge by the researchers who conducted it. For example, in a 2012 research article published in *Nature*, Church et al. demonstrated a method for encoding digital information into DNA, reporting the successful storage and retrieval of 700 terabytes per gram (Church, Gao, & Kosuri, 2012).
- **The primary literature** refers to the broader collection of research articles published in peer-reviewed journals, forming a nonfiction genre called *primary literature*. For example, the above DNA storage study is part of the primary literature on synthetic biology, which includes thousands of peer-reviewed studies spanning topics like CRISPR gene editing, artificial chromosomes, and bioinformatics platforms.

Box 1.2 Knowledge Check: Question 1.1 Which factor contributes most to the "Hidden Curriculum" around critical reading of primary literature?

A. Instructors often believe critical reading strategies require little formal instruction.
B. Journal publishers hide critical reading guidelines behind paywalls to restrict readers' access.
C. Most peer-reviewed journals have replaced reading with videos and webinars.
D. Scientific societies discourage active reading techniques in higher education.

(Check your understanding using the Knowledge Check Key at the end of the book.)

1.3 Reading Primary Literature Compared to Other Texts

Reading primary literature is a unique kind of academic work. It is not a genre built for enjoyment or comfort. Unlike a novel read front to back, or a textbook consulted for clear, structured explanations, research articles ask us to piece together a puzzle that the author has not yet finished – and maybe does not know where it is going (Lie et al., 2016). Comparing primary literature to other academic texts invites a reflection on the different reading strategies we need to use. Understanding these contrasts helps clarify what makes primary literature uniquely challenging – and uniquely valuable.

- **Fiction and narrative nonfiction** invite immersive reading. The goal is to follow characters, narratives, and themes, often in linear order. Emotions guide the pace. Literary devices like metaphors and foreshadowing do the heavy lifting. Readers are encouraged to feel, imagine, and intuit rather than critique methods or evaluate empirical claims.
- **Magazine articles** aim for broad accessibility. They often summarize scientific ideas or social trends in engaging, simplified language. These texts typically omit methodological details and references, which means they can be read quickly, often skimmed for takeaways. The reader's role is passive – receiving information rather than interrogating it.
- **Textbooks** are purposely dense texts written at a frustration level. Their job is to organize established knowledge into digestible chapters, usually structured from basic to advanced concepts. Readers often dip in selectively, looking up definitions, scanning examples, or reviewing diagrams. Textbooks rarely include original data. Instead, they translate research into foundational knowledge and disciplinary concepts.
- **Web content** varies wildly in quality and intent. Some web articles summarize peer-reviewed research with care; others are casual opinion pieces or curated content with no clear source. Readers must become gatekeepers who evaluate credibility, authorship, and purpose. While this invites a form of critical reading, it is usually aimed at assessing the source quality, not the internal logic of a scientific argument.

- **Primary literature** is none of these. It is a collection of original research reports, often dense, and demands a reading approach that is both analytical and nonlinear. Readers must understand how the authors designed their study, what they found, and – most importantly – whether their reasoning holds up. Indeed, most readers new to a field scan the abstract and read sequentially, but what most do not know is that this genre has the density of a textbook with the structure similar to a magazine article, and that research articles are never meant to be read from start to finish.

Scientific articles are dense for a reason: They are built to document, document, and document some more. This means loads of specialized vocabulary, method-heavy sections, various analyses, and other technical details that may be unfamiliar even to experienced readers. Understanding these articles often requires extensive background knowledge in the field and the ability to critically evaluate each element of the study – from the framing of the research question to the justification of the conclusions. As readers, we must reconstruct and evaluate the authors' logic in relation to the evidence and decide if the claims are reliable. This kind of reading is really not reading at all as we typically know it, but an active, iterative investigation.

1.4 Common Challenges in Reading Primary Literature

Reading research articles can feel like being dropped into the middle of someone else's weird dream – there's no guide or key to understanding the symbols or deeper meaning. Even advanced critical readers can find the experience disorienting, especially when reading in a new field. As the authors of this book on critical reading, we still experience this disorientation each time we read in a new field, like when we start a new collaboration and need to read a lot of new research or critique research in fields in which we are *not* experts, as we did to develop many of the worked examples in the Student Supplement.

One of the reasons behind this experience of disorientation is that research articles rarely signal their argument points clearly. The main claim may be buried in the abstract and discussion; the evidence is scattered across the methods, findings, and discussion; and the reasoning may be implied rather than stated. Add to that a heavy dose of jargon and loads of required background knowledge, and it is easy to feel overwhelmed and adrift, uncertain of what matters most or how to connect the pieces.

However, we do not let discomfort stop us from engaging with new or challenging research articles. We use the methods we lay out in this book, and they work. We hope that by the end of this book, you will feel more equipped to work through feelings of discomfort and overwhelm.

One important caveat is that sometimes research articles are just very, very difficult. Some great research articles were cut from this book and moved to the supplement as worked examples because they were just too dense for most of us to understand. For example, there is an article about Hermitian by Rüter et al. (2010). It reports something about a mathematical matrix used in quantum mechanics, but beyond that, who knows? Give it a try if you feel bold.

1.4.1 The Abstract Leaves Out A Lot

Many readers start by skimming abstracts to decide whether an article is worth reading. This approach is highly practical. We use this approach to decide if an article is worth more time and effort. But abstracts, by design, are compressed summaries. They often skip the fine-grained details that make or break an argument, such as how the data were collected, why the methods were chosen, and what reasoning ties it all together. While abstracts are enormously useful for assessing if an article is relevant to a project, they rarely capture the study's full arc of evidence and reasoning or its broader implications.

Learning to read past the abstract deeper into the main text – systematically and critically – is essential. This does not mean reading every word in order (Lie et al., 2016). Instead, it means developing strategies for uncovering and evaluating what has been left unsaid in the abstract. Figuring out which strategies work best for you can reduce overwhelm and foster curiosity, especially when a deeper dive into an article's main text reveals unexpected twists and turns.

Because of space limits in this book, we often quote an article's abstract when modeling a specific reading skill. An abstract is compact enough to illustrate a strategy on a single page, yet still rich enough to show how the argument fits together. Please use our demonstrations as only a starting point only. Genuine critical reading means engaging with the full article whenever it is relevant to your project.

1.4.2 Jargon and the Curse of Expertise

A neuroscience major described her first encounter with a CRISPR article as, "I felt like I'd opened a cookbook written in another alphabet" (Bjorn, 2023). The barrier she described is the dense shorthand expert research readers use among themselves. Unfamiliar terminology is the single strongest predictor of student disengagement (Nelms & Segura-Totten, 2019). A quick fix is to encourage novice research readers name the unknowns – literally listing unknown terms in the margin – then crowd-sourcing definitions in class and research groups. The act of flagging gaps turns confusion into a question, and questions drive learning.

1.4.3 Finding the Argument in a Sea of Results

Beyond the traditional section headers like Introduction and Methods, published articles do not come with flags for the essential elements of the article's argument from evidence. Rarely will you see headings labeled "Claim," "Evidence," or "Reasoning." However, this is precisely the information we, as readers, must extract to evaluate the study's validity. Readers are expected to reverse-engineer the logic: Figure out what the authors are arguing, why it matters, whether the data support it, and whether the reasoning is sound. This guessing game is shockingly opaque for a civilization built on scientific advancement. Many of our students have asked us: Why not just make the argument clear? That is an important question that we will dive into in section 1.8.

For now, let's play a quick prediction game. Do not worry if microalgae engineering is not your field; it is not ours either. Consider the following research article title:

"Engineered chlorophyll catabolism conferring predator resistance for microalgal biomass production" (Kashiyama et al., 2021)

Now, predict the article's main claim and the likely evidence the authors developed. Jot it down somewhere if you can…

How did you do? Here's a quick check of our thinking:

Claim: "Engineered chlorophyll catabolism" … Somehow, tweaking the microalga's chlorophyll breakdown pathway "conferring predator resistance" … makes the cells harder for grazers to eat, "microalgal biomass production" … so the culture ends up growing more total biomass.

Evidence: Although the title itself does not spell out data, we can use it to guess some possible types of evidence:

- **Evidence bucket #1**: Predator-resistance tests to see if engineered microalgae get grazed a lot less than the control microalgae. Common micro-grazers could be small filter feeders, rotifers, microzooplankton, and other microscopic ocean creatures.
- **Evidence bucket #2:** Biomass comparision to see if, over time, the engineered microalgae cultures rack up more weight or cell density than the control microalgae growing in the same light and nutrient conditions.

There are many more possible lines of evidence, and we went with the two that first came to mind. If you listed something else, like genetics or respiration, it is probably correct.

Next, let's predict what kind of reasoning the authors might use to logically move from the evidence to the claim. Again, take about a minute to predict the authors' reasoning and jot it down somewhere…

How did you do? Here's a quick check of our thinking:

Reasoning: Changing how chlorophyll is broken down probably tweaks the release of compounds or pigment chemistry and those changes may put off grazers because of bad taste, wrong color, or toxicity, etc. With predators backing off, more microalgal cells survive each day, so the total biomass piles up faster.

Notice that we focus on the logic of what would have to be true for this evidence-claim pair to make sense: Changing how chlorophyll breaks down makes the microalgae taste or look bad, so predators leave them alone and they grow more. We do not cite any preexisting theories about predator–prey relationships or other articles with similar findings, although we certainly could if we knew more about this field.

Predicting in plain language an article's main argument of claim, evidence, and reasoning improves our predictive understanding of what the article is about. Then, we can check our understanding as we critically read. This common sense approach holds true even if we are new to the topic. *Claim–evidence–reasoning* give us a strategic approach to critical reading and thinking that we can use every time we encounter a research article.

With claim–evidence–reasoning as our starting point, our practice in this book moves us from a simple prediction game to strategies for writing a full literature review, ultimately deepening and strengthening analytical reading skills. Once you can spot how scientific arguments are built, it is possible to understand how even the most complex research develops – one claim, one dataset of evidence, and one line of reasoning at a time.

1.5 Types of Primary Literature

Primary literature is where scientific knowledge enters the world for the first time. But not all research articles serve the same function. There are two main types: research articles and review articles. Figure 1.1 shows how these types relate. A single review article might synthesize dozens – or even hundreds – of research articles. Both types are essential, but they offer different insights and require different reading strategies.

1.5.1 Research Articles

Research articles are the initial reports of a scientific discovery, like our previous example of a plastic-eating microbe. They present new findings generated directly by the researchers conducting the study. These articles contribute original evidence to the field and are the building blocks of the scientific record. There are two common types of research articles:

- **Experimental articles** report studies in which researchers observe or manipulate variables to test hypotheses, explore causal relationships, or evaluate interventions. These studies typically take place in controlled settings. For instance, a biology article might investigate how sleep deprivation affects immune performance by comparing test results of well-rested and sleep-deprived participants.
- **Theoretical articles** propose, refine, or challenge frameworks that explain observed phenomena. Instead of collecting new data, these articles build arguments using logic, mathematics, simulations, or reinterpretations of existing studies. For example, a theoretical article in astrophysics might critique quantum models of spacetime and propose a classical alternative, grounded in simulation results and previous observations.

The boundary between these types is often porous. Theories inspire experiments. Experiments test and reshape theories. Scientific progress depends on this interplay.

1.5.2 Review Articles

Review articles serve a different purpose: they pull together what is already known. Instead of reporting new findings, they analyze and synthesize the results of many

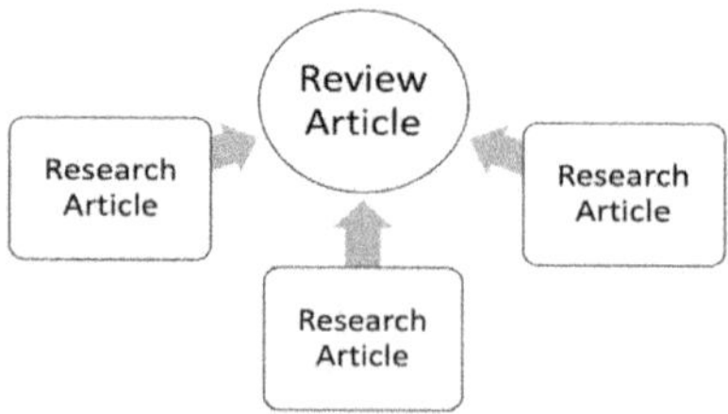

Figure 1.1 Research and review articles.
Review articles synthesize multiple studies, while research papers present original data.

Box 1.3 Knowledge Check: Question 1.2 Which statement best describes a research article?

A. A secondary source of summarized research findings compiled by professional editors.
B. An original research publication that claims new knowledge based on data.
C. A classroom handout that compiles excerpts from multiple journal articles.
D. A peer-reviewed editorial focusing exclusively on theoretical debates without data.

(Check your understanding using the Knowledge Check Key at the end of the book.)

studies on a shared topic, method, or tool. Reviews are especially helpful for getting up to speed in a field, making them a go-to resource for beginning students, interdisciplinary researchers, or anyone scouting a topic's current landscape. There are two advanced forms of review articles worth noting:

- **Meta-analysis** focuses on quantitative research. It combines numerical results from multiple studies using statistical techniques. This allows researchers to detect patterns, weigh evidence across studies, and evaluate the strength of effects better than any single study can. For example medical research commonly uses meta-analysis, such as Cochrane reviews, to synthesize clinical trial results.
- **Meta-synthesis** focuses on qualitative research. It gathers themes, metaphors, or conceptual insights from interview-based or observational studies. This allows researchers to weave the findings into a larger narrative or theoretical model and to surface testable hypotheses. For example, behavioral research uses meta-synthesis to surface new research questions and hypotheses.

Review articles are easy to spot: many journals label them prominently near the title. While they do not provide new data, they offer something just as valuable – a broader view of where a body of evidence points. Reading reviews helps readers zoom out, see connections and gaps between studies, and identify promising directions for further investigation.

1.6 Overview of Peer Review

Peer review is one of the oldest quality control systems in science – and one of the least standardized. Since 1665, when the *Philosophical Transactions of the Royal Society* first employed expert judgment to vet submissions, peer review has grown into the dominant model for assessing scientific credibility (Gregory & Denniss, 2019). By 2020, scholars published over four million peer-reviewed journal articles annually, with estimates of annual growth between 5% and 6.5% (STM, 2021). These articles are cataloged in large commercial databases, such as Scopus and Web of Science, which cover almost every discipline. Scopus lists 25,000 active journals and Web of Science lists 18,000. However, the actual number of active journals is higher because no single platform captures the full scope of global scholarship. In addition, there are dozens of academic databases that cover specialty topics, such as PubMed for biology and medicine, and IEEE for engineering.

Despite this enormous publication output, there is no universal training for peer reviewers, and typically, individual publishers provide a patchwork of guidance (e.g., Elsevier, Inc., 2025). In many cases, peer reviewers are expected to intuit how to evaluate manuscripts without explicit guidance or shared criteria (Glonti et al., 2019). Editors have described peer review instructions from publishers as vague, unspecific, or rough. This ambiguity raises real concerns, such as:

- How do we assess the rigor of a system with no consistent standards?
- How transparent is the process if its rules are unwritten?

As we will explore further in Chapter 10, these questions do not undermine the importance of peer review. Instead, they highlight the need to approach it critically and with great concern.

1.6.1 Why Peer Review Matters

Despite inconsistencies, peer review serves two essential purposes in scientific communication. First, it acts as a filter. Manuscripts are vetted to ensure that the knowledge they present is new, significant, and grounded in sound methods and reasoning. Second, it works as a refinement process, helping authors revise and strengthen their work before publication (e.g., Glonti et al., 2019; Kelly et al., 2014).

Peer review remains one of the key mechanisms by which scientific claims gain credibility. The process is flawed, but it adds friction to the publication pipeline, slowing things down just enough to include critique. Reviewers can challenge methods, interrogate assumptions, and assess whether conclusions follow from the evidence (The American Physiological Society, 2025). These challenges are what make peer-reviewed research more than just information; it adds a layer of trustworthiness.

1.6.2 Peer Review and Scientific Credibility

When done well, peer review can improve a study's published report. Ideally, reviewers help catch errors, question overreach, and raise new possibilities. This critical dialogue between reviewer and author can make research stronger and support the integrity of the primary literature as a whole. A 2015 survey of researchers showed that 82% of respondents agreed that "without peer review there is no control in scientific communication" (Publishing Research Consortium, 2016).

Peer review also shapes behavior. Knowing that peer experts will scrutinize their work, researchers are more likely to follow ethical norms, use rigorous methods, and communicate clearly. In this way, peer review functions as a gatekeeping mechanism and as an incentive for scientific improvement.

However, peer review is not a guarantee. Like any human system, peer review is shaped by norms and assumptions. Understanding its value – and its limitations – is part of becoming a critical reader of scientific research.

Box 1.4 Knowledge Check: Question 1.3 What is one of the primary goals of the peer review process in scientific publishing?

A. To promote competition among journals
B. To screen and improve the quality of research articles before publication
C. To increase the number of published journals annually
D. To provide standardized training for peer reviewers

(Check your understanding using the Knowledge Check Key at the end of the book.)

1.7 Citing Research Articles

Citing other research is one of the defining practices of research articles. It shows where ideas come from, how evidence accumulates, and how knowledge builds across time. When you read research articles or reviews, including this book, you will notice citations scattered throughout. Each citation is a record of knowledge and an invitation for readers to trace the point back to its original publication.

Different scientific fields use different citation styles, but the goal is always the same: to give credit, build context, and make sources easy to find. Figure 1.2 shows the anatomy of a journal article.

Next, let's look at the various citation styles used by disciplines in the natural sciences. Note that there are a lot of variations by sub-discipline and geography, and these are some broad usage patterns to orient us to citation practices. Below, each field name is followed by the style name, then an example underneath, and followed by the style's source citation (note: this is a *fictitious citation* for illustration purposes).

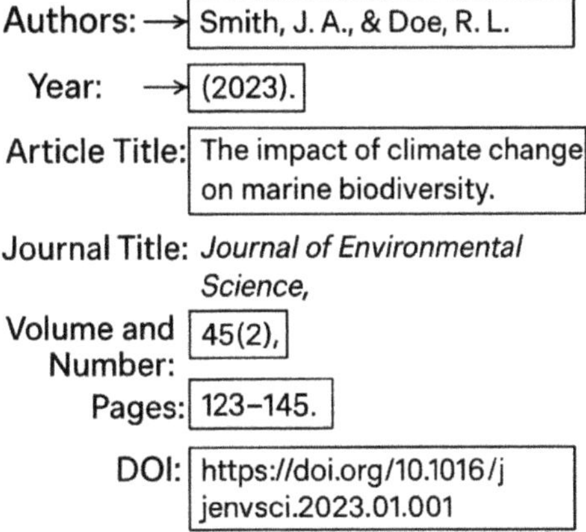

Figure 1.2 Anatomy of a citation.
Citation anatomy – author, year, title, journal, volume, pages – is essential for article retrieval.

Astronomy: AAS (ApJ/AASTeX) – Author-Year
Smith, J. A., & Doe, R. L. 2023, *J. Environ. Sci.*, 45, 123–145. https://doi.org/10.1016/j.jenvsci.2023.01.001
Source: AAS (2025, June 14)

Biology: CSE – Name-Year
Smith JA, Doe RL. 2023. The impact of climate change on marine biodiversity. *Journal of Environmental Science*. 45(2):123–145. https://doi.org/10.1016/j.jenvsci.2023.01.001
Source: Council of Science Editors (2006)

Chemistry: ACS – Numeric/Superscript
Smith, J. A.; Doe, R. L. The impact of climate change on marine biodiversity. *J. Environ. Sci.* **2023**, 45(2), 123–145. https://doi.org/10.1016/j.jenvsci.2023.01.001
Source: ACS Style Quick Guide (2020)

Earth Sciences: AGU – Author-Year
Smith, J. A., and R. L. Doe (2023), The impact of climate change on marine biodiversity, *J. Environ. Sci.*, 45(2), 123–145. https://doi.org/10.1016/j.jenvsci.2023.01.001
Source: AGU Grammar and Style Guide (2025)

Physics: AIP – Numbered
[1] J. A. Smith and R. L. Doe, *J. Environ. Sci.* 45, 123 (2023). https://doi.org/10.1016/j.jenvsci.2023.01.001
Source: AIP Publishing LLC (2025, July 1)

Combination Fields: APA 7 (or latest edition)
Smith, J. A., & Doe, R. L. (2023). The impact of climate change on marine biodiversity. *J. Environ. Sci.*, 45(2), 123–145. https://doi.org/10.1016/j.jenvsci.2023.01.001
Source: American Psychological Association (2020)

The last example of the American Psychological Association (APA 7) is used in fields that combine natural and social sciences – such as biology, psychology, and environmental science – especially when the research involves human behavior or interdisciplinary themes. We use APA 7 in this book because we are writing about *human behavior* – namely, critical reading and thinking.

Let's review the key points about citations and their use.

- **Citations are a shorthand way to point you to a published research article, and each part of it has a specific function**. They allow readers to quickly locate and evaluate sources based on the search criteria of authors, year, title, journal, volume, issue, page numbers, and DOI (a permanent Digital Object Identifier that links directly to the article online). Unlike URLs, which can break, DOIs are built to last.
- **Citations let us follow the argument back to its source, locate the study, and see who else has cited the work**. If you're unsure which citation style to use, start by checking the references in a published article from your field or consult the journal's submission guidelines. You can also check with instructors, as many also include preferred styles in assignment prompts or syllabi.

- **Citing prior work is more than academic courtesy; it is how researchers join the ongoing conversation of their field.** Citations reveal where ideas come from, how they evolve, and what questions remain unanswered. When done well, they position a study within the larger landscape of scientific knowledge, making its contribution clear and its relevance visible.

In these ways, citations act as the connective tissue of research, linking today's questions to yesterday's answers – and paving the way for tomorrow's discoveries.

1.8 Research Article Structure

Grab your nearest research article and look at its overall structure. The title, authors, and all-important abstract appear at the top of a research article. Next comes the main body of the text, usually accompanied by illustrative figures and data-rich tables. Appendices and acknowledgments might follow this, and then a long list of references. Now, go back to the main body of the text and ignore everything except the subject headings. These are probably laid out in the following order: Introduction, Methods, Results, and Discussion (IMRaD) or with slight variations. The remainder of this section dives into why we have this structure and how we can work with it.

1.8.1 IMRaD

The standard organizational structure of scientific articles did not begin that way. In the eighteenth century, researchers shared their findings in letter form, weaving context, methods, and reflections into narrative prose. These letters were personal and compelling, but they were also inconsistent. Key details were often missing. Reproducibility was an afterthought. Science was a conversation, but not always a clear or repeatable one.

As the scientific enterprise grew in the nineteenth and early twentieth centuries, so did the need for clarity and consistency. Journals began to formalize what good reporting looked like. Readers needed more than a story: they needed a blueprint. The result was a shift toward structured communication, especially around methods and results, which laid the groundwork for the format we now know as *IMRaD*, as shown in Figure 1.3.

By the 1940s, that format – Introduction, Methods, Results, and Discussion – began to take hold in fields such as biology, chemistry, and medicine (Sollaci & Pereira, 2004). *IMRaD* tidied up research reports and standardized how science was communicated. Instead of letting each author decide how to organize their work, *IMRaD* offered a shared framework. Readers could quickly scan for methods, evaluate evidence, and follow conclusions with greater ease. This standardization helped science become more transparent, efficient, and cumulative. In the 1970s, the American National Standards Institute formally recommended the *IMRaD* format for scientific manuscripts, first in 1972 (ANSI, 1972) and again in 1979 (ANSI, 1979; Day, 1989).

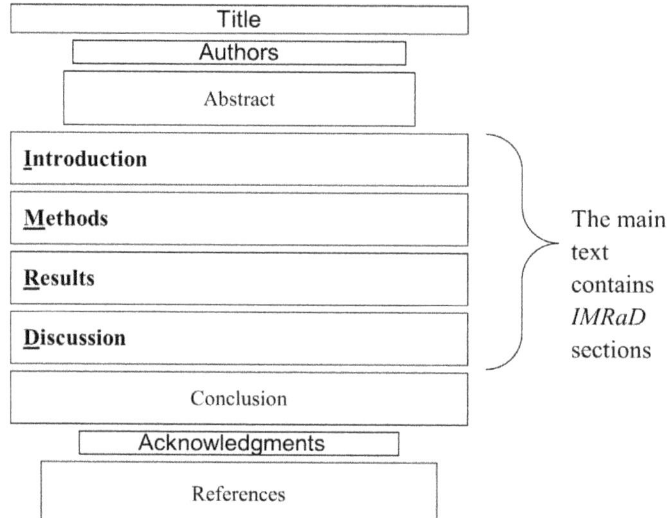

Figure 1.3 Typical *IMRaD* structure of a research article.
IMRaD stack from title to references, showing the standard publication format of Introduction, Methods, Results, and Discussion.

With that endorsement, *IMRaD* became the dominant model for research writing – a kind of common language for scientists across disciplines that endures to this day. It endures, in part, because *IMRaD* is a reliable structure that answers four essential questions:

- Why did the authors do the study? (Introduction)
- How did they do it? (Methods)
- What did they find? (Results) and
- What does it mean, and why does it matter? (Discussion)

This publication order reflects the classic arc of discovery. Tracing this arc is the first step toward unpacking research. Each *IMRaD* section has specific functions described in detail by Nair & Nair (2014) and Wu (2011) and summarized below:

- **Introduction:** This section provides the background and context for the research. It consists of a review of relevant literature, the research question or hypothesis, and the study's objectives.
- **Methods:** This section details the procedures and techniques used to conduct the research. It includes information on the study design, data collection methods, and analysis techniques, allowing other researchers to expand or replicate the study.
- **Results:** This section presents the study's findings, often using tables, figures, and descriptive text. It focuses on the data collected and the analysis without interpretation.
- **Discussion:** This section interprets the results, explaining their significance and implications. It relates the findings to the original research question, discusses limitations, and suggests areas for future research.

IMRaD's standardized format is one reason many university writing centers teach it as a structured approach for writing research articles (e.g., *Scientific Writing: IMRAD format; BYU Writing Center*, 2025).

However, science rarely follows a script, nor do journals. The *IMRaD* structure is common across research articles, grant proposals, and technical reports, but many fields must adapt it to meet disciplinary practices. Some journals include additional elements such as abstracts, keywords, acknowledgments, and appendices to help readers navigate or contextualize a study (Nair & Nair, 2014). Others customize the core structure itself, for example:

- *Nature* moves the Methods section to the end of the article and prints it in a smaller font, subtly signaling its role as a technical appendix rather than part of the story arc (Wu, 2011).
- *Science* often omits formal Introductions altogether, embedding background in the opening paragraphs.

Hartley (1999) proposed nesting topical subheadings within each *IMRaD* section to improve clarity and readability. This idea echoes what many journals already do: tailor structure to suit the discipline, the data, and the audience.

Despite these variations, *IMRaD* persists for good reasons. It supports the values that make science credible and shareable: transparency, reproducibility, and a common language for collaboration and replication. Like any reliable framework, it is designed to organize knowledge presentation and to make it possible for others to find and build on it.

1.8.2 Hero's Journey Analogy

At first glance, the *IMRaD* structure might look like another outline. But look deeper, and you'll see something much more familiar. The structure of a scientific article closely mirrors the arc of the Hero's Journey: a storytelling pattern elucidated by Joseph Campbell found in myths, novels, and films across cultures. This analogy can help demystify why research articles are structured the way they are, and how each part serves a specific purpose in the logic of discovery. Let's break it down:

- The **Introduction** section is the call to adventure. Here, the research question is posed, the gap in knowledge revealed, and the stakes made clear. The hero – the study – emerges from the ordinary world of established knowledge to face a challenge that no one has fully solved.
- The **Methods** section is the quest itself. This is where the path is chosen, the tools are assembled, and the journey unfolds step by step. It details the planning, procedures, and techniques used to navigate uncertainty and gather evidence.
- The **Results** section is the moment of truth. The hero faces tests and trials, and we find out what was discovered. This is where the data appear, unembellished – just the raw outcomes of the journey, whether triumphant, surprising, or inconclusive.
- **The Discussion** section is the return and reflection. The findings are interpreted, their meaning unpacked, and their implications weighed. This is where the hero

shares hard-won wisdom with the scientific community, linking back to the original problem and pointing toward new horizons.
- **The Conclusion**, when included, serves as the epilogue. It distills the key lessons of the journey, acknowledges the limits of what was learned, and gestures toward the next adventure – whether by proposing future studies or revisiting old assumptions with new insight.

Framing *IMRaD* as a journey humanizes research. It can help us as readers recognize that scientific research, like storytelling, unfolds with purpose. There is a beginning, a middle, and an end. There are challenges, tools, discoveries, and transformations. And when we learn to see research this way – not just as data, but as a structured quest for understanding – we read to critique its meaning. We argue in this book that an effective critique must center on the argument from evidence.

1.8.3 Reproducibility of the Primary Literature

These patterns in literature publication highlight a central point. While *IMRaD* gives research a tidy scaffold, the structure alone cannot guarantee that findings are reliable, verifiable, or even meaningful. The lasting value of a research article depends on what sits inside that framework, namely, transparent methods and reproducible results. Reproducibility – the capacity for independent teams to repeat a study's procedures and arrive at comparable outcomes – improves scientific credibility because it lets the community confirm that claims survive beyond the originating researchers.

However, reproducibility has significant problems. Beginning in 2013, the *Reproducibility Project: Cancer Biology* (Center for Open Science, 2025) set out to replicate 50 pivotal experiments drawn from highly cited cancer articles. Several mouse-tumor replications were conducted, yet the aggregate analysis later showed median effect sizes that were about 85% smaller than the originals (Errington et al., 2021). That shortfall is not confined to cancer research. An audit of 441 biomedical articles published between 2000 and 2014 found that only 1.6% shared raw data and 8% included a full protocol (Iqbal et al., 2016). A follow-up survey of 149 articles from 2015 to 2017 showed that data-sharing statements climb to 18% and detailed methods to 21%, but comprehensive transparency was still the exception (Wallach et al., 2018). Similarly, in psychology, the Open Science Collaboration's effort to replicate 100 studies found that only about 36% reproduced statistically significant results consistent with the original findings in psychology (Aarts et al., 2015).

Figure 1.4 visualizes these trends, showing at a glance how much work remains to achieve routine reproducibility. It is a sobering and nuanced picture. In cancer biology, for instance, the median effect sizes reproduced by the Reproducibility Project landed at roughly 15% of those reported initially. Psychology fared somewhat better on direct replications, yet only 36% of studies produced statistically significant results consistent with their originals. Across biomedical research more broadly, raw-data availability and full-protocol disclosure were vanishingly rare between 2000 and 2014, though both indicators improved modestly – if still inadequately – by 2015–2017. Audits across biomedicine

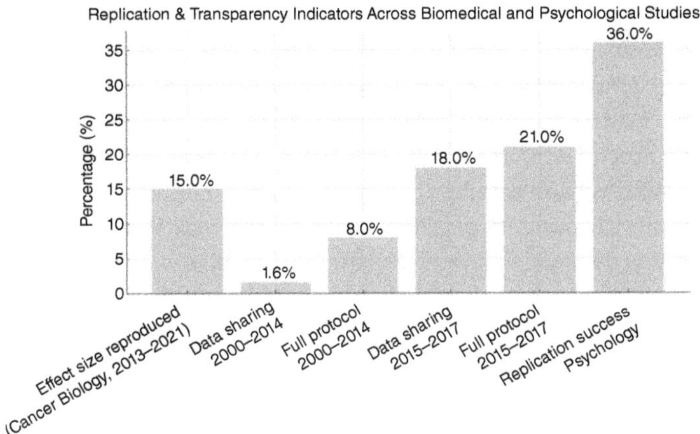

Figure 1.4 Reproducibility trends.
Each bar shows some reproducibility trends, with the largest replication success in psychology (36 %), underscoring the need for transparent methods.

(Iqbal et al., 2016; Wallach et al., 2018) and large-scale replication efforts in cancer biology and psychology (Ioannidis, 2017; Errington et al., 2021; Nosek & Errington, 2017) show how essential details for study replication are rarely, if ever provided.

Closing this reproducibility gap demands ambitious change, such as open-science frameworks, preregistered protocols, and public repositories for data and analysis scripts. As journals and funders adopt these changes as new requirements, transparency could become the default and replication a feasible follow-up rather than a heroic, six-figure exception. For the time being, reproducibility remains a worthy, if largely unmet aim.

1.8.4 Validity through Evidence, Reasoning, and Limitations

While *IMRaD* offers a clear structure for organizing research, and reproducibility ensures that findings can be independently verified, neither guarantees that a study's conclusions are trustworthy. For that, we need to look deeper at the quality of the evidence, the logic of the reasoning, and the transparency with which researchers acknowledge the study's limitations. These elements form the core of scientific validity. Let's unpack these ideas.

- **Evidence**: In the natural sciences, empirical evidence forms the bedrock of credible research (Krajcik & McNeill, 2015). This evidence is derived from systematic observations, controlled experiments, and meticulous data collection. The credibility of such evidence is enhanced when it can be independently replicated and verified by other researchers. Detailed documentation of methodologies, data analyses, and experimental protocols is essential, as it allows for reproducibility and critical evaluation within the scientific community.
- **Reasoning**: Transforming raw data into meaningful conclusions requires logical reasoning. This involves connecting evidence to hypotheses or research questions

through structured argumentation (Allen & Baker, 2016). Scientific reasoning often employs both inductive and deductive approaches. Inductive reasoning involves drawing general conclusions from specific observations, while deductive reasoning tests hypotheses based on established theories. The clarity and coherence of this reasoning process are vital for the validity of the research findings.

- **Limitations**: Acknowledging the limitations of a study is a cornerstone of scientific integrity (Dawson et al., 2024). Limitations may stem from various sources, such as study designs, sample size constraints, methodological challenges, or external factors influencing the results. By transparently discussing these limitations, researchers provide context for their findings and guide future investigations. This practice fosters a culture of openness and continuous improvement within the scientific community.

In summary, the validity of scientific research is underpinned by the rigorous collection of evidence, the application of rigorous logical reasoning, and the transparent acknowledgment of study limitations. These components collectively ensure that scientific findings are credible, logical, and reproducible.

1.9 Approaches to Reading Primary Literature

Reading primary literature is a skill that evolves over time and benefits from structured guidance. Educators have developed various strategies to help readers engage with scientific texts each tailored to different learning objectives and educational levels. These methods range from annotation and scaffolding to inquiry-based learning and argumentation frameworks (Goudsouzian & Hsu, 2023). Let's break them down by approach.

1.9.1 Analysis and Annotation

Annotation serves as a powerful tool to demystify complex scientific articles. The American Association for the Advancement of Science (AAAS) offers the "Science in the Classroom" initiative, providing annotated research articles to support student understanding. Similarly, Kararo and McCartney (2019) advocated for annotated primary literature as a pedagogical tool that scaffolds undergraduate engagement with scientific texts. Building on these approaches, Bjorn (2024) introduced a method that focuses on the elements of a scientific argument, the CERIC method. This method is flexible, yet structured analysis that can integrate other approaches, including annotation and social collaboration to enhance critical reading skills. Here is an example:

- **Situation:** You have been assigned to read a recent article on CRISPR diagnostics. You may work with a group, but if so, you must show how you read together.
- **Individual Steps:**
  - Download a PDF of the article.

  - Open it in Adobe Acrobat Preview, or Perusall and use five highlighter colors keyed to the CERIC argument categories (yellow = Claim, blue = Evidence, green = Reasoning, pink = Implications, orange = Context).
- **Group Steps:**
  - Add margin notes that translate jargon, for example, write in the margin, "LOD = limit of detection, for example, the quietest sound a microphone can pick up."
  - When you meet with your lab or study group, share the annotated file so the colors guide discussion without anyone re-reading the whole article aloud.
  - If the group decides to adopt this approach, consolidate similarities and differences in your margin notes.

Anecdotally, our students who have run this approach report to us feeling more interested in the research and confident in their findings.

1.9.2 Inquiry-Based and Interdisciplinary Methods

Inquiry-based learning encourages students to ask questions and explore scientific literature actively. Ness (2016) highlights the cognitive and motivational benefits of question generation in fostering reading comprehension. In an interdisciplinary context, Liotta and Almeida (2005) describe a seminar that combines organic chemistry and cell biology, using primary literature to tie together concepts from both fields. Here is an example:

- **Situation**: You are in a joint biology–chemistry seminar and have been assigned to form a group with at least one biology major and one chemistry major. You have been assigned to critically read an article on testing athletes for illegal doping.
- **Individual Steps:**
  - Scan the abstract and main text, then brainstorm five "how/why" questions that neither discipline alone can answer (e.g., "How do natural metabolic products limit testing applications?").
  - Each person researches the question for 30 minutes.
- **Group Steps:**
  - Reconvene to pool answers and decide whether the article's methods and findings really bridge both fields.

Anecdotally, our students who have run this exercise report a much better understanding of what makes and breaks interdisciplinary research.

1.9.3 Longitudinal and Intensive Programs

Long-term engagement with primary literature can lead to significant improvements in students' comprehension and critical-thinking abilities. Sato et al. (2014) found that repeated exposure to scientific articles resulted in longitudinal gains in subsequent lab courses. Kozeracki et al. (2006) describe an intensive, literature-based teaching program that prepares undergraduate science majors for advanced studies, facilitating their transition to doctoral programs. Here is an example:

- **Situation:** You have been assigned to keep a weekly reading log for an entire term.
- **Steps:**
  - Each Friday, pick one core article in your thesis or dissertation area, complete a one-page notes summary, and brainstorm two follow-up experiments you might run.
  - At week 12, submit the log plus a 500-word reflection on how your questions evolved.

Anecdotally, our students who have run this exercise report sharper qualifying exam answers because they see the growth trajectory of their own thinking.

1.9.4 Structured Frameworks and Argumentation

Structured frameworks help students dissect scientific arguments effectively. The Scientific Argumentation Model (SAM) is a heuristic that guides students through the rhetorical structure of research articles (van Lacum et al., 2026). Similarly, Koeneman et al. (2013) used an argumentation analysis framework to teach students how to evaluate claims and evidence. The CERIC method builds on these frameworks by making analytical categories explicit, empowering a deeper understanding of the argument's elements and how they interrelate. Here is an example:

- **Situation**: You have been assigned a research article on microplastic uptake in fish to read critically in a journal club.
- **Steps**:
  - You choose the SAM and use a worksheet.
  - After reading an article on you, fill in the SAM grid: Research question → Claim → Supporting Evidence → Underlying Reasoning → Possible Limitations.
  - In the journal club meeting where you are leading, you focus the debate on the reasoning and limitations columns.

Anecdotally, our students who have run this exercise report that it forces the group to scrutinize and debate the logic instead of re-summarizing results.

1.9.5 Selecting an Appropriate Approach

Choosing the right approach to critical reading of research articles depends on where you are in your learning journey. Figure 1.5 shows a decision tree to help choose the best reading approach for the situation: when to pick annotation, inquiry, structured frameworks, longitudinal reading – or simply mix and match. If you're just beginning to read primary literature, start with annotation-based tools or guided frameworks like CERIC to help you break down unfamiliar texts into manageable parts. As you build confidence, try summarization or inquiry-driven reading to practice identifying main ideas and asking meaningful questions.

Advanced readers – especially those preparing for individual research projects, qualifying or comprehensive exams, or a graduate-level literature review – might benefit most from structured argumentation models or intensive critical reading programs that

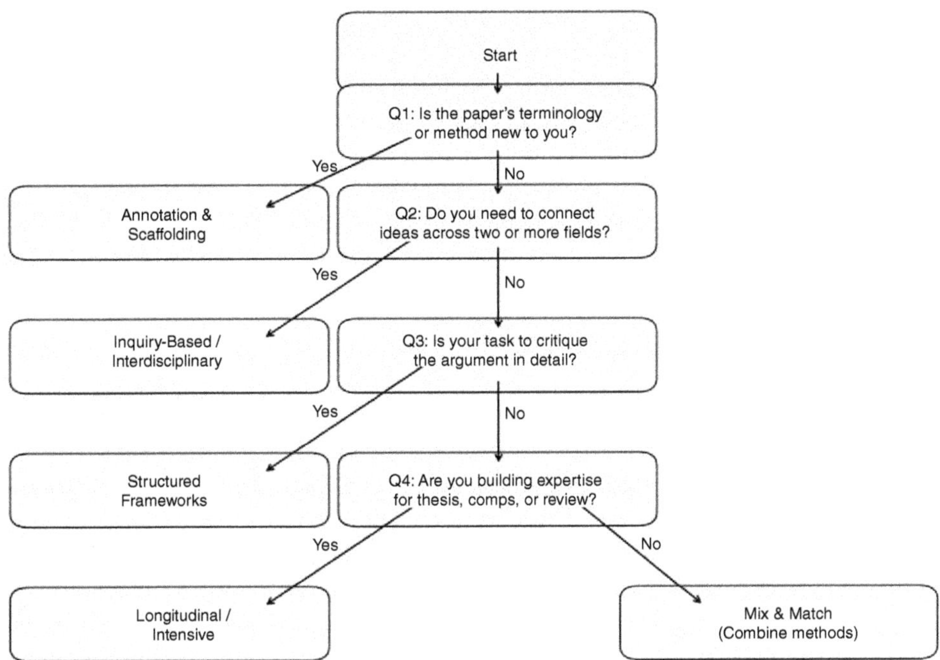

Figure 1.5 Decision tree for reading approaches.
Decision tree highlights how readers might skim, scan, or deep-read based on goal and time.

emphasize synthesis across articles. There's no one-size-fits-all strategy; the key is to select a method that challenges you just enough to grow your skills without overwhelming your understanding. Do not be afraid to mix and match strategies. Some of the best reading habits emerge through a blend of structured support and self-directed exploration.

For example, annotation-based methods are suitable for introductory courses, providing more conceptual guidance for novice research readers. Structured frameworks like SAM and CERIC are beneficial for both novice and advanced readers, promoting critical analysis and argumentation skills. Inquiry-based and interdisciplinary approaches can stimulate curiosity and integrate knowledge across fields at any reading level. The key point here is that ongoing and critical engagement with research articles is necessary to develop the foundational skills necessary for research-intensive careers. Find a method or mix of methods that works for you and stick with it.

1.10 Conclusion and Looking Ahead

Understanding these foundational aspects of primary literature – its importance, types, credibility, the *IMRaD* structure, challenges with reproducibility, and reading strategies – prepares readers for the essential work of learning how to conduct a deeper critical analysis of research articles. Many approaches and methods exist for analyzing primary literature. This book focuses on the CERIC method as a guide to help advanced undergraduate and graduate students through a structured analysis of the essential elements of a scientific argument. CERIC can be easily combined with many other methods. The explicit CERIC categories of information are what we will hunt for and analyze in each article and in each chapter ahead, starting with Claims in Chapter 3. This method also provides instructors with a framework to assess and address critical reading challenges and growth.

1.11 Chapter Key Takeaways

1. **Primary literature is a cornerstone of scientific research**: Primary literature comprises original research articles that present new knowledge. These articles are essential for advancing scientific knowledge. The structured formats of *IMRaD* enhance clarity and consistency in publication.
2. **Critical reading is a cornerstone skill:** Critical reading is fundamental for academic success, particularly in graduate education in the natural sciences. Despite its importance, critical reading is often overlooked in formal instruction and coursework, leaving many learners underprepared to evaluate arguments, evidence, and methodologies. This gap hinders our abilities to analyze and synthesize knowledge.
3. **Many barriers to understanding primary literature:** Novice critical readers face many challenges, such as dense jargon, assumed background knowledge, and the nonlinear nature of the research process. Abstracts often omit critical details, requiring readers to engage deeply with the entire article to grasp its argument, methodology, and implications.
4. **Effective teaching and learning:** Effective methods to teach critical reading include annotation and summarization, peer discussions, interdisciplinary and inquiry-based approaches, and categorical frameworks like SAM and CERIC. These methods help readers locate and evaluate essential elements, such as claims, evidence, and reasoning.

2 The CERIC Method

Reading scientific articles is like trying to decipher a code written in a foreign language with the codebook missing.

—2nd Year Doctoral Student in Cognitive Science

2.1 Overview

Chapter 2 introduces the CERIC Method, an evidence-based form of categorical reading that allows readers to break down a research article into its five main components of scientific argumentation: Claim, Evidence, Reasoning, Implications, and Context. This chapter starts with the historical development of modern scientific argumentation in the Toulmin et al. (1984) model and considers its strengths and weaknesses. Next, this chapter provides an overview of traditional and evidence-based reading strategies. Then, this chapter introduces the CERIC elements, demonstrating how to strategically identify them in a research article. This chapter closes by summarizing the empirical evidence demonstrating that CERIC improves reading comprehension, research self-efficacy, and reading apprehension, providing a structured approach to engaging with and analyzing research literature.

2.2 Arguing from Evidence

In the 1970s and 80s, Stephen Toulmin, Richard Rieke, and Allan Janik set out to update the centuries-old study of argumentation for a modern scientific audience. Their landmark text, *An Introduction to Reasoning* (Toulmin, Rieke, & Janik, 1984), shows that a persuasive scientific claim rests on more than a stack of facts. It must also link those facts to a defensible conclusion through transparent reasoning. This deceptively simple insight reshaped how scholars teach scientific communication and how reviewers assess manuscripts. When instructors introduce beginning researchers to the Toulmin et al. model, they are giving students a durable mental model for distinguishing well-founded science from shaky speculation. So, let's get into it.

Toulmin and colleagues distilled scientific argument into three interlocking elements, illustrated in Figure 2.1:

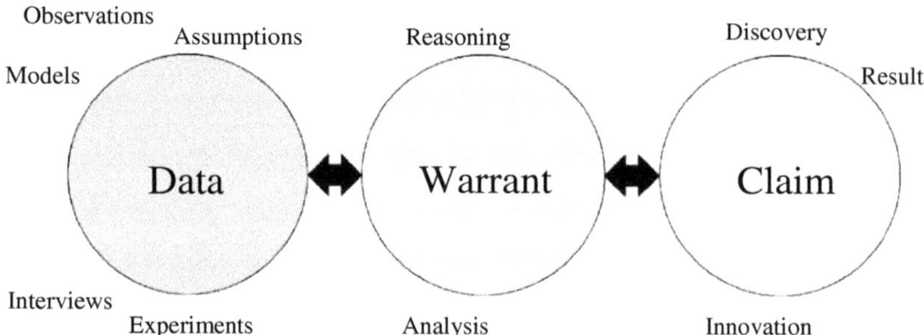

Figure 2.1 Summary of the Toulmin et al. (1984) model of scientific argumentation.

(1) **Data**: The observations, measurements, or model outputs that ground the discussion.
(2) **Warrant**: The chain of reasoning that connects those data to a broader inference or assertion.
(3) **Claim**: The conclusion that follows when data and warrant work in concert.

This anatomy of argumentation is powerful because it emphasizes reasoning – often the most elusive part of a research article. Yet, like all models, it has limits. Real articles rarely present a single, tidy trio of data, warrant, and claim. Instead, they weave nested sub-arguments that branch into comparative models, sensitivity analyses, or tests of instrument calibrations. In such cases, a "warrant" can morph into a subordinate claim (or subclaim), and data for one research question can become context for the next.

The use of the term *warrant* – a word that seems more at home in formal logic textbooks but seldom appears in modern research writing – can be misleading. Seasoned scholars sometimes confuse warrants with rationales (the motivation for a study), data, or conclusions (the logical consequences of the study). A practical remedy is to substitute the more familiar word *reasoning* and to assess the quality of that logic against alternative explanations.

Another challenge can emerge when readers catalogue every data point yet stop short of asking whether those data are adequate to support the claim. The Toulmin et al. model labels evidence but does not structure its evaluation. To cultivate that practice, readers can rate evidence on multiple dimensions relevant to the discipline before tracing reasoning. For instance, readers of an observational astronomy article might first find and rate the calibrations, instrumentation checks, and model assumptions before delving into evidence and reasoning.

In addition, scientific arguments live in social as well as logical space. Editorial gatekeepers, funding priorities, and public values all influence which claims reach print. Because the Toulmin et al. model does not directly capture these external forces, we need another approach to situate an argument within its disciplinary conversation, including probing the future implications of a claim and contextualizing the research and social ecosystems that shaped a study's research questions.

Box 2.1 Knowledge Check: Question 2.1 Why does the Toulmin et al. model potentially overlook complexities in scientific discourse?

A. Because it focuses too heavily on the strength of data rather than the claim itself.
B. Because it reduces arguments to three components, while real research often has sub-arguments and multiple lines of evidence and reasoning.
C. Because it only applies to ethics and philosophy, and not to other scientific fields.
D. Because it requires specialized statistical knowledge to evaluate each argument.

(Check your understanding using the Knowledge Check Key at the end of the book.)

2.3 Reading and Writing Out of Order

Research rarely follows a straight-line path from scientific question to publication. A project may begin with an observing proposal or a survey design, only to run into clouded skies or a thin response rate. Fresh data sometimes contradict the hunch that inspired the study, or a newly published article may reframe the main question that originally motivated the study. These surprises do not derail the scientific narrative; they simply shift its point of entry. Much like a novelist who pens the plot twist before the first chapter, scientists often draft the middle of a manuscript – methods, preliminary findings, even a figure or two – long before they craft the abstract or introduction that eventually headline the published work.

Writing in segments creates a mosaic that authors later polish into a seamless story. One week the team captures the experimental setup; the next, they attach raw measurements and some initial analysis results. Literature review paragraphs accumulate in a background file until they coalesce into an introduction. When inspiration strikes, coauthors sketch implications and future directions. Only after these pieces fall into place does the crown – title and abstract – lock the sections together, ready for final revision and organization into the *IMRaD* format.

A reading strategy that accounts for this messy process can help readers navigate the arguments that underlie a research study. **Categorical reading**, rooted in the Information processing theory of learning (summarized in Schunk, 2012), invites readers to enter an article wherever the claim pulls them, instead of marching from the title page to the methods and the findings, to the conclusions. Unlike the passive skim that precedes most university note-taking, categorical reading is an *active hunt* for five key classes of scientific information – claims, evidence, reasoning, implications, and context. By sampling sections in a purposeful manner, readers turn a linear *IMRaD* structure into a flexible menu that centers what matters most in the actual research.

Information processing theory, with multiple contributors and facets, illuminates why this categorical approach works so well. Four concepts are especially helpful, as shown in Figure 2.2:

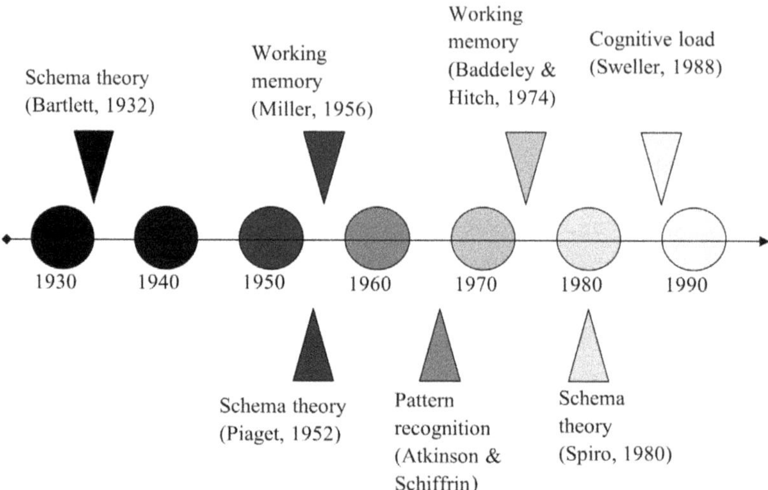

Figure 2.2 Chronology of Learning Theories Supporting Categorical Reading.

- **Pattern recognition** allows readers to spot familiar textual landmarks – abstracts, subsections, and captions – and label them quickly as evidence, reasoning, or claim (Atkinson & Schiffrin, 1968).
- **Working memory** then keeps those labelled pieces in mind while screening out details that do not serve the immediate goal (Miller, 1956; Baddeley & Hitch, 1974).
- **Long-term memory** weaves the extracted pieces into evolving mental models, also known as *schemas*, making each new article easier to parse than the last (Bartlett, 1932; Piaget, 1952; Spiro, 1980). Providing categories in advance seeds the schema formation.
- **Cognitive load theory** reminds us that working memory has limits; categorical reading lightens the load by grouping information into pre-grouped packets (Sweller, 1988).

When we pair categorical reading with the Toulmin et al. model, we can locate each argument element quickly and evaluate its reliability before cognitive overload sets in. The result is improved comprehension with more durable knowledge that migrates into lab meetings, literature reviews, and article-writing sessions. In other words, our strategic reading actions coalesce into other essential learning tasks, turning reading scientific literature into an interactive brainteaser where we read, think, and write, out of the *IMRaD* order to focus on what is most important to the actual research process.

2.4 Evidence-Based Methods of Reading

Type "how to read a journal article" into your browser and dozens of listicles appear. Some are from seasoned researchers on YouTube, and others are from publishers that host marquee journals. Most share a familiar origin story: I struggled, experimented,

Box 2.2 Common Confusion: Quick-Read Shortcuts ≠ Critical Reading

- **What shortcuts do:** Skimming the abstract, scanning headings, or jumping straight to the figures can tell us what the authors studied, and roughly what they found.
- **What shortcuts miss:** They rarely reveal how the evidence supports the claim or whether the reasoning holds up. Without that logic chain, we cannot gauge reliability, spot hidden assumptions, or identify follow-up questions.
- **Graduate-level goal:** Move from being a passive consumer of facts to an active evaluator of arguments. That shift requires slowing down, mapping each CERIC element, and asking, "Does this evidence truly support the claim?"

Takeaway: Use skimming for orientation and relevance, and reserve argument analysis for any article you plan to review, critique, or build upon in your own research.

and finally hit on a trick that worked for me, so it might work for you. The advice typically goes like this:

- Skim the title and abstract.
- Skip the introduction and methods.
- Focus on the figures and discussion.
- Mine the reference list for related articles.
- Reflect on what you read and form your own opinion.
- Reread the article front-to-back if you need more details, worry you have missed key points, or simply feel afraid of looking stupid.

These steps offer a brisk tour of any article, yet they seldom help us understand or critique its main finding or the arguments supporting that finding. Glancing at the figures shows what the authors measured, but says little about how the data were obtained, analyzed, or why they support the claim, if at all. Rapid skimming can also create an illusion of comprehension. Skimming is fast, so we feel good about making progress, but it can trick us into thinking we understand more than we do about an article, especially its reasoning, which is often the most challenging part of an article to assess. Reasoning frequently hides in the fine print of inferences where we cannot skim.

Like many classroom shortcuts, these "quick look" strategies endure primarily by tradition; there is no formal research supporting this kind of trial-and-error approach. They are simply handy shortcuts passed down to us, and, in turn, they are often what we teach as instructors and research mentors. The goal of this book is to transform reading from a quick look to an active reading critique that effectively and efficiently allows us to analyze any research article's main arguments.

2.5 Introducing the CERIC Method

CERIC – rhymes with *cleric* – stands for Claim, Evidence, Reasoning, Implications, and Context. Together, these five elements map the complete argumentative arc of a scientific research article. We define these elements as follows:

- **Claim** is a plain language, declarative answer to a research question representing new knowledge. A Claim can take many forms, such as a quantifiable relationship between variables, a discovery, or a new methodology. Valid Claims are well supported by Evidence and Reasoning. Articles often feature one primary Claim supported by smaller, secondary Claims (also called subclaims). Claims are usually found in the title, the abstract, and the conclusion section of an article.
- **Evidence** is the pieces of information that are collected, computed, or measured and must be analyzed to generate the Claim. Evidence can be quantitative, qualitative, or both. It can include empirical measurements, theoretical models, computational simulations, analysis methodologies, survey data, interview data, focus group data, and assumptions used to create a body of information that will lead to the Claim. Evidence is specifically those data that are relevant to the study's motivating research question and the resulting Claim of an article. It is often found in the figures, the tables, and the methods or results section of an article.
- **Reasoning** is the logic that directly leads from the Evidence to the Claim. Reasoning can include identifying patterns, statistical analyses, modeling and interpretation, arguments that exclude alternative hypotheses, coding and thematic analysis of qualitative data, and other logical links that support the Claim's viability. Reasoning is easily confused for a study's rationale or motivation (i.e., why the study was conducted in the first place). Rationale and motivation existed before the study began, whereas Reasoning is part of the work reported within the article itself. Reasoning is often found in the results and methods, and the first part of the discussion sections of an article.

To sum up, Claim, Evidence, and Reasoning form the article's core argument.

- **Implications** encompass the forward-looking narrative that emerges from the article's Claim. Implications usually answer some or all of the following questions: How is this result significant beyond the immediate findings? What was learned? What new predictions emerge from this study that require further investigation to refute or confirm the Claim? What are the ethical and social dimensions for consideration and broader discourse? Implications are often found in the discussion and conclusion sections of an article.
- **Context** is the backward-looking narrative that places the work within the broader scope of the field of study, providing the rationale or motivation for the investigation. The Context should identify "the gap" in the current literature that necessitates this study, whether it be conflicting observations or theories, novel phenomena, or limitations in current approaches. A strong Context clearly states the rationale, which explains why this specific study needed to be done and is often found in the introduction and early discussion sections of an article.

To sum up, Implications and Context form the article's future and past aspects that frame the core argument.

Box 2.3 Common Confusion: Alternate CERIC Terms

Many readers confuse the definitions of the CERIC elements with other terms, because our current research-based language does not always reflect common research terms we may have learned before. Here are some examples of synonyms or similar concepts to the CERIC elements that are worth distinguishing:

- **Claim vs. hypothesis:** A hypothesis launches a study, and a Claim concludes it. A hypothesis is introduced early in a scientific investigation and might be the initial "guess" to answer a research question. However, a hypothesis need not be supported by Evidence and Reasoning, which makes it functionally different from a Claim. (See more in Chapter 3.)
- **Evidence vs. data:** Evidence is the relevant slice of the larger data pie that relates directly to the Claim. In other words, Evidence is a subset of the full body of a study's data, relevant only to addressing a particular scientific question. Curating data down to the relevant information reduces the universe of possible information present in a research study and makes it easier to process. (See more in Chapter 4.)
- **Reasoning vs. analysis:** The specific purpose of Reasoning is to connect the Evidence to a Claim. Analysis encompasses a broader set of "work" done in the study, including processes such as data reduction, sample cleaning, or statistical testing. Not all analysis is necessary to support the Claim. (See more in Chapter 5.)
- **Reasoning vs. rationale:** Reasoning is often confused with rationale, which is the original motivation for a research study. Since an article's rationale is based on the existing body of research, it precedes the study and is part of the Context. Reasoning, in contrast, is the logical framework within the article, bridging from Evidence to Claim. (See more in Chapter 5.)
- **Implications vs. future work:** Future work on the specific topic of an article is certainly one kind of implication, but Implications can expand into other areas, such as ethical, policy, and social implications. (See more in Chapter 6).
- **Context vs. introduction:** While Context is often found within the introduction section of an article, it is more than just background information. Context distills the broader research ecosystem to the specific rationale for the current study. It can also include past measurements or theories that are used to validate or contextualize the present study. (See more in Chapter 7.)

If you are still feeling confused about these terms, it might help to have a study group conversation about them.

The list of CERIC elements above might be surprising, since it does not strictly align with the *IMRaD* order of elements we see in a typical research article. As shown in Figure 2.3, CERIC begins with the Claim because the entire investigation of an article pivots on the assertion of new knowledge. Evidence and Reasoning are next because the Claim-Evidence-Reasoning sequence is the core argument of every

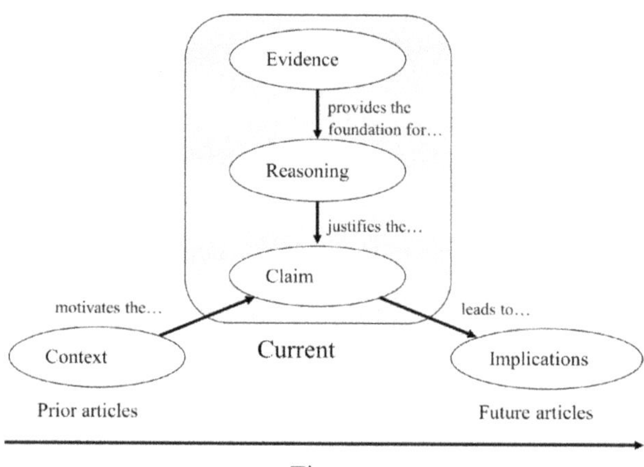

Figure 2.3 Conceptual Diagram of the CERIC Elements in an Article.

research article. Evidence grounds the Claim, and Reasoning backs it. A Claim without Evidence and Reasoning is just an assertion, opinion, or speculation, not something that would stand up to scientific scrutiny. Implications and Context come into play when we consider the scientific ecosystem in which a research investigation occurred. Implications follow naturally from Claim-Evidence-Reasoning as future research directions and new study directions. While Implications are about the future, Context is about the past. Context reflects how the published study was motivated by past work; in particular, what gap in knowledge necessitated the study (also known as the rationale). Context gives us the big starting picture about the whole Claim-Evidence-Reasoning-Implications sequence in the article we are reading.

We deliberately place Context at the end of the CERIC sequence, even though this order departs from the familiar *IMRaD* structure, and many of our former students have asked why it does not come first like the traditional Introduction. The answer is pragmatic and rooted in our classroom experience. When novice critical readers start with an article's introduction, many often become mired in unfamiliar jargon and dense background information – the rhetorical "quicksand" that can swallow attention before anyone ever reaches the authors' main argument. By postponing Context, we help readers engage first with what matters most: the article's Claim, the supporting Evidence, the Reasoning that links the two, and the Implications that follow. Once these elements are clear, readers can circle back to the introduction with a purpose, using it as a resource for the rationale rather than an obstacle.

Remember that *IMRaD* (Introduction → Methods → Results → Discussion) was designed for experts communicating chronologically with other experts. CERIC (Claim → Evidence → Reasoning → Implications → Context) was designed for readers at any level who need to decode expert reports by following the logic of an argument. Chapters 11 and 12 will show you how, once you are comfortable using CERIC, to rearrange its elements to mirror *IMRaD* when you synthesize and present findings from multiple articles.

Box 2.4 Knowledge Check: Question 2.2 Which statement best clarifies the difference between rationale and reasoning?

A. Rationale deals with how the study was funded, whereas reasoning explains the results.
B. Rationale includes the past motivation for conducting a study, whereas reasoning connects the evidence to the claim in the current study.
C. Rationale is part of the implications, whereas reasoning is part of the data.
D. They are functionally the same in a research article and are used interchangeably.

(Check your understanding using the Knowledge Check Key at the end of the book.)

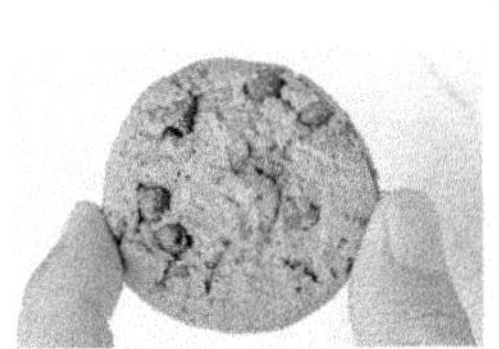
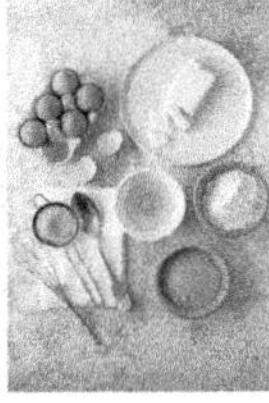

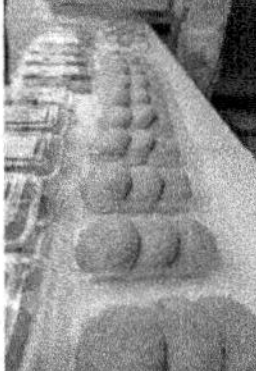

Figure 2.4 Cookie analogy.

2.5.1 Cookie Analogy for the CERIC Elements

Sometimes the best way to introduce a complex framework is through a warm, gooey metaphor – preferably one packed with chocolate chips. For many readers, CERIC can feel abstract the first time they encounter it. So, we offer an analogy that is almost universally relatable: making cookies, as shown in Figure 2.4.

Let's say the **Claim** is, "I made the most delicious chocolate chip cookie." That's a bold statement; one we should be skeptical of. What follows is a familiar scientific instinct – question the claim! How do you know it's the most delicious? Compared to what? Who said so? And what even makes a cookie delicious?

These questions point straight to the CERIC elements. To support the "I made a chocolate chip cookie" portion of the Claim, we begin with **Evidence**. This includes: the raw ingredients – flour, eggs, sugar, chocolate chips – and the tools – bowls, spatulas, oven. Maybe you used almond flour instead of wheat or opted for a solar oven over electric. Every variation in method or material can affect the result, and so the Evidence matters.

Ingredients and tools alone do not bake a cookie; there needs to be a recipe, some **Reasoning**. In what order and how much of the ingredients were mixed? Were measurements precise or casual? Did you bake at 350°F or 375°F? How long? Did the cookies rest for 10 minutes after coming out of the oven? These procedural details, also part of the Evidence are necessary to explain, because the reader, like the skeptical taster, cannot assess the quality of the cookie – or the research – without knowing how it was made.

"I made a cookie" was not the Claim, however; recall that it was "I made the most delicious cookie." That's a high bar. Do you have data to back it up? Did you run a taste test against competing cookies? Or is this a personal opinion wrapped in a scientific-sounding assertion? Here, we need to look for **Reasoning** that supports the argument that we do not just have a cookie but the most delicious cookie. Reasoning could include the criteria for deliciousness, independent cookie tasters' previous reports (hopefully not just the baker), the comparison cookies, and possibly a counterfactual (which factors, if absent, make a truly bad cookie). A strong cookie Claim, like a strong scientific Claim, requires both clear Evidence and sound Reasoning.

Once we have accepted the Claim, **Implications** enter the scene. Now what? Should you eat the cookie? Share it? Sell it? Could the recipe be improved – double the chocolate chips, perhaps? Implications invite broader questions: How might others benefit? Could the cookie's carbon footprint be reduced? This is the realm of future work, societal impact, and follow-up studies, whether you're publishing in *The Journal of Cookie Science* or *Nature Cookies*.

Finally, we need to remember that there is a **Context** to this work. Why did you make these cookies in the first place? Maybe store-bought cookies were disappointing. Maybe this was Grandma's secret recipe. Maybe it is a startup venture. Context explains why you pursued this delicious cookie quest in the first place, identifying the gap in current cookie offerings and providing the rationale for your culinary experiment. If your goal was just to enjoy a snack while grading lab reports, the motivation might be simple. If your aim was to dethrone Oreos as the nation's favorite, that calls for a much deeper dive into market research.

2.6 Finding the CERIC Elements in a Research Article

Recall the *IMRaD* structure of a research article (Figure 1.3). You may notice that the CERIC elements do not follow this publication order. So, we need to follow a **categorical reading** approach to find these elements. Where do we look?

Table 2.1 summarizes where CERIC elements are found in the *IMRaD* research article structure. The abstract is the most CERIC-rich part of the article, potentially collecting all the CERIC elements in one place. This richness is one reason the abstract is essential critical reading for an article, and why even anecdotal advice emphasizes reading the abstract completely. However, we are also likely to find expanded versions of the CERIC elements concentrated in other sections: Evidence in the methods section and in figures and tables; Reasoning in the results and discussion sections, and often in key figures; Implications in the discussion and conclusions sections; and Context in the introduction. A well-written article will have the Claim in the title. By focusing on these sections in a hunt-in-seek process, we can efficiently collect the main CERIC elements of a research study.

The CERIC elements also provide a useful scaffold for taking consistent notes while reading critically, facilitating better understanding, discussions, and comparisons between articles (Bjorn et al., 2022). We will explore some examples of notetaking with CERIC reviews in Chapters 8 and 9, and we will demonstrate how these reviews become essential when the reading starts to pile up for a literature review (Chapter 11).

CERIC Element	...may be **(is likely to be)** found in...
Claim	**Title, Abstract,** Introduction, **Results,** Discussion, **Conclusions**
Evidence	Abstract, Introduction, **Methods**
Reasoning	Abstract, **Methods, Results,** Discussion
Implications	Abstract, **Discussion, Conclusions**
Context	Abstract, **Introduction**

Article Section	...may **(is likely to)** contain...
Title	**Claim**
Abstract	**Claim**, Evidence, Reasoning, Implications, Context
Introduction	Claim, Evidence, **Context**
Methods	**Evidence, Reasoning**
Results	**Claim, Reasoning**
Discussion	Claim, Reasoning, **Implications**
Conclusions	**Claim, Implications**

Table 2.1 Mapping CERIC elements and *IMRaD* sections.

Box 2.5 Knowledge Check: Question 2.3 Where is a researcher most likely to find all five CERIC elements (Claim, Evidence, Reasoning, Implications, and Context) condensed in one section of a journal article?

A. Title
B. Methods
C. Discussion
D. Abstract

(Check your understanding using the Knowledge Check Key at the end of the book.)

2.7 How the CERIC Method Works

Using CERIC as a support can help novice readers feel less overwhelmed by the process and stay more focused and engaged with the article. Indeed, when the Toulmin et al. model is modernized and expanded with a categorial method, like CERIC, we gain a portable and durable strategy that we can use in many contexts, such as coursework, a journal club, internships, research group meetings, a peer-review checklist, and a literature review draft – topics we will return to in Part II of this book.

As scientists, we value learning approaches that are supported by evidence. To that end, we provide a worked example below. It is a CERIC review of Dr. Bjorn's article about the CERIC Method (play on words intended). This example will give you a clearer idea about what each element is, where it is located, and how they function together to form an argument.

- Title: "The CERIC method plus social collaborative annotation improves critical reading of the primary literature in an interdisciplinary graduate course."

- Citation: Bjorn, G. (2024). *Frontiers in Education, 9*, 1257747. https://doi.org/10.3389/feduc.2024.1257747
- Abstract:

  "**Background**: Innovative approaches to graduate education that foster interdisciplinary learning are necessary, given the expansion of interdisciplinary research (IDR) and its ability to explore intricate issues and cutting-edge technology.

  Purpose: This study examines an intervention to develop critical reading skills of the primary literature (CRPL), which are often assumed and unaided by formal instruction in graduate education (GE), yet are crucial for academic success and adapting to new research fields.

  Methods: This study applied mixed methods and a pre-post design to assess the effectiveness of a CRPL intervention among 24 doctoral students from diverse fields engaging in the interdisciplinary field of science policy research. The intervention was a 4-week online course with explicit instruction in a categorical reading approach, the CERIC method (claim, evidence, reasoning, implications, and context), combined with social collaborative annotation (SCA) to facilitate low-stakes, peer-based discourse practice. It examined how participation changed participants' CRPL skills and self-perceptions.

  Results: The intervention significantly improved CRPL, $t(23) = 13.6$, $p < 0.0001$; research self-efficacy, $t(23) = 4.9$, $p < 0.0001$; and reading apprehension, $t(23) = 4.3$, $p < 0.0001$. Qualitative findings corroborated these findings and highlighted the importance of explicit CRPL instruction and the value of reading methods applicable to IDR. These results aligned with sociocultural and social cognitive theories and underscored the role of discourse and social engagement in learning critical reading, which is traditionally viewed as a solitary activity.

  Conclusion: The findings present a valid and innovative model for developing CRPL skills in interdisciplinary GE. This approach provides a model for scaffolding CRPL that can be adapted to IDR contexts more broadly.

  Implications: The study findings call for revising graduate curricula to incorporate explicit CRPL instruction with peer-based discourse, emphasizing integrations in higher education, anywhere students encounter primary literature. The findings advocate for formal and informal adoption of the reviewed methods, offering a significant contribution to interdisciplinary GE pedagogy."

Identifying the Claim: We are looking for a declaration of new knowledge, and we find it up top in the title as:

> The CERIC method plus social collaborative annotation improves critical reading of the primary literature in an interdisciplinary graduate course.

The new knowledge here is that CERIC + SCA improve critical reading skills at the graduate level. In addition, articles often have secondary Claims that build off the main Claim. One appears in the conclusion section of the abstract as:

> The findings present a valid and innovative model for developing CRPL skills in interdisciplinary GE.

This secondary Claim points to CERIC as a model for skills development in graduate education. We would need to read into the main text for more supporting Evidence.

Identifying the Evidence: Next, we are looking for the methods and data that support this Claim. In the Methods part of the abstract, we first find the study design that lays out the conditions of data collection:

> This study applied mixed methods and a pre-post design to assess the effectiveness of a CRPL intervention among 24 doctoral students from diverse fields engaging in the interdisciplinary field of science policy research.

So, this explains the study design, sample size, and population. This would be considered preliminary Evidence (testing a novel idea), as opposed to confirmatory evidence (replicating a result).

Next, we look for information about how they collected the data. Again, the Methods section of the abstract gives us more details:

> The intervention was a 4-week online course with explicit instruction in a categorical reading approach, the CERIC method (claim, evidence, reasoning, implications, and context), combined with social collaborative annotation (SCA) … (). It examined how participation changed participants' CRPL skills and self-perceptions.

This information gives us an overview of the 4-week online intervention study and adds key details about SCA, CRPL skills, and self-perceptions.

Identifying the Reasoning: Reasoning is the logic that points from the Evidence to the Claim. Because this was an interventional study using mixed methods, we expect to see the big-picture approach of deductive reasoning (e.g., using data to support an existing learning theory). This deductive approach probably also includes statistical analysis of pre-post test scores (e.g., mathematical reasoning) plus an analysis of interview data showing practical significance of the findings compared against existing learning theories. The Results section of the abstract is where we find some of the expected Reasoning:

> Pre-post testing indicated quantitative findings significant for improvement in reading comprehension ($t(23) = 13.6$, $p < 0.0001$), research self-efficacy ($t(23) = 4.9$, $p < 0.0001$), and reading apprehension ($t(23) = 4.3$, $p < 0.0001$).

This is mathematical reasoning in the form of t-tests (e.g., $t(23)$) of the pre-post skills tests. We see levels of quantitative significance reported as $p < 0.0001$ values. We also see Reasoning about qualitative data:

> Qualitative findings indicated that novice research readers needed and benefited from explicit instruction in a strategic method in combination with a low-stakes discussion group. Participants indicated that they wished they had learned a CRPL method earlier in their academic careers instead of the trial-and-error process they experienced.

This qualitative information gives us reasoning in terms of the practical significance of the quantitative results, mainly that participants benefitted from explicit instruction in critical reading and many wished they had learned it earlier in their academic careers. Finally, we get the alignment of the findings with existing learning theories:

> These results aligned with sociocultural and social cognitive theories and underscored the role of discourse and social engagement in learning critical reading, which is traditionally viewed as a solitary activity.

This line of reasoning brings in broader alignment with sociocultural and social cognitive theories, which confirms our initial prediction that this study does indeed use deductive reasoning.

Identifying the Implications: The future next steps are found in the Implications section of the abstract:

> The study findings call for revising graduate curricula to incorporate explicit CRPL instruction with peer-based discourse, emphasizing integrations in higher education anywhere students encounter primary literature. The findings advocate for formal and informal adoption of the reviewed methods, offering a significant contribution to interdisciplinary GE pedagogy.

They call for revising graduate education and whenever primary literature is used. One additional implication is in the Conclusion section:

> This approach provides a model for scaffolding CRPL that can be adapted to IDR contexts more broadly.

This is a broader implication about adapting and using this critical reading method in interdisciplinary research. In addition, limitations of study design were discussed in the main text and indicate larger trials in different contexts are needed.

Identifying the Context: The previous research supporting doing the study is found in the abstract's background section:

> Innovative approaches to graduate education that foster interdisciplinary learning are necessary, given the expansion of interdisciplinary research (IDR) and its ability to explore intricate issues and cutting-edge technology.

Graduate education and interdisciplinary research frame the study. Next, the reason for conducting the study appears in the purpose:

> This study examines an intervention to develop critical reading skills of the primary literature (CRPL), which are often assumed and unaided by formal instruction in graduate education (GE) yet are crucial for academic success and adapting to new research fields.

Developing essential, but often unaided, critical reading skills is the rationale or motivation for the study. This rationale, plus the past Context, explain the need to do the study. In addition, the study's evidence logically supports the Claim and is supported by Reasoning. The future Implications are relevant to the Claim and Evidence. Together, these elements form a cohesive argument.

Notice how the CERIC structure has allowed us to capture the essential finding of the study (CERIC improves reading comprehension), the primary argument supporting that finding (reading self-efficacy survey before and after an online course), and the broader scope of the study in both addressing an underlying gap (the need to teach effective critical reading for graduate success) and the further work needed to validate and build on the finding (a more extensive trial, integration into higher education pedagogy). We obtain a comprehensive picture of this article that allows us to both understand its main results and evaluate whether they are robust.

We will see more worked examples of CERIC applied to articles in natural science fields across Chapters 3–7, as we dive into each of the CERIC elements. Additional worked examples of each of the natural science disciplines appear in the Student Supplement. Further guidance for classroom facilitators appears in the Instructor Supplement.

2.8 Chapter Key Takeaways

1. **Arguing from Evidence**: This chapter frames the primary elements of a scientific argument through the conceptual model of Toulmin et al. (1984), which connects evidence to a claim through a warrant or reasoning. It then highlights some of the weaknesses of this model when applied to scientific research, and the misalignment with the traditional *IMRaD* structure (Introduction, Methods, Results, and Discussion).
2. **The value of categorical reading:** This chapter describes the value of the hunt-and-seek approach of categorical reading, and its alignment with the main elements of the Information Processing Theory of Learning: pattern recognition, working memory, long-term memory, and cognitive load. It also reviews the shortcomings of current recommendations for efficient reading which are often anecdotal or unsupported by research.
3. **The CERIC method:** The main focus of this chapter is describing the elements and structure of the CERIC method (Claim, Evidence, Reasoning, Implications, and Context) as a systematic approach to reading and understanding research articles. This method is designed to help readers identify and categorize key elements within a research article, facilitating a more comprehensive and critical understanding of its findings, supporting arguments, and broader scope. The CERIC method addresses the challenges discussed in this chapter with other reading methods and offers a structured approach to engaging with and analyzing scientific literature.

3 Claim

The problem is not when someone is in love, it's when they claim to be in love.
—paraphrasing findings by Horan & Booth-Butterfield (2013)

3.1 Overview

Chapter 3 focuses on the essential element of research articles, the Claim, a declarative response to a research question signifying new knowledge. Claims are plain language, concise statements supported by evidence and reasoning, which makes them distinct from hypotheses, assumptions, or opinions. The Claim is often located in the title, the abstract, and the conclusion section. Depending on the nature of the study, the Claim may fit into one of several different categories, and there may be one or more main Claims and multiple subclaims in a given article. This chapter provides several worked examples to illustrate how to identify the Claim in research articles.

3.2 What Is a Claim?

The fundamental element of a research article is its **Claim**: *a plain language, declarative answer to a research question that presents new knowledge*. Let's pick apart this definition:

- **A Claim should use plain language:** A Claim needs to be clear, concise, and understandable without being embedded in jargon. A good Claim can be communicated to nonexperts in a single sentence. For example, remdemesvir given within the first five days of SARS-COV2 infection reduces morbidity and mortality.
- **A Claim is declarative:** It declares a finding or conclusion that logically arises from research. The Sun is made of hydrogen. Genetic inheritance is encoded in DNA. Folic acid supplementation during early pregnancy reduces neural tube defects. These statements state the findings clearly.
- **A Claim is an answer to a research question:** Claims do not arise spontaneously; they emerge from a gap in our knowledge – the rationale – and requires research to answer it. A clear research question is what separates Claims from basic facts, observations, hunches, assertions, or assumptions. The question, "Why is the sky blue?" leads to research into the interaction of light and matter, which ultimately leads to the Claim,

"because of the Rayleigh scattering of sunlight off of air molecules." Ultimately, the Claim is what the research concludes based on Evidence and Reasoning.

- **A Claim represents new knowledge:** Research moves forward Claim by Claim. A research article establishes a new Claim that is supported by research. Some types of research articles, such as literature reviews or institutional reports, summarize, analyze, or synthesize previously established knowledge (i.e., prior Claims) and are distinct from the primary research articles that generate and support new Claims in the first place.

Box 3.1 Knowledge Check: Question 3.1 Which characteristic best defines a claim in a research article?

A. It must be extensively detailed with technical jargon.
B. It is the same as a hypothesis, only written in the past tense.
C. It is a declarative statement representing new knowledge supported by evidence and reasoning.
D. It is a statement of personal opinion unrelated to the research question.

(Check your understanding using the Knowledge Check Key at the end of the book.)

The phrase "new knowledge" deserves some attention here, as how we define knowledge depends on the research discipline we work in. In the physical sciences, knowledge is often seen as an objective fact. It is a piece of information about how some aspect of the Universe works or provides new insight into nature, like the speed of light or Newton's laws. This knowledge is constructed from measurements obtained through controlled experiments or observation, mathematical models, and logical or statistical analyses. Rarely do the researchers' identities, opinions, or biases factor directly into the research.

In contrast, in the social sciences and any field that deals with human behavior, knowledge can include objective measurements (e.g., numerical data) and subjective insight (e.g., narrative data). Researcher positionality, opinions, and personal experience are all vital building blocks of knowledge when human behavior is the subject of investigation. When evaluating studies across disciplines, it is crucial to recognize these different forms of knowledge when assessing the studies' Claims and what they can be. A good example is the field of physics, which recognizes positivist forms of knowledge (i.e., "objectively verifiable" information like laws and equations). In contrast, physics education considers both a positivist viewpoint and the subjective experiences and behavior people have learning physics.

No matter which kind of knowledge is valid, it is important to emphasize that an empirical Claim is *always* supported by Evidence and Reasoning, which distinguishes Claims from hypotheses – these are always best guesses about the research gap that motivated the study. In addition, a research article can have more than one Claim, although generally, there is one primary Claim with perhaps secondary claims or subclaims, as discussed in Section 3.3.

3.3 Types of Claims

The type of Claim a research article establishes depends on the nature of the study, its research goals, and the findings that emerge. Here are some common empirical Claim types (i.e., based on Evidence) found in the natural sciences' primary literature:

- **Correlation:** This Claim describes a quantifiable relationship between two or more variables. Correlation Claims are associations that do not rise to the level of cause and effect. Examples of Correlation Claims include students' improved performance on tests after writing self-affirmation statements or the observation that the frequency of urban violence is higher in warmer weather.
- **Cause–effect**: This Claim specifically demonstrates that one phenomenon causes another, possibly rising to the level of an empirical law. The Cause–Effect Claim is stronger than the Correlation Claim and requires the highest levels of Evidence and Reasoning to establish that a relationship is causative – that one directly and necessarily leads to the other. Examples of Cause–Effect Claims include validating the greenhouse effect by demonstrating that a glass filled with carbon dioxide is hotter than the one filled with regular air when exposed to sunlight or a double-blind, placebo-controlled randomized control trial that validates the efficacy of a new vaccine to increase immunity to a disease.
- **Confirmation or refutation of a previous claim**: This Claim declares that a new set of Evidence or a new line of Reasoning confirms or refutes a previous Claim. These types of Claims are iterative, building incrementally off prior research. Confirmation Claims are often found in serial articles and may be published by the same research group. An example of a Confirmation Claim is the confirmation of a new particle in a high-energy collision experiment, originally marginally indicated (2-sigma level) but now detected at high significance (10-sigma level). Refutation Claims are often found among competing research groups. An example of a Refutation Claim is the rejection of a preliminary measurement of drug efficacy, initially indicated as significant in a small exploratory sample, that is ruled out as insignificant in a large clinical trial.
- **Description of a phenomenon**: Phenomenon Claims often address observable natural events or objects, human behavior, and social interactions. Some examples of Phenomenon Claims include the spatial distribution of the cosmic microwave background or molecular interactions observed in specific temperature–pressure conditions. Thus, this Claim describes an existing phenomenon's qualitative or quantitative properties, providing further insight into its nature.
- **Development of a new resource, tool, or instrument:** This Claim focuses on research tools and is distinct from Methodological Claims that describe how such tools might be used to conduct the research. Resource Claims focus on mechanical, computational, or organizational entities that enable new approaches to existing research questions. Examples of Resource Claims include the description of a new mass spectrometer or a publicly available compilation of college admissions data.

Box 3.2 Knowledge Check: Question 3.2 Which of the following examples best illustrates a cause–effect claim (and *not* a correlation claim)?

A. Students who attend extra tutoring sessions tend to have higher exam scores.
B. A new chemical reagent was found to dissolve solid ionic salts faster.
C. Exposure to ultraviolet light directly increases mutation rates in bacterial DNA.
D. A dataset analyzing employment rates in different zip codes reveals a positive association between location and job availability.

(Check your understanding using the Knowledge Check Key at the end of the book.)

- **Discovery**: Discovery Claims are often enabled by new technologies, instrumentation, or methodologies that allow the exploration of previously inaccessible areas or research. Examples of Discovery Claims include detecting a new exoplanet with an advanced imaging system or identifying a novel avian flu variant from the genetic sampling of a large bird population. Thus, this Claim reports the discovery of a new object, phenomenon, or process previously undocumented in the literature.
- **Identification of a pattern or relationship (non correlated)**: This Claim describes a qualitative relationship between two or more conditions, factors, situations, living beings, or people. Examples of Pattern or Relationship Claims include aggression between captive chimpanzees when treats are given unfairly or first-year students' feelings of belonging to a science community. Importantly, Pattern/Relationship Claims *never* explain cause and effect.
- **Methodological**: This Claim presents an advance in methodology that enables a new line or area of investigation. Methodological Claims can be highly specific and technical and may require significant background knowledge in the field to understand and apply. Examples of Methodological Claims include a new procedure for analyzing stellar spectra to increase radial velocity measurement precision or a new approach for coding qualitative survey data using artificial intelligence tools.
- **New theory or update to existing theory**: Theory Claims present a new theory or framework, or they add or modify existing theories or frameworks. Examples of Theory Claims include Einstein's theory of general relativity and the theory of plate tectonics that explains how earth's crust is divided into several large, moveable layers, which can interact. In addition, this Claim may draw on mathematical or conceptual formalisms to craft a perspective that explains an entire class of observations of phenomena.

3.4 Finding the Claim

Given its importance, the Claim is often (or should be) the most straightforward CERIC element to find in a research article. A clear research article headlines its Claim right in the title. (Did your future self catch that? It is an important tip for when you are writing a research article.)

3.4.1 Claims in the Title of an Article

One of the most highly cited articles in astronomy by Abbott et al. (2016) is entitled "Observation of Gravitational Waves from a Binary Black Hole Merger." This title reports the Claim clearly and upfront, without ambiguity about what the reader can expect. On the other hand, a title such as "On the Nature of Educational Research" by Soltis (1984) tells us the article's broad topic but does not convey a Claim. Looking deeper into the abstract, this article is a review article and, as such, is not a primary source for which a novel Claim would be expected.

3.4.2 Claims in the Abstract of an Article

The Claim is almost always found in the abstract of an article and can be identified by key phrases or ***sentence builders***, such as: "we show that," "we find," "we reject," "there is," "can be explained by," or "the discovery of." Below, we work a couple more examples of Claims in astronomy abstracts, where we have highlighted the **Claim** as boldface type and the sentence builders as underlined text.

- Title: "**An explanation for the gap in the Gaia HRD for M dwarfs**."
- Citation: MacDonald, J., & Gizis, J. (2018). *Monthly Notices of the Royal Astronomical Society, 480(*2), 1711–1714. https://doi.org/10.1093/mnras/sty1888
- Abstract:

> **We show that the recently discovered narrow gap in the Gaia Hertzsprung-Russell Diagram near $M_G = 10$ can be explained by standard stellar evolution models** and results from a dip in the luminosity function associated with mixing of ^{3}He during merger of envelope and core convection zones that occurs for a narrow range of masses.

This is a clear example of a Cause–Effect Claim, indicated by the phrase "can be explained by," which is a statement of causal relation. This Claim is also hinted at (although not clearly stated) in the title. Let's consider another astronomy example.

- Title: "**GW170817: Observation of gravitational waves from a binary neutron star inspiral**."
- Citation: Abbott, B. P., Abbott, R., Abbott, T., … & Cahillane, C. (2017). *Physical Review Letters, 119*(16), 161101. https://doi.org/10.1103/PhysRevLett.119.161101
- Abstract:

> On August 17, 2017, at 12:41:04 UTC **the Advanced LIGO and Advanced Virgo gravitational-wave detectors made their first observation of a binary neutron star inspiral**. The signal, GW170817, was detected with a combined signal-to-noise ratio of 32.4 and a false-alarm-rate estimate of less than one per 8.0×10^4 years. We infer the component masses of the binary to be between 0.86 and 2.26 $M_\odot$, in agreement with masses of known neutron stars. Restricting the component spins to the range inferred in binary neutron stars, we find the component masses to be in the range 1.17 – 1.60 $M_\odot$, with the total mass of the system $2.74_{-0.01}^{+0.04} M_\odot$. The source was localized within a sky region of 28 deg^2 (90% probability) and had a luminosity distance of 40 – 14 + 8 Mpc, the closest and most precisely localized gravitational-wave signal yet.

> The association with the γ-ray burst GRB 170817A, detected by Fermi-GBM 1.7 s after the coalescence, corroborates the hypothesis of a neutron star merger and **provides the first direct evidence of a link between these mergers and short γ-ray bursts**. Subsequent identification of transient counterparts across the electromagnetic spectrum in the same location further supports the interpretation of this event as a neutron star merger. This unprecedented joint gravitational and electromagnetic observation provides insight into astrophysics, dense matter, gravitation, and cosmology.

Here, we have a case of multiple Claims. The first is a Discovery Claim about the "first observation" of gravitational waves from a merging pair of neutron stars, also indicated in the title. The second is a Correlation Claim that "links" two phenomena: neutron star mergers and short γ-ray bursts. The wording here distinguishes this as a Correlation Claim (i.e., "evidence of a link") versus a Cause–Effect Claim (i.e., one thing directly and necessarily leads to the other). However, the authors make a case for a causal connection in the main text of the article based on the location and timing of the two events, and additional research is needed to confirm that cause-effect case.

3.4.3 Claims throughout an Article

Claims are often repeated in the main text, typically in the Introduction or Analysis sections, and later in the Conclusions/Summary section. For example, in Abbott et al. (2017):

> Section I. Introduction:
> On August 17, 2017, the LIGO-Virgo detector network **observed a gravitational-wave signal from the inspiral of two low-mass compact objects consistent with a binary neutron star (BNS) merger**.
>
> Section III. Detection:
> **... a highly significant detection of a binary neutron star signal in coincidence with the independently observed gamma-ray burst GRB 170817A**.
>
> Section VI. Conclusions:
> In this Letter **we have presented the first detection of gravitational waves from the inspiral of a binary neutron star system**.

We find first the Discovery Claim in the Introduction, followed later by the second Correlation Claim, and finally repeating the Discovery Claim in the Conclusions. By repeating these Claims, the authors establish these as the article's main findings.

Box 3.3 Common Confusion: Claims vs. Subclaims

Confusion over an article's Claim is common for novice readers and even experienced researchers. It can be difficult to separate Claims, subclaims, and findings from a quick skim of an abstract or an article. A research article often has several declarative statements throughout, and determining which is the main Claim (or Claims) can be unclear, particularly for research articles with ambiguous titles or long abstracts. After checking all the conditions given in this chapter – plain language, declarative, answers a research question, represents new knowledge, and follows from Evidence and Reasoning – a question you might pose to yourself to find the main Claim is: **"Would this Claim be sufficient to publish an article?" The answer is: it depends!**

Here's an example. Dr. Burgasser studies low-temperature stars using the technique of spectroscopy, in which starlight is spread out in wavelength, just like sunlight is spread out by raindrops to form a rainbow (yes, he studies star rainbows!). A spectrum allows astronomers to measure all kinds of physical properties of a star, which involves several steps of calibration, analysis, and interpretation. In the end, he might be able to state that a star's spectrum indicates that it is 7.6 ± 2.2 billion years old. Is this a Claim?

In the article "On the Age of the TRAPPIST-1 System" by Burgasser & Mamajek (2017), this value is precisely what is reported for the age of a star called TRAPPIST-1, which has at least seven Earth-sized planets orbiting it – an unusually rich planetary system. Determining the age of TRAPPIST-1 was new knowledge that helped researchers understand how long these planets have existed and whether life had time to evolve on any of the planets, possibly into advanced civilizations we could talk to. Given its importance, the age of TRAPPIST-1 warranted its own publication and was the article's primary Claim.

In subsequent studies of exoplanet systems, Dr. Burgasser and his colleagues have conducted similar analyses for other planet-hosting stars, where the age was not a Claim. For these studies, the research focused primarily on reporting the discovery of the exoplanets (i.e., Discovery Claims) with an assessment of their nature (e.g., size, mass, and habitability generating a Phenomenon Claim). The host star's age is a finding that provides part of the Reasoning used to constrain the size of the planets, assess their orbital stability, or conjecture on the existence of life. For these systems, determining a star's age is insufficient to warrant its own publication and, therefore, is not a Claim.

Whether a declarative statement comprises a Claim or a broader piece of a research puzzle can depend on the scope of the study and its contribution to the broader field of research.

3.5 Multiple Claims, Subclaims, and Findings

The Abbott et al. (2017) example discussed earlier illustrated a case of two Claims of different types in the same article. The presence of multiple Claims is common for articles exploring a dataset (e.g., a catalog of galaxies or a meta-analysis of clinical trial results) or looking at phenomena from multiple perspectives (e.g., a binary neutron star merger in gravitational waves and gamma rays). In addition, there are often secondary Claims or subclaims that are of less importance as compared to the main Claim.

Box 3.4 Knowledge Check: Question 3.3 Which statement correctly distinguishes a main claim from an intermediate finding or subclaim?

A. An intermediate finding is listed only in the introduction, while the main Claim is always in the abstract.
B. A subclaim or intermediate finding never appears in an article's results section.
C. The main Claim addresses the motivating research question and represents publishable new knowledge; subclaims or findings often support it.
D. An intermediate finding is known before the study begins, whereas a main Claim is proven only after the study is concluded.

(Check your understanding using the Knowledge Check Key at the end of the book.)

The Abbott et al. (2017) abstract we worked on previously provides examples of both of these cases:

> The source was localized within a sky region of 28 deg^2 (90% probability) and had a luminosity distance of 40 – 14 + 8 Mpc, **the closest and most precisely localized gravitational-wave signal yet**.

This is a Discovery Claim. It is a "superlative" discovery as indicated by the qualifiers "closest and most precisely." This Claim is independent of and secondary to the main Discovery Claim of detecting a binary neutron star merger, but it nevertheless warrants mention given its exceptional nature.

Some statements that sound like Claims may, in fact, be intermediate findings that build the logical Reasoning leading to the primary Claim. Consider this section of the Abbott et al. (2017) abstract:

> The association with the gamma-ray burst GRB 170817A, detected by Fermi-GBM 1.7s after the coalescence, corroborates the hypothesis of a neutron star merger … ().

This phrase sounds like a Confirmation Claim. However, because the "hypothesis of a neutron star merger" is one of the article's main Claims, this statement is simply a step in the Reasoning that allows the authors to establish their main Claim.

Here's another example from the same abstract of something that sounds like a Claim but is not:

> We find the component masses to be in the range 1.17 – 1.60 $M_\odot$, with the total mass of the system $2.74_{-0.01}^{+0.04} M_\odot$.

Here, the phrase "we find" suggests this to be a Discovery Claim, but again, the mass measurements ultimately support the main Claim of the discovery of a binary neutron star merger. The masses are a finding that should be considered as either Evidence (if the masses emerge directly from the measurements) or Reasoning (if the masses are part of the analytical or logical argument chain) in support of the Claim.

3.6 Worked Example

The astronomy article by Abbott et al. (2016) reported the first detection of gravitational waves with the Laser Interferometer Gravitational-Wave Observatory (LIGO) and led to the awarding of the 2017 Nobel Prize in Physics to Rainer Weiss, Barry C. Barish, and Kip S. Thorne "for decisive contributions to the LIGO detector and the observation of gravitational waves." This is a classic example of a Discovery Claim in action. Let's analyze it.

- Title: "Observation of **Gravitational Waves from a Binary Black Hole Merger**"
- Citation: Abbott, B. P., Abbott, R., Abbott, T., … & Cavalieri, R. (2016). *Physical Review Letters, 116*(6), 061102. https://doi.org/10.1103/PhysRevLett.116.061102
- Abstract: "On September 14, 2015, at 09:50:45 UTC the two detectors of the Laser Interferometer Gravitational-Wave Observatory simultaneously observed a transient gravitational-wave signal. The signal sweeps upwards in frequency from 35 to 250 Hz with a peak gravitational-wave strain of 1.0×10^{-21}. It matches the waveform predicted by general relativity for the inspiral and merger of a pair of black holes and the ringdown of the resulting single black hole. The signal was observed with a matched-filter signal-to-noise ratio of 24 and a false alarm rate estimated to be less than 1 event per 203 000 years, equivalent to a significance greater than 5.1 σ. The source lies at a luminosity distance of 410_{-180}^{+160} Mpc corresponding to a redshift $z = 0.09_{-0.04}^{+0.03}$. In the source frame, the initial black hole masses are $36_{-4}^{+5}\ M_{\odot}$ and $29_{-4}^{+4}\ M_{\odot}$, and the final black hole mass is $62_{-4}^{+4}\ M_{\odot}$, with $3.0_{-0.5}^{+0.5}\ M_{\odot}c^2$ radiated in gravitational waves. All uncertainties define 90% credible intervals. **These observations demonstrate the existence of binary stellar-mass black hole systems. This is the first direct detection of gravitational waves and the first observation of a binary black hole merger.**"

There are two main Claims in this article:

(1) **"This is the first direct detection of gravitational waves and the first observation of a binary black hole merger."** This is a classic Discovery Claim, indicating the first detection of two phenomena, gravitational waves and a binary black hole merger. In fact, one could argue these as being two separate Claims that are tightly linked, as it is possible to detect gravitational waves from systems other than binary black holes (e.g., Abbott et al., 2017). Still, we could only detect a black hole binary merger by detecting their gravitational waves since such systems do not emit light, a detail noted in the article. In any case, this combined discovery is clearly the article's primary Claim.

(2) **"These observations demonstrate the existence of binary stellar-mass black hole systems."** This is also a Discovery Claim, although of secondary importance to detecting gravitational waves. Before this article, there was evidence of individual stellar-mass black holes (most discovered through a technique called microlensing) and stellar-mass black holes in binary systems with another normal star (known as X-ray binaries because the black hole accretes matter from the

other star, and the matter emits X-rays as it falls into the black hole). However, binary black holes do not emit light and can only be detected by the ripples they produce in spacetime as they orbit each other; these are the gravitational waves. Hence, this article makes this secondary Claim, which represents new knowledge linked to a research question (Do binary stellar-mass black hole systems exist?) and based on the Evidence (gravitational wave detection) and Reasoning (consistent with binary stellar-mass black hole merger, which must have been preceded by an unmerged binary system).

It is tempting to interpret the second Claim as an Implication. The existence of a binary stellar-mass black hole merger would seem to imply the existence of binary stellar-mass black hole systems. However, Implications are always forward-looking, such as: What can be done next? Where do we go from here? In this article, the phrase "demonstrate the existence" emerges from the article's Evidence and Reasoning. Thus, it is another Claim in the article.

3.7 Chapter Key Takeaways

1. **A Claim is a declarative statement of new knowledge:** A Claim responds to a research question; is clear, concise, and in plain language; and importantly, is supported by Evidence and Reasoning, making it distinct from opinions, assumptions, assertions, or hypotheses. Claims are also found up front – almost always in the abstract, often in the title, and repeated elsewhere in an article's Introduction and Conclusion sections.
2. **Claim types vary:** Claims can be about associations, relationships, discoveries, methodologies, theories, cause–effect, or descriptions of phenomena. What they share is that they are always supported by Evidence and Reasoning in the study.
3. **Articles can have more than one Claim:** Research articles often contain a primary Claim and secondary Claims or subclaims. The relative importance of these Claims can often be inferred by their placement within the abstract or main text. Intermediate findings that are used to support a primary Claim are better categorized as Evidence or Reasoning.
4. **Claims evolve with research:** Today's Claim is tomorrow's Evidence or Reasoning. What was once a significant Claim in the past might become a standard observation or analysis technique in the present, demonstrating how research progresses.

4 Evidence

The evidence does not support your claim.

—Attorney James Kildunne, June 20, 2025

4.1 Overview

This chapter focuses on the second CERIC element, the Evidence. Evidence is the main "input" of a research study. It can include data, test results, measurements, and observations that provide the foundation for the Claim. Evidence can also include theoretical inputs, such as models and simulations, that may be compared to data. Evidence can be quantitative or qualitative, and it is typically found in the methods and results sections of a research article, as well as in tables and charts. This chapter provides several examples to illustrate how to identify the Evidence in research articles.

4.2 What Is Evidence?

Establishing a Claim requires two important elements: Evidence and Reasoning. Evidence can be considered the inputs for a research study that form the basis of a Claim, while Reasoning (discussed in Chapter 5) is the logical framework that points from the Evidence to establish the Claim. Evidence can take on many different forms depending on the field and specific type of research study. In experimental studies, Evidence is most often data acquired through measurements, test results, or other data collection methods. It includes both quantitative (numerical) and qualitative (descriptive) data. Finally, Evidence can also include theoretical inputs, such as models and simulations, that may be compared to collected or preexisting data to establish a Claim. Here are a few general examples of Evidence:

- *In-vitro* measurements of neurons were made in a controlled laboratory experiment.
- Observations obtained with an astronomical instrument.
- Blood test results from rats given an experimental drug.
- Patients' responses to a validated survey instrument.
- A set of galactic evolution models compared to the measured properties of a sample of galaxies.
- Simulations of protein folding compared to experimental folding data.

The validity of Evidence is a key consideration when assessing the viability of a Claim; hence, a sufficient description of Evidence must include the study design, methodology, samples, and assumptions that go into the data collection or model generation process. If Evidence is described too tersely, it may be insufficient to assess its reliability. Let's consider this example.

- Title: "Testing for the presence of planets with CRIRES infrared radial velocities"
- Citation: Trifonov, T., Reffert, S., Zechmeister, M., Reiners, A., & Quirrenbach, A. (2015). Precise radial velocities of giant stars – VIII. Testing for the presence of planets with CRIRES infrared radial velocities *Astronomy & Astrophysics, 582*, A54. https://doi.org/10.1051/0004-6361/201526196
- Abstract:

  "*Context*. We have been monitoring 373 very bright (V ≤ 6 mag) G and K giants with high precision optical Doppler spectroscopy for more than a decade at Lick Observatory. Our goal was to discover planetary companions around those stars and to better understand planet formation and evolution around intermediate-mass stars. However, in principle, long-term, g-mode nonradial stellar pulsations or rotating stellar features, such as spots, could effectively mimic a planetary signal in the radial velocity data.

  Aims. Our goal is to compare optical and infrared radial velocities for those stars with periodic radial velocity patterns and to test for consistency of their fitted radial velocity semiamplitudes. Thereby, we distinguish processes intrinsic to the star from orbiting companions as reason for the radial velocity periodicity observed in the optical.

  Methods. **Stellar spectra with high spectral resolution have been taken in the H-band with the CRIRES near-infrared spectrograph at ESO's VLT for 20 stars of our Lick survey**. Radial velocities are derived using many deep and stable telluric CO2 lines for precise wavelength calibration.

  Results. We find that the optical and near-infrared radial velocities of the giant stars in our sample are consistent. We present detailed results for eight stars in our sample previously reported to have planets or brown dwarf companions. All eight stars passed the infrared test.

  Conclusions. We conclude that the planet hypothesis provides the best explanation for the periodic radial velocity patterns observed for these giant stars."

In addition to the deeper into the main text, the authors report additional information:

> We have observed our subsample of 20 stars with CRIRES over four semesters in an attempt to provide more evidence in favor of the companion hypothesis. (p. A55)

Together, these statements give a general description of the sample, what instrument was used to collect the data, how long the data was collected, and why that data was collected. However, it raises questions about the quality of the data, such as:

- What stars were actually measured?
- How were they selected?
- What is the quality of the data used to make these measurements?

Box 4.1 Knowledge Check: Question 4.1 How is Evidence best distinguished from Context in a research article?

A. Any numerical data, regardless of its purpose, is always Evidence.
B. Evidence directly supports the Claim, whereas Context situates or motivates the study.
C. Context always appears in the Methods section, while Evidence appears only in the Introduction.
D. Only newly acquired measurements can be considered Evidence, while archival data is Context.

(Check your understanding using the Knowledge Check Key at the end of the book.)

- With what precision were the velocities measured?
- What calibration steps were used to verify the accuracy of the measurements?

Without sufficient description of the data collection and calibration process, we cannot be certain that the Evidence is of sufficient quality to support the Claim.

In addition to requiring sufficient detail, Evidence is distinct from past knowledge or data collected in a different study, which provides Context and is more often a motivating or situating factor for the research (see Chapter 7). Nonetheless, preexisting data may constitute Evidence if it is reexamined using a new theoretical framework or reanalyzed using a new technique or algorithm that leads to a new Claim. For example, a researcher could draw on a sample of archival data to put new measurements into historical perspective. In this case, the archival data serves as Context, not Evidence. Conversely, if the archival data is reanalyzed to improve its precision or accuracy, then these data become new Evidence for a Claim.

Finally, Evidence is not limited to experimental measurements or observations. A set of theoretical models, when compared to new or preexisting data, would constitute Evidence, particularly if the comparison leads directly to a new Claim. Other theoretical or computational inputs, including simulations, can be Evidence if their inclusion is essential to establishing a Claim.

4.3 Types of Evidence

The types of Evidence present in a research article depend largely on the Claim that is being established. There are typically many forms of Evidence that can establish a single Claim. Here are some common Evidence types found in primary sources.

- **Archival data**: Archival data are past measurements or observations that are reevaluated in a manner that establishes a new Claim. Research studies that apply a new analysis technique to improve precision or accuracy, or synthesize past data from various sources in a novel manner, have archival data as their Evidence. In contrast, if archival data serve merely to motivate or contextualize other Evidence, then these data provide Context for the study.

- **Experimental measurements**: The most common form of Evidence is measurements obtained in an experimental setting. Experimental measurements are often obtained in a controlled environment, with few (often one) "free parameters" allowed to vary in that environment. Measurements are characterized by both precision (the statistical uncertainties of the measurements) and accuracy (how well the measurements reflect the "true" nature of a system). These qualities, in turn, are established by the experimental procedures used to make the measurements, including repeatability and calibration. Measurements are generally quantitative, but can also include qualitative information, such as the extent of morbidity in a medical study, or a set of classification categories.
- **Models**: A model is the mathematical representation of a theory and provides a means of directly comparing the predictions of a theory with empirical data. When the inclusion of models is necessary to establish a Claim, the models form part of a research study's Evidence. The method by which a set of models is constructed, including the assumptions that go into the models, is essential for an adequate description of Evidence to ensure that incompleteness or bias are not responsible for the Claim. Models and samples are similar in these ways.
- **Observations**: Observations are distinct from experimental measurements in that they are not obtained in a controlled environment; rather, they are collected from a system or event as it appears at a particular time or in uncontrolled conditions. Like measurements, observations are characterized by their precision and accuracy, although establishing these in the case of nonrepeatable observations (e.g., a supernova or a particular earthquake) requires care. Observations can also be quantitative (the brightness of a supernova) or qualitative (the color of a reagent).
- **Processes and procedures**: In studies in which the Claim focuses on a new resource, tool, or instrument, part of the Evidence may be devoted to describing the process or procedure in which the tool or instrument was designed or constructed. This thinking also applies to Methodological Claims. Examples include a description of the steps involved in preparing a sample for analysis or the design specifications and calibration procedures for a new measurement device.
- **Sample(s):** A key consideration of an experimental, observational, or survey study is the sample being measured, whether it is a collection of stars or a group of patients. An accurate description of a sample is necessary to assess whether the resulting measurements could be biased by an incomplete or skewed sample. Descriptions of how the sample was constructed and how inherent biases were addressed are essential components of Evidence, similar to how models must be described.
- **Simulations**: For a complex system, a model may take the form of a simulation, a computational structure that aims to recreate the measurable components of that system. For example, a binary star system can be simply modeled analytically using Kepler's Laws or orbital motion, whereas a dense cluster of mutually interacting stars may require a more complex simulation. As for models, the methodology and assumptions that go into a set of simulations must be included for a complete description of Evidence.
- **Survey data**: For studies in which experimental data is collected from human subjects, surveys may be a key component of the Evidence. These differ from measurements and observations in that they are often self-reported and depend on an

individual person's perceptions of what is being measured. Survey data can be quantitative, often in the form of Likert scales (e.g., "rate your pain on a scale of 1 to 10 ..."), whereas qualitative survey data in the form of personal narratives or interviews may contain more depth, nuance, and insight. There are well-established methods for conducting analysis on purely qualitative and mixed quantitative and qualitative ("mixed methods") survey data.

- **Theoretical and conceptual frameworks**: Researchers distinguish between conceptual and theoretical frameworks, each offering a complementary vantage point for research design and interpretation. A conceptual framework organizes empirical regularities; in gas chemistry, this role belongs to the suite of gas laws – Boyle's, Charles's, Gay Lussac's, Avogadro's, and the combined or ideal-gas law – which together map the predictable covariation of pressure, volume, temperature, and amount of gas. These relationships provide an operational schema that chemists use to plan experiments, generate quantitative predictions, and interpret results. The kinetic-molecular theory (KMT), sometimes called the kinetic theory or ideal-gas model, supplies the corresponding theoretical framework. KMT begins with first-principles postulates: gas molecules behave as point particles in constant, random, elastic motion; intermolecular forces are negligible; and collisions with container walls transfer momentum while conserving energy. From these microscopic assumptions, one derives the macroscopic gas laws listed above (Atkins & de Paula, 2023; Maeng & Bell, 2013). In short, the conceptual framework (gas laws) summarizes what is observed, whereas the theoretical framework (KMT) explains why those regularities arise, jointly guiding hypothesis formation, study design, and data interpretation in research on gas behavior.

Box 4.2 Common Confusion: Evidence versus Data

In every study, data comprise the full set of measurements, observations, or recordings the researchers collect – the whole data "pie" that includes everything from sensor readings to interview transcripts. In contrast, Evidence is the curated "slice" of that data pie that speaks directly to the study's research question, and therefore, underpins the article's main Claim. Another way to think of Evidence is as *signal to the Claim*, and data as *all signals plus background*. This view can help us appreciate why not every datum earns equal billing in the Results section. Researchers often handle this distinction between Evidence and data pragmatically, such as:

- **Main text:** Present only the data points that answer the focal research question (the evidence) and shows how it supports the Claim.
- **Supplementary files:** Archive of the complete data set so peers can reconstruct and replicate the analysis, explore alternative interpretations, or reuse the material in meta-studies.

Takeaway: Evidence appears up front while the rest of the data wait in the wings because publication space is limited, and argumentation depends on highlighting only the subset of data that moves the scientific story forward toward the Claim.

4.4 Distinguishing Evidence from Other Sources of Information

Scientific articles often overflow with numbers, images, and references, yet only a subset of that material counts as Evidence in the CERIC sense. Evidence is limited to those data that actively do the argumentative work of supporting the article's Claim. Everything else – however interesting or meticulously presented – plays a different rhetorical role. The following guidelines help us sort one category from another.

4.4.1 Does This Datum Carry the Weight of the Claim?

A study may cite stellar coordinates, reference genome sequences, or list boiling points of solvents. These pieces of information become Evidence only if the authors must rely on them to persuade the reader of the Claim. If the coordinates of a star simply orient the reader to a patch of sky, they form part of the Context. If, instead, the coordinates demonstrate that the star lies inside a tightly bound cluster – thereby substantiating a gravitational-binding Claim – they graduate to Evidence.

4.4.2 Treat Archival or "Borrowed" Data with the Same Test

Researchers frequently import measurements from public repositories or prior publications. Those legacy data serve as Evidence when the new article's Claim fails without them. If, however, the prior measurements merely frame a new observation historically, they belong in Context. In practice, we can annotate every dataset in the methods and results sections and ask whether or not the article's main Claim would collapse logically if that dataset disappeared. If it does, then that is Evidence.

4.4.3 Locate the Boundary between Evidence and Reasoning

Calibration, cleaning, and validation procedures often confuse readers because they sit at the interface of collecting data and thinking with data. A pair of simple rules is useful:

- **Rule 1:** If the paper's main advance is a new calibration that unlocks the discovery, treat the raw data as Evidence and the calibration work as Reasoning – the logic that links data to the Claim.
- **Rule 2:** If the calibration is already standard and simply prepares data for later tests, count both the raw and calibrated data as Evidence; the statistical analyses that follow constitute the Reasoning.

Let's apply these rules and consider how they play out in a set of related studies. Here is an example of the rules applied to some paleoclimatology ice core data, examining isotopes of argon (Ar), nitrogen (N), and oxygen (O).

- **Case 1:** Orsi et al. (2014) developed a new high-resolution temperature reconstruction that combines $\delta^{15}N$ and $\delta^{40}Ar/^{36}Ar$ with firn-diffusion modeling. The innovation is the calibration itself, which refines the speed and magnitude of the Dansgaard–Oeschger event 8 warming. **Evidence** is the original gas-isotope profiles extracted from the ice. **Reasoning** is the newly devised calibration procedure – forward-model tests and sensitivity analyses – that transforms those profiles into temperature and underpins the Claim that the Dansgaard–Oeschger event 8 warmed the Earth ~9.4°C in ~20 years.
- **Case 2:** Johnsen et al. (1997) provided the standard spatial slope of $\approx 0.67‰°C^{-1}$ that linked $\delta^{18}O$ values to surface temperature. A later study (Capron et al., 2021) applied that routine calibration to new EastGRIP $\delta^{18}O$ data, then ran changepoint statistics to show Greenland warmed ≈9°C–12°C in a few decades during Dansgaard–Oeschger event 8. **Evidence** is the raw $\delta^{18}O$ series plus the calibrated temperature curve, because the conversion step is conventional. **Reasoning** is the statistical changepoint test that links the Evidence to the Claim of abrupt warming.

Jargon aside, these paired studies show the rules in action. When the conversion from proxy to temperature is standard, it travels with the data as part of the Evidence (Case 1). When the conversion method is the paper's advance, that procedure becomes the logical bridge – Reasoning that links Evidence to Claim (Case 2). The key difference is that when conventions in a field render certain calibration routines as routine processes or procedures, those merge into the data-collection narrative that produces the initial Evidence. Only when the authors need to defend the calibration itself as part of the persuasive arc do those steps migrate into Reasoning.

4.4.4 Watch for Figures that Mix Categories

Many research figures intertwine data (Evidence) with statistical overlays or model fits (Reasoning). It can be very challenging to separate them. We can practice "de-layering" such visuals to try to isolate each element. Here are some tips:

(1) Isolate the plotted measurements.
(2) Identify the analytical overlay.
(3) Map each component to its CERIC element.

It may be necessary to check the article's Supplemental data to get the figure's values. An important boundary with de-layering is that it can become time-consuming, so it is important to focus on only the most important figures.

In summary, by keeping the Claim firmly in view and interrogating every data point for its argumentative necessity, readers develop a sharper eye for the architecture of scientific persuasion. Such precision equips us to read, teach, and ultimately produce research that makes its logical scaffolding transparent (Toulmin, Rieke, & Janik, 1984). Finally, this precision of thinking about Evidence lays the foundation for training the mind in more advanced evidentiary skills, like detecting data that should be there but is not (omissions) and when data might be falsified (cheating).

Box 4.3 Knowledge Check: Question 4.2 Which scenario best illustrates archival data being used as Evidence rather than merely as Context?

A. Researchers republish weather data from a satellite without any new interpretation.
B. A team references a past article's population statistics only to explain why they repeated the same analysis.
C. Scientists apply a new calibration to older telescope images, achieving more precise star distances that support a new Claim.
D. An author briefly cites historical literacy rates in the introduction to show how reading habits have changed since 1950.

(Check your understanding using the Knowledge Check Key at the end of the book.)

Box 4.4 Knowledge Check: Question 4.3 Which statement correctly distinguishes Evidence from Reasoning in a research article?

A. Reasoning refers strictly to textual citations, while Evidence refers to any numeric data.
B. Evidence is data or models supporting the Claim, and Reasoning is the logical process connecting that data to the Claim.
C. Reasoning is optional in specific disciplines, whereas Evidence is mandatory in all research.
D. Both Evidence and Reasoning only appear in the Discussion section of an article.

(Check your understanding using the Knowledge Check Key at the end of the book.)

4.5 Finding the Evidence

In most articles, the Evidence is largely found in the "Methods" (sometimes called "Methodology") and "Results" sections of an article, the "M" and "R" in *IMRaD*. For studies that include experimental or observational data, this section will typically include description of the sample and measurement procedures, including aspects of data calibration. For studies that include theoretical models or simulations, these may be included in the Methodology or the subsequent Results section (sometimes labeled as "analysis"), and typically include description of the modeling approach and underlying assumptions. Sentence builders that signal Evidence in the text are often centered around active verbs such as "we measured," "we observed," "we modeled," or "we calculated."

Evidence is also commonly found in an article's Tables and Figures. For example, an article may include a table summarizing the (previously measured) properties of an experimental sample, or specifics of the acquired data, such as observation date or measurements and their uncertainties. An article may also use figures to visualize the measurements or outline the main steps for data calibration.

Box 4.5 Better Practice: The Ethics of Evidence Sharing

Why share? In one word: transparency. Open-data policies let outside teams rerun the analyses that support "headline" claims – whether in oncology trials or room-temperature superconductors (Nosek et al., 2015). Successful replications reinforce confidence; failed ones surface errors or misconduct, protecting science's self-correcting engine (Allison et al., 2016).

- **Bonus payoff:** Archived datasets can spawn entirely new findings. Deep-learning re-scans of Kepler files revealed previously missed exoplanets. (Shallue & Vanderburg, 2018), and NASA's Space Cloud Watch project (NASA, 2025) is crowdsourcing noctilucent-cloud images for future mesospheric studies, crediting citizen scientists along the way.
- **Real-world frictions:** There are so many of these that we will suggest only some of the major sources. Petabyte storage is costly. Human-subject files demand deidentification and sometimes re-consent. Industry partners guard intellectual property. Researchers need brief embargoes to finish primary analyses.
- **A workable standard:** Funding agencies now promote the FAIR principles – Findable, Accessible, Interoperable, Reusable – and encourage tiered access calibrated to data sensitivity (Wilkinson et al., 2016).

Takeaway: Transparency is not all-or-nothing. Responsible Evidence sharing means balancing scientific benefit against ethical, legal, and logistical constraints – a judgment call every researcher must learn to make.

In some high-impact journals (those with the highest numbers of submissions and citations), Evidence may be detailed more fully in a Supplementary section or an Appendix following the main article, often linked separately on the journal's website. This practice allows readers to focus on the Context, Claims, and Implications of a study in a shorter format, while the more knowledgeable (or skeptical) reader can examine the experimental setup, assumptions or models used, and logical development of the Claim in greater detail.

4.6 Worked Example

In biology, Long et al. (2020) were working in the early days of the COVID-19 pandemic. Medical practitioners did not know why some people exposed to the COVID virus experienced significant and even life-threatening symptoms, including severe respiratory illness, fever, and cough, while others showed mild or no symptoms whatsoever. Understanding the variance in response to virus exposure was an important step in planning treatment for those infected, as well as predicting the spread of the virus among so-called silent spreaders. This study found important differences between asymptomatic and symptomatic patients, including that the former shed virus longer and had a weaker immune response than the latter. The study focused on a group of people who were exposed to the SARS-CoV-2 virus but did not exhibit any clinical symptoms related to infection. Let's examine the Evidence, highlighted in **bold**, in the following example. We also underline relevant sentence builders.

- Title: "Clinical and immunological assessment of asymptomatic SARS-CoV-2 infections"
- Citation: Long, Q. X., Tang, X. J., Shi, Q. L., … & Huang, A. L. (2020). *Nature Medicine, 26*(8), 1200–1204. https://doi.org/10.1038/s41591-020-0965-6
- Abstract:
  "The clinical features and immune responses of asymptomatic individuals infected with severe acute respiratory syndrome coronavirus 2 (SARS-CoV-2) have not been well described. **We studied 37 asymptomatic individuals in the Wanzhou District who were diagnosed with RT–PCR-confirmed SARS-CoV-2 infections but without any relevant clinical symptoms in the preceding 14 d and during hospitalization. Asymptomatic individuals were admitted to the government-designated Wanzhou People's Hospital for centralized isolation in accordance with policy. The median duration of viral shedding in the asymptomatic group was 19 d** (interquartile range (IQR), 15–26 d). The asymptomatic group had a significantly longer duration of **viral shedding** than the symptomatic group (log-rank P = 0.028). **The virus-specific IgG levels** in the asymptomatic group (median S/CO, 3.4; IQR, 1.6–10.7) were significantly lower (P = 0.005) relative to the symptomatic group (median S/CO, 20.5; IQR, 5.8–38.2) in the acute phase. Of asymptomatic individuals, 93.3% (28/30) and 81.1% (30/37) had reduction in **IgG and neutralizing antibody levels**, respectively, during the early convalescent phase, as compared to 96.8% (30/31) and 62.2% (23/37) of symptomatic patients. Forty percent of asymptomatic individuals became **seronegative** and 12.9% of the symptomatic group became **negative for IgG in the early convalescent phase**. In addition, asymptomatic individuals exhibited lower levels of **18 pro- and anti-inflammatory cytokines**. These data suggest that asymptomatic individuals had a weaker immune response to SARS-CoV-2 infection. The reduction in IgG and neutralizing antibody levels in the early convalescent phase might have implications for immunity strategy and serological surveys."

The main text then goes on to detail some of the tests undertaken to track the infection, immune response, and symptoms of this group:

> **A complete blood count, blood biochemistry, coagulation function, liver and renal function and infection biomarkers were measured** upon admission (Supplementary Table 2) to monitor the potential disease progression.

The study collected four categories of data to capture both infection severity and immune response:

- Chest CT scans to visualize lung scarring characteristic of COVID-19.
- Serial nasal swabs to chart viral load over time.

- Serum immunoglobulins (IgG and IgM) were measured during the acute phase and again eight weeks post-discharge.
- A panel of cytokines and chemokines to monitor systemic inflammation at the same two time-points.

Because the authors aimed to test a Correlation Claim – whether asymptomatic and symptomatic patients differ immunologically – the sampling design was critical. They enrolled one cohort of asymptomatic, PCR-confirmed cases and an equal-sized symptomatic cohort that was sex-, age-, and comorbidity-matched, thereby limiting demographic bias. Both groups were compared with a third, similarly matched, uninfected control group to detect possible false-positive signals. All participants came from the same Wanzhou District and were studied contemporaneously, controlling for healthcare-access differences and circulating viral variants. Thus, the study's Evidence comprises the careful construction of comparable samples and multiple laboratory measurements.

The technical details – such as patient demographics, the complete list of tests, and assay protocols – appear in the article's online Methods supplement. Placing these specifics outside the main narrative allows the text to focus on the other CERIC elements (Claim, Reasoning, and Implications) while still providing readers with all the information needed to evaluate the Evidence in depth. When we examined the Evidence in the abstract, main text, and supplemental Methods, we conclude that the sampling design, data collection methods, and data collected are sufficient to make their Correlation Claim.

4.7 Chapter Key Takeaways

1. **Evidence encompasses the inputs developed in a research study** that form the basis of a Claim, and can include quantitative (numerical) and qualitative (descriptive) data, theoretical inputs, or descriptions of samples and processes that are specifically relevant to the Claim.
2. **Specific forms of Evidence** include experimental measurements, observations, survey data, sample information, archival data, models, simulations, processes & procedures, and theoretical frameworks.
3. **Evidence is a subset of the entirety of data** that could, in principle, be brought to bear on a particular scientific study. It includes only those data inputs that directly support the Claim, *not* the entire universe of collected data.
4. **Evidence may encompass some degree of processing or analysis on "raw" data** for calibration and validation, depending on whether the analysis procedures are standard to the field. If the validity of the Evidence is central to establishing the Claim, or a nonstandard analysis is applied, these procedures may be considered as Reasoning.
5. **Evidence is typically found in the Methods section of an article**, and occasionally in the Results (or analysis) section. Evidence can also be summarized in an article's Tables and Figures, and in some journals may be detailed in a Supplementary section following the main article text. Sentence builders that signal Evidence often use active verbs such as "we measured," "we observed," "we modeled," "we studied," or "we calculated."

5 Reasoning

People are generally better persuaded by the reasons which they have themselves discovered than by those which have come into the mind of others.

—Blaise Pascal, Pensées I.10, 1670

5.1 Overview

Chapter 5 explores the CERIC element of Reasoning, the logical chain of argumentation that connects Evidence to Claim. This chapter begins by defining Reasoning, its significance in scientific research, and the various categories of reasoning and strategies used across disciplines. Through real-world examples in fields like healthcare diagnostics and climate science, this chapter clarifies deductive, inductive, and abductive approaches, emphasizing their roles in different research contexts. This chapter also discusses how Evidence, models, accuracy, precision, and flexibility interact with Reasoning, underscoring the importance of transparent and well-structured logic in scientific discourse. It concludes with examples of Reasoning styles from diverse research articles, reinforcing logical and analytical thinking in advancing scientific knowledge.

5.2 What Is Reasoning?

Reasoning is the process by which we justify our ideas, decisions, or beliefs. Sound reasoning involves linking past knowledge, evidence, observation, or data to conclusions through a structured and logical process of argumentation. At its core, reasoning seeks to make findings, conclusions, or decisions understandable and acceptable to others by offering a robust argument that is clear, relevant, and sufficient for the context and can withstand scrutiny, challenges, and counterarguments.

Let's consider the following example: We want to persuade a friend who's skeptical about your belief that exercise improves mental health. How might we use reasonable arguments to do this?

We begin by noting that there's a substantial body of scientific evidence supporting this idea. For instance, a study of 1.2 million people, published in *Lancet*

Psychiatry by Checkroud et al. (2018), found that individuals who exercise regularly report fewer days of poor mental health compared to those who do not. Note that by stating the source of the research, both the citation and the journal it was published in, lends credibility to this finding, as compared to "something I read online."

Next, you might explain the causative reasons why exercise reduces symptoms of depression and anxiety by triggering the release of endorphins – the natural "feel-good" chemicals in the brain – and lowering stress hormones, such as cortisol. Exercise also improves blood flow to the brain, which enhances focus and emotional regulation.

To strengthen your point, you could mention that organizations such as the American Psychological Association and the Centers for Disease Control recommend regular exercise as a vital part of mental health care.

At the same time, it may be helpful to acknowledge that exercise might not work equally well for everyone. If someone is experiencing a severe mental health challenge, therapy or medication may still be necessary alongside physical activity.

By framing physical exercise as a simple, accessible tool that helps many people feel better – but not presenting it as a cure-all – you hopefully have offered a balanced and persuasive argument that convinces your friend.

5.2.1 How Reasoning Is Used in Scientific Research

Reasoning is a cornerstone of scientific inquiry. **Scientific reasoning** is the systematic process of constructing arguments where Claims are justified by Evidence and Reasons, and where reasons are linked to the Claims through explanations of why the Evidence supports the conclusion (Toulmin, Rieke, & Janik, 1984). Thus, Reasoning in scientific arguments must draw directly from the relevant Evidence and clearly show why that Evidence substantiates the Claim.

Understanding the chains of Reasoning is essential for evaluating their quality. When researchers present new scientific Claims, their primary goal is often to persuade others to reconsider or revise their beliefs. This persuasion, however, cannot rest on assumptions or unfounded assertions. Researchers must offer clear and well-justified reasons to support their Claims, and these reasons need to be scrutinized to assess their strength, relevance, and significance.

Even after a rigorous defense, an argument may still fail to persuade its intended audience if it lacks sufficient clarity or rigor. Objections to Reasoning can then prompt revisions or refinements to the original assertions. This iterative process, where arguments evolve through a sequence of ever-refined Reasoning, is central to scientific discourse and is one of the most challenging skills to learn as a scientist in training. When people speak vaguely of reading or thinking "like a scientist," this complex skill is often what they mean.

Additionally, the purposes of scientific arguments vary. In some cases, researchers will propose a Claim and use Reasoning-based arguments to defend it. In other cases, they may begin with questions or problems for which answers are not immediately apparent. Here, Reasoning serves as a tool for exploration and discovery. This duality highlights two key functions of scientific Reasoning: **advocacy**, where

Reasoning is used to support an established Claim, and **inquiry**, where Reasoning is used to uncover discoveries. Understanding this distinction is vital for appreciating the diverse roles Reasoning plays in advancing scientific knowledge.

5.2.2 Typical Forms of Reasoning

Three main categories of Reasoning are typically found in research articles: deductive, inductive, and abductive reasoning. The arguments connecting Evidence to Claim may comprise one or more of these categories, depending on the nature of the research and the standards of a given field.

Deductive Reasoning begins with a general theory or hypothesis, and tests its validity through specific observations or experiments. It is the most direct chain of logical Reasoning. The following syllogism illustrates how it works:

All mammals have lungs. All dolphins are mammals. Therefore, all dolphins have lungs.

This type of Reasoning is commonly used in research to validate or refute theoretical frameworks. A research example is a clinical trial testing the hypothesis that a new drug reduces blood pressure; this is a form of deductive Reasoning because the researchers are evaluating whether the experimental outcomes align with the initial theory (Diaz Quezada & Sepúlveda Albornoz, 2023).

Inductive Reasoning works the opposite way and draws general conclusions from specific observations or experimental results. For example:

The sun has risen in the east every day in recorded history. Therefore, the sun will rise in the east every day, continuing to do so into the future.

Researchers make inductive arguments by identifying patterns or trends in their data, then apply these patterns to broader contexts, often to make predictions. An example of this in a research study is observing that a specific teaching method consistently improves student performance in an individual classroom, then concluding that it is likely to be effective in other classrooms teaching the same topic (Rost & Knuuttila, 2022).

Abductive Reasoning is used to infer the most likely explanation for a set of evidence, or **inference to the best explanation**. This approach is particularly valuable in contexts where the evidence is incomplete or the cause of an observed phenomenon is uncertain. Abductive arguments are frequently used in medical diagnostics. For example:

- **Observed symptoms:** A patient presents with fever, productive cough, and shortness of breath.
- **Abductive Reasoning:** These symptoms could indicate pneumonia, bronchitis, or a viral infection. However, chest X-ray findings and an elevated white blood cell count suggest pneumonia is the most likely cause.
- **Conclusion:** The clinician infers pneumonia as the most probable diagnosis and initiates treatment accordingly.

With pneumonia as the most likely explanation, the clinician would initiate treatment while monitoring for new symptoms or signs of treatment failure. If either of those occurred, the other diagnoses would be considered more likely.

5.2.3 Abduction and Bayesian Probability

When abductive Reasoning is combined with **Bayesian probability**, the approach becomes quantitative. Bayesian probability combines prior knowledge with new observations or data to infer an updated (or "posterior") probability (Macagno & Rapanta, 2019). Abduction then selects the hypothesis with the highest posterior probability as the **inference of the best explanation**, while acknowledging uncertainty – unlike deduction (proof) or induction (broad generalization).

Bayes' theorem (Bayes & Price, 1763) provides the mechanism for updating our prior knowledge to posterior probability based on new evidence. If you have taken a statistics course, you may recall this theorem:

$$P(A|B) = \frac{P(B|A) \times P(A)}{P(B)}.$$

In this equation:

- **A** = hypothesis (e.g., the patient has pneumonia).
- **B** = evidence (e.g., symptoms and lab results).
- **P(A)** = prior probability or baseline prevalence (e.g., how common pneumonia is in the general population).
- **P(B|A)** = likelihood of the evidence given the hypothesis (e.g., probability of the symptoms or lab results for someone with pneumonia).
- **P(B)** = marginal probability or the overall chance of seeing the evidence for any reason (e.g., how common the symptoms are generally).
- **P(A|B)** = posterior probability or how likely the diagnosis is *after* seeing the evidence (e.g., the likelihood of the diagnosis of pneumonia).

The inclusion of uncertainty in the decision-making process makes abduction distinct from deduction (proof) or induction (broad generalization). Toulmin et al. (1984) described this process as a practical strategy for dealing with uncertainty:

> When we are faced with a set of puzzling facts, we look for the explanation that, if correct, would best make sense of them. (p. 149)

It is especially common in medical science, theoretical modeling, and diagnostic Reasoning, where direct experimentation is limited.

Let's see how to apply Bayesian reasoning to diagnose a disease, using the following *hypothetical* clinical case study.

Case presentation: A 75-year-old patient presents with fever, productive cough, and shortness of breath. A chest X-ray shows patchy dark areas (consolidation), and the white blood cell (WBC) count is high, indicating an infection. But what kind of infection?

Plausible diagnoses are:

1. X: Bacterial pneumonia
2. Y: Bacterial bronchitis
3. Z: Viral infection

Step 1. Set priors, P(A):

Assume seasonal prevalence (P) in this age group.

1. P(X) = 0.30
2. P(X) = 0.30
3. P(Y) = 0.20

Step 2. Add likelihoods, P(B|A):

From published studies:

1. 80% of pneumonias show consolidation + leukocytosis → P(B|X) = 0.80
2. 40% of bronchitis cases → P(B|Y) = 0.40
3. 10% of viral infections → P(B|Z) = 0.10

We see that the Evidence pattern of consolidation plus high WBC is most common in pneumonia (0.80), less common in bronchitis (0.40), and uncommon in viral illness (0.10).

Step 3. Compute the evidence term or marginal probability, P(B):

$$P(B) = 0.80(0.30) + 0.40(0.20) + 0.10(0.50) = 0.37$$

This represents the probability of experiencing symptoms (B) for any type of disease.

Step 4. Update each disease hypothesis with the evidence term to calculate the posterior probability:

Table 5.1 Bayesian updating of diagnostic hypotheses. Each prior probability is multiplied by the likelihood of observing the evidence, then normalized by the marginal probability to yield the posterior probability (i.e., case likelihood) for pneumonia, bronchitis, and viral infection.

Diagnosis	(Prior prob. × Likelihood)	/Marginal prob.	= Case likelihood
Pneumonia	(0.80 × 0.30)	0.37	= 0.65 (65%)
Bronchitis	(0.40 × 0.20)	0.37	= 0.22 (22%)
Viral Infection	(0.10 × 0.50)	0.37	= 0.14 (14%)

Step 5. Interpretation: A 65% case likelihood (i.e., posterior probability) makes pneumonia the leading explanation, so first-line antibiotics are justified. Bronchitis (22%) and a viral illness (14%) remain backup possibilities if the patient fails to improve. Understanding the abductive plus probabilistic reasoning behind the numbers helps readers evaluate and critique whether the inputs – the priors and likelihoods – best explain the clinical context.

To gain more experience with this type of logical Reasoning, we recommend experimenting by changing one prior, for example, increasing the prevalence of viral infections during flu season. Then recalculate the posteriors. You will see immediately how assumptions shape conclusions, an insight applicable to any research article.

In reality, physicians are rarely able to make these calculations manually in a busy clinical setting. Instead, they rely on probabilistic software for diagnostic decision-making, such as the Bayesian Diagnostic Uncertainty tools developed by Chatzimichail and Hatjimihail (2023, 2024), among others.

5.2.4 Sharpening Causal Arguments

New researchers often blur the distinction between **sufficient** and **necessary** conditions when explaining why an outcome occurs. Sorting the two apart strengthens Causal Claims, reveals hidden variables, and guards against the perennial "correlation versus causation" trap. Because reviewers occasionally refer to these ideas in logical shorthand, we introduce the symbols here – even though you will rarely need to write them yourself.

5.2.4.1 What Counts as Sufficient?

A logical condition (or set of conditions) is sufficient when its presence guarantees the outcome, but the outcome may also arise in other ways. For example, consider summer temperatures in a temperate climate (Figure 5.1).

Everyday Wording	Logic Shorthand
If it is summer (A), it is warm (B	A ⇒ B

Figure 5.1 Illustration of sufficiency using everyday language and logical shorthand: in a temperate climate, summer (A) is sufficient for warmth (B), expressed as A ⇒ B (where "⇒" the rightwards double arrow is read as *implies*).

A ensures B, yet B could still happen without A – an unseasonably warm spring day can feel like summer. The arrow points one way, and warmth does not point back to only summer.

5.2.4.2 What Counts as Necessary?

A logical condition is necessary when the outcome cannot happen in its absence. In a way, this reverses the logic of sufficiency, as the observation of the outcome implies the existence of the condition. For example, the nature of tropical plants can allow us to infer something about the climate of a region (Figure 5.2).

Everyday Wording	Logic Shorthand
Tropical plants (B) require abundant sun and water (A). I see tropical plants, hence it must be sunny and wet here.	A ⇐ B

Figure 5.2 Illustration of necessity using everyday language and logical shorthand: the presence of tropical plants (B) implies the necessary conditions of abundant sun and water (A), expressed as A ⇐ B (where "⇐" the leftwards double arrow is read as *implied by*).

5.2.4.3 What Counts as Sufficient and Necessary?

Sufficient and necessary logical conditions can be combined in a way that two situations can only occur together, often linked by the phrase *"if and only if."* An example in astronomy is that a full Moon requires a specific Sun–Earth–Moon alignment (Figure 5.3).

Everyday Wording	Logic Shorthand
A full Moon (B) occurs *if and only if* the Earth lies between the Sun and the Moon (A).	$A \Leftrightarrow B$

Figure 5.3 Illustration of sufficiency and necessity using everyday language and logical shorthand: a full Moon (B) occurs if and only if the Earth lies between the Sun and the Moon (A), expressed as $A \Leftrightarrow B$ (where "$\Leftrightarrow$" the left right double arrow is read as *if and only if*).

The alignment is both sufficient and necessary for a full Moon. The double arrow means each implies the other.

5.2.4.4 Using Counterfactuals

Causal arguments wobble when a condition that is merely sufficient is treated as necessary, or vice versa. A quick way to vet a Claim is the **counterfactual test**, defined as mentally removing or altering the "necessary" condition and asking whether the outcome still occurs (Gomes, 2023). Although difficult to perform experimentally, the mental exercise highlights hidden assumptions that statistics can miss.

To maintain transparency in the logic, the following example employs a few symbols as visual shortcuts. Think of them as road signs. You may need to recognize them in reviewer comments or journal articles, yet your work will rarely require them.

Case: Counterfactual in Action for Catching a Cold

Let's step through a counterfactual test and use logical shorthand to explore the relationship between exposure and catching a cold.

Claim: Exposure to people who visibly have colds always leads to catching a cold.

Logic shorthand:

- E = exposure to people who visibly have colds
- C = catching a cold

Interpreting "always leads to" as necessary and sufficient, this Claim can be logically represented as $E \Leftrightarrow C$ or *E if and only if C.*

Now let's examine the conditions for sufficiency and necessity, exploring the counterfactuals.

Sufficiency: If I am exposed to people who visibly have colds, I will catch a cold.

Logic shorthand: $E \Rightarrow C$

Read aloud as: If E is true, then C is also true.

Exploring the counterfactual: I could be exposed to people who visibly have colds but do not catch one myself.

Logic shorthand: $E \Rightarrow \sim C$

Read aloud as: If E is true, then C could be not true.

Necessity: If I have a cold, I must have been exposed to people who visibly have colds.

Logic shorthand: $E \Leftarrow C$

Read aloud as: If C is true, then E must also be true.

Exploring the counterfactual: If I am not exposed to people with visible cold symptoms or exposed to people without any visible cold symptoms, a cold would still follow.

Logic shorthand: $\sim E \Leftarrow C$

Read aloud as: *If C is true, E could be not true.*

Interpretation: Let's consider the validity of our counterfactuals. It is certainly possible we could be in contact with people who are visibly sick but not catch a cold. Perhaps we have a strong immune system, or the ill people are not contagious. However, these cases are probably rare; we generally catch colds if we are around sick people. So, we can consider our argument to be likely sufficient.

On the other hand, we can certainly catch a cold even if we are in contact with people who are not sick (e.g., we may have contracted the cold virus from a door knob) or those we contact are asymptomatic but still contagious. Either case can produce the same outcome: catching a cold. Because this situation is plausible, we rule out the necessity aspect of our argument.

Revised Claim: Exposure to people who visibly have colds is likely sufficient – though not necessary – for catching a cold.

5.2.5 Modeling as a Form of Reasoning

In Chapter 4, we introduced models as an element of Evidence that may lead to establishing a Claim. We can also include the act of modeling or the critical comparison of models to data as part of the Reasoning that establishes a Claim. Models are simplified representations of complex real-world systems or phenomena, serving as essential tools for Reasoning, simulation, and exploration (Frigg & Hartmann, 2025). By abstracting the key components and relationships within a system, models help researchers understand, test, and predict phenomena and their underlying causes.

Models can take various forms, each suited to specific disciplines and research purposes. Some common types of models include:

- **Animal models:** These models use living, nonhuman organisms to replicate key aspects of human biology or disease. They are valuable for probing physiological mechanisms and testing interventions when direct experimentation in people is impractical, as in mouse models of cancer or zebrafish models of neural development.
- **Computational models**: These models utilize algorithms and computational frameworks to simulate real-world systems or processes. They are beneficial for studying complex systems where direct experimentation is not feasible, such as climate models or astrophysical simulations.
- **Conceptual and symbolic models:** These models are abstract frameworks that outline the relationships between different variables or concepts. They are often used to guide investigations. Examples include ball-and-stick models of molecular structures or reaction equations that symbolize chemical interactions.

- **Mathematical models:** These models use mathematical equations to represent quantitative relationships between variables. They are widely used in fields like astronomy, economics, engineering, and physics. Examples include predicting the motion of objects moving under gravity or predicting the response of consumers to a change in supply.
- **Physical models:** These models are tangible, often scaled-down representations of real-world objects or systems. They are commonly used in environmental, engineering, and architecture studies. Examples include wind tunnels that allow analysis of drag forces on aircraft or shake tables that test the structural integrity of a support beam during an earthquake.
- **Statistical models** utilize statistical techniques to analyze data and draw inferences about the relationships between variables. They are commonly used in social sciences and medical research, for example, as regression models that predict health outcomes based on income.

For example, one efficient application of models is in public health surveillance, where researchers use mathematical models to assess the safety of large-scale interventions, such as the distribution of the SARS-CoV-2 vaccine. A common approach is the **observed versus expected (OE) analytic framework, also known as OE analysis,** a form of complex Reasoning in which models are used to represent the "expected" outcome, which is then compared to the "observed" data (Mahaux, Bachau, & Van Holle, 2016). Here is a brief summary of how it works:

- OE models can estimate the *expected* rate of adverse events (e.g., deaths) that occur by chance in a population, assuming no causal relationship between the COVID-19 vaccine and these events.
- The *expected* rate is then compared to the *observed* rate of adverse events over a comparable period.
- A significant discrepancy between the *expected* and *observed* rates can indicate a potential safety concern, such as the need to recall a particular batch of vaccines.

OE analysis provides a structured approach for differentiating genuine safety concerns from random occurrences and serves as a framework for reasoning for hypothesis-driven studies.

5.2.6 How Much Reasoning Is Enough?

All arguments from Evidence must ultimately lead to the Claim, but knowing when an argument is sufficient to establish a Claim can be one of the most challenging questions for researchers-in-training to answer. Standards of sufficient Reasoning depend on the study design, the nature of the Claim, and the standards of Evidence and Reasoning expected in the field of study. For example, efficacy studies of drug treatments require a new treatment to perform better than a placebo at a statistically significant level. The level of statistical significance needed to establish a Claim can also depend on the nature of the Claim itself. Suppose a result that supports an existing paradigm may require only a "3-sigma" level of confidence (99.7% confidence or a p-value of 3×10^{-3}), while a result that contradicts the existing paradigm may need a "5-sigma" level of confidence (99.99997% confidence, or a p-value of 3×10^{-7}) or higher to be persuasive.

The standards of proof are set by the scientific community and can take on both quantitative and qualitative forms, such as:

- **Iterative validation**: A resolution often serves as part of an iterative cycle of inquiry, in which findings address the current question but can also generate additional hypotheses or lead to further investigations. For example, confirming the efficacy of a drug at a given statistical significance may lead to follow-up studies on longer-term effects or broader applicability.
- **Peer review and consensus:** In modern scientific studies, the gold standard for resolution is when a result withstands peer review and validation studies and achieves consensus within a scientific community.
- **Practical significance**: Another quantitative, yet less rigorous, resolution may be the demonstration of a clear practical improvement in a methodology, such as enhanced efficiency or predictive accuracy.
- **Statistical significance:** This common form of quantitative resolution demonstrates that the Evidence meets a given statistical threshold (e.g., a p-value < 0.05) to validate a prediction or differentiate two hypotheses.

Let's look at some examples in the natural sciences of how much Reasoning is typically enough to resolve an argument:

- **Astronomy**: In stellar astrophysics, the resolution of an argument may involve proving that a particular theory explains the observed properties of an object. This proof can be established by comparing the measurements of the masses and radii of objects to computational or mathematical models for different theories, then determining which are supported and which are ruled out.
- **Biology:** In evolutionary biology, resolving an argument may involve demonstrating that a particular trait confers a survival advantage. For example, researchers might compare the reproductive success of organisms with and without the trait, using statistical analyses to show that the trait significantly increases fitness.
- **Chemistry:** In synthetic chemistry, the resolution of an argument may involve proving that a new synthetic method produces a desired compound more efficiently than existing methods. This proof can be established by comparing yields, reaction times, and purity levels, with the resolution being the demonstration of statistically significant or practical improvements in the production of a compound.
- **Earth Sciences:** In geology, resolving an argument may involve demonstrating that a particular geological process, such as plate tectonics, can explain the formation of specific landforms. This proof can be established by comparing geological data from different regions and using computational or physical models to demonstrate that the proposed process accurately predicts the observed features.
- **Physics:** In physics, resolving an argument often means showing that a model's quantitative forecasts line up with experimental data inside carefully stated error bars. Take the 2012 headline discovery of the Higgs boson (Aad et al., 2012). The

> Standard Model predicted a neutral scalar particle with specific production rates and decay-channel ratios. ATLAS and CMS collected proton-collision data at 7 – 8 TeV, reconstructed invariant-mass spectra for $\gamma\gamma$, ZZ, and WW final states, and then asked a purely statistical question: Does any known background process reproduce these excesses? Global-fit likelihoods and "blind" cross-checks showed a 5-sigma signal near 125 GeV – far beyond the chance-fluctuation threshold. Because the observed branching fractions and production cross-sections matched Standard-Model calculations within combined experimental and theoretical uncertainties, the Reasoning (model → prediction → data agreement → robustness tests) was sufficient to settle the decades-long debate over the Higgs mechanism.

Resolutions in arguments, and specifically knowing when the Reasoning is enough, are field-specific and provisional, relying on a balance of evidence, evaluation, and corroboration to contribute to practical understanding and a broader scientific consensus.

5.3 From Reasoning to Argument

The types and elements of Reasoning described above can be considered part of the building blocks of constructing an **argument from evidence** that justifies and supports the Claim, while addressing alternative interpretations that may counter the Claim. Arguments are the logical foundation upon which a Claim is established and are also the focus of studies that evaluate, critique, or overturn prior Claims. As such, it is worth spending some time reviewing the elements and structure of a well-constructed argument, with a focus on Reasoning.

5.3.1 Revisiting the Toulmin et al. (1984) Model

In Chapter 1, we introduced the concept of argumentation from evidence by considering the Toulmin et al. (1984) model. Let's take a closer look at this model, focusing on the central role of reasoning.

Figure 5.4 illustrates an adaptation of the Toulmin et al. (1984) model applied to scientific research. The main throughline of the argument is its **logical development** or the "Warrant." This line of logic leads us directly from Evidence to the Claim, but this Reasoning element is not sufficient on its own. We also must consider the **supporting reasons** for our argument, or the "Backing," represented as a bridge's foundations that support logical development.

Figure 5.4 also highlights some of the argument elements that apply directly to the Evidence and the Claim. For instance, qualifiers influence the Evidence through the experimental **uncertainties**, and the Claim through **qualifiers** based on variation in outcomes or unaccounted-for factors. Similarly, **limitations** in our Reasoning influence both the Evidence (what can be measured) and the Claim (the new knowledge from the findings). Finally, our Claim can be constrained by **counterfactuals** and supported by **prior confirmation** studies. A well-reasoned study addresses all these aspects of Reasoning.

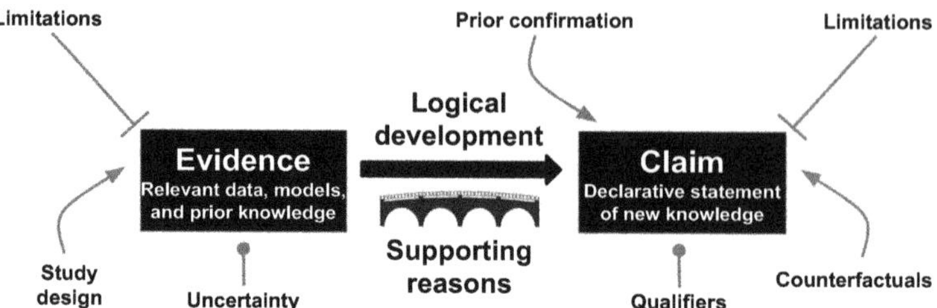

Figure 5.4 Detailed model of how reasoning connects evidence to the claim.

Let's return to our conversation with a skeptical friend about the benefits of physical exercise on mental health at the beginning of this chapter. We can frame our discussion within the Toulmin et al. (1984) model by identifying all the parts of a complete argument within that example:

(1) **Claim**: "*Regular physical exercise leads to improved mental health.*"
This is a clear, declarative statement of new knowledge (to the friend).

(2) **Data (Evidence)**: "*Numerous studies, such as a large one of 1.2 million people published in the Lancet Psychiatry by* Checkroud et al. (2018), *show that individuals who exercise regularly report fewer days of poor mental health compared to those who do not exercise. Exercise was associated with reductions in symptoms of depression and anxiety.*"
This is the specific data that supports our Claim. In our example, it is the prior research by Checkroud et al. (2018) that makes the association (or correlation) between regular exercise and depression/anxiety symptom reduction. Here, "association" refers to a standard statistical test, such as a Pearson correlation or Spearman's rank correlation.

(3) **Warrant (Reasons):** "*Exercise is known to release endorphins, which are chemicals in the brain that act as natural mood elevators. Additionally, physical activity reduces levels of stress hormones such as cortisol and increases blood flow to the brain, supporting cognitive function and emotional regulation.*"
As the famous saying goes, "correlation does not imply causation," so we need to establish the **why or how** this connection between exercise and mood works. Toulmin et al. call this the "warrant," or in plain language, the reasons for the correlation. In CERIC, we call this **"Reasoning"** when it refers to the logical thought process of connecting Evidence to the Claim.

(4) **Backing (Support for Reasons):** "*Professional health organizations like the American Psychological Association (APA) and the Centers for Disease Control and Prevention (CDC) recommend regular physical activity as part of a holistic approach to mental well-being.*"
What if the skeptical friend challenges this study as invalid or untrustworthy? As noted, this kind of challenge is an essential part of scientific discourse, leading us to the second part of Reasoning, in which we must **support our reasons** with additional lines of logic or evidence. Supporting our reasons strengthens their validity.
Here, we draw support from two widely recognized health agencies, the APA and CDC, which recommend exercise as part of a comprehensive approach to

improved mental health. Presumably, scientists at these health agencies analyzed the body of relevant research literature, drew reliable conclusions, and then made reasonable recommendations to the public.

If we want more than just advice from regulatory agencies, there are other lines of support or logic we can use to bolster our reasons. Evidence from meta-analyses or longitudinal studies would demonstrate repeated validation of the link between exercise and mental health. Cross-cultural studies would demonstrate the global applicability of exercise for improving mental health. Neuroscience research, such as studies on how exercise stimulates neurogenesis in the hippocampus (associated with memory and emotion regulation) or increases brain-derived neurotrophic factor, would explain the process by which exercise promotes brain health.

(5) **Qualifier (Uncertainty and Counterfactuals):** "*While exercise benefits many, its effects may vary depending on individual circumstances or other health conditions.*"
In our example, we qualified the argument by acknowledging that exercise effects may vary depending on individual circumstances, such as the severity of mental health conditions or physical limitations. This statement of uncertainty signals ambiguity or variability in the evidence or the argument. Uncertainty can take the form of statistical uncertainty in the Evidence, or other factors (known or unknown) that could modify or blur predicted outcomes. Uncertainty functions as a form of **oppositional Reasoning**, which recognizes that the Claim may not be universally valid, adding nuance to the argument by clarifying where it does not hold. Recognizing the boundaries of the research strengthens the argument because experienced scientists know that uncertainties are inherent in every dataset, and failure to discuss qualifiers is often a major red flag that Reasoning is incomplete.

(6) **Rebuttal (Limitations and Counterfactuals):** "*For individuals with severe mental health conditions, exercise alone may not suffice and should be complemented by other treatments, such as therapy or medication.*"
In addition to uncertainty, findings may be limited to a specific set of conditions or criteria (limitations) or there may be conditions or situations where the Claim may not hold (counterfactuals). In this case, the authors indicate that the severity of mental health could limit the efficacy of exercise and would require additional treatments (therapy or medication) to address. Alternatively, the authors might have found that individuals with severe mental health issues did not respond to exercise alone, a counterfactual that narrows the scope of the findings. These kinds of statements acknowledge how the experimental design or the nature of the system under investigation constrain the scope of the Claim. Clearly stating limitations and counterfactuals strengthens the Reasoning by anticipating challenges and providing a clear domain of applicability, and it improves the trustworthiness of the author.

This structure provides a means of evaluating the quality of the argument we are making to support exercise as a means of improving mental health. However, it does leave out a few key elements that are particularly important for experimental studies that rely on quantitative and qualitative evidence, often framed within a large body of prior studies.

5.3.2 Regular and Critical Arguments

Figure 5.4 illustrates that scientific arguments can be approached from two perspectives: by working within an accepted framework or by challenging it. Let's explore these classes of argumentation in more detail.

- **Regular arguments** operate within an accepted framework, relying on the field's established standards, patterns of reasoning, and concepts, to draw conclusions.
- **Critical arguments**, in contrast, challenge the framework itself, seeking to rethink or reevaluate existing standards and norms.

The choice between these approaches shapes the kinds of Reasoning that are applied. Regular arguments emphasize reliability and consistency by leveraging well-established reasons and patterns of logic within the field. Conversely, critical arguments require rethinking and reevaluating, possibly drawing on Reasoning from other fields or perspectives to disrupt norms and enable paradigm shifts.

Regular arguments often draw from scientific laws, standardized methodologies, or statistical Reasoning that can be repeatedly used without needing to revalidate their dependability or relevance. Mastering established frameworks is a crucial aspect of doctoral training in scientific disciplines, laying the foundation for conducting reliable and meaningful research. Some examples of regular arguments include:

- **Astronomy:** Using high-precision transit photometry and radial-velocity measurements to test Keplerian orbital solutions and confirm the mass, radius, and orbital period of newly detected exoplanets.
- **Biology:** Applying Mendelian genetics to predict inheritance patterns in controlled breeding experiments.
- **Chemistry:** Applying density-functional theory calculations, then validating predicted reaction pathways and activation energies with kinetic and spectroscopic experiments.
- **Earth Science:** Combining stable-isotope geochemistry with climate-system models to reconstruct past temperatures from ice cores and verify model skill against independent proxy records.
- **Physics:** Applying the laws of thermodynamics to optimize energy transfer in engineering systems.

The main takeaway is that **regular arguments apply standard reasons and rules to produce reliable results**, making them a fundamental approach for conducting research.

Critical arguments are the opposite of regular arguments: They challenge the standards, rules, or reasons to modify how arguments are made. These types of arguments, if well-justified, can be paradigm-shifting, changing how a field operates or creating entirely new areas of inquiry. Some examples of critical arguments include:

- **Astronomy:** Arguing that the flat rotation curves of spiral galaxies require large amounts of non-luminous dark matter, not just the visible stars and gas.
- **Biology:** Contending that extensive horizontal gene transfer among microbes invalidates a single, tree-like "great chain of life" and demands a network view of evolution.
- **Chemistry:** Critiquing the rigid application of the octet rule in chemical bonding, enabling the exploration of hypervalent molecules and bonding phenomena that defy traditional rules, such as the bonding seen in sulfur hexafluoride (SF_6).
- **Earth Science:** Proposing continental drift – later formalized as plate tectonics – against the long-held belief in fixed continents.
- **Physics:** Questioning the completeness of classical mechanics by introducing quantum mechanics, radically redefining our understanding of the atom and the nature of determinism.

Importantly, critical arguments in one field may be closely related to the regular arguments in another field, illustrating the interconnectedness of Reasoning across scientific domains. Here are a few examples:

1. Ecology and Economics
   a. **Ecology:** Abrams (2000) and Cortez & Ellner (2010) showed that predator–prey cycles cannot be fully explained without allowing prey and predator traits to evolve during the interaction – a critique that expanded classical Lotka–Volterra theory.
   b. **Economics analog:** In differential game theory (e.g., Dockner, 2000), companies update strategies continuously in response to rivals; adaptation is a built-in function, rather than a response to a challenge.
2. Environmental Science and Economics
   a. **Environmental science:** Daly's (1996) steady-state economics rejects GDP maximization as ecologically untenable – a critical stance that redefines the concept of "progress."
   b. **Economics analog:** Standard Ramsey-type optimal-growth models with environmental constraints (Nordhaus, 2017) treat carbon abatement as a routine resource-allocation problem; sustainability enters as an internal cost rather than a paradigm shift.
3. Genetics and Computer Science
   a. **Genetics:** The discovery of reverse transcriptase (Coffin, 2021, citing Baltimore, 1970 and Temin & Mizutani, 1970) and later evidence that epigenetic markers can be inherited (Jaenisch & Bird, 2003) compelled biologists to expand Crick's central dogma to encompass feedback pathways from RNA (or chromatin state) back to DNA expression.

b. **Computer-science analog:** In back-propagation learning (Rumelhart et al., 1986), error signals propagate from output to earlier network layers; feedback is standard, not controversial.

In summary, what counts as a "critical" argument in one discipline (e.g., adding trait evolution to predator–prey models) may already be business-as-usual in another (e.g., dynamic game theory). This paradox is particularly important to remember with interdisciplinary research.

5.3.3 Elements of Sound Argumentation

Sound scientific arguments do more than list facts: they guide readers through a clear, logical structure supported by precise language, consistent Reasoning, and credible evidence. Researchers build trust and clarity by avoiding logical fallacies, using well-placed supports and grounding Claims in accurate and precise data. Understanding these components strengthens your ability to evaluate arguments and craft your own with scientific integrity.

5.3.3.1 Establishing the Logical Flow of an Argument

A well-written argument builds logically, step by step, to guide the reader seamlessly through the chain of Reasoning. Organizing arguments using a logical structure enhances the readability and persuasiveness of scientific writing. For example, in an article on deforestation, a logical flow might begin with the causes of deforestation, followed by its ecological impacts, and conclude with mitigation strategies. Barroga and Matanguihan (2021) emphasize that structuring arguments cohesively enables readers to follow the progression of ideas, thereby strengthening the argument's impact.

Logical flow can be achieved through precise word choice, clear sentences, and well-structured paragraphs, often relying on the effective use of transitions. Transitional words and phrases, such as *therefore*, *however*, and *in addition*, play a vital role in maintaining clarity and coherence in academic writing. Let's consider a few ways that transition words (in bold italic) are used in the following example that Dr. Bjorn drafted for an upcoming Context section:

> Ocean acidification, primarily driven by increasing atmospheric CO_2 levels, poses a significant threat to marine ecosystems. ***For example***, coral reefs are particularly vulnerable to changes in ocean chemistry. As CO_2 dissolves in seawater, it lowers the pH, resulting in ocean acidification. ***Consequently***, this reduces the availability of carbonate ions, which corals need to build their calcium carbonate skeletons. ***Furthermore***, the weakening of coral skeletons makes reefs more susceptible to erosion and less able to provide habitats for marine species. ***However***, mitigating ocean acidification requires coordinated global efforts to reduce CO_2 emissions and protect vulnerable ecosystems.

Here is a breakdown of the transition words and their function in scientific articles:

- **Introducing the topic:** *"For example,"* presents a specific instance (*"coral reefs"*) to illustrate the broader threat of ocean acidification to marine ecosystems.

- **Explaining the cause–effect relationship:** *"Consequently"* indicates a direct cause–effect connection, in which *"acidification"* leads to reduced *"availability of carbonate ions."*
- **Describing the broader impact:** *"Furthermore"* indicates an additional consequence, *"weakened coral skeletons and ecosystem impacts,"* which expands the implications of ocean acidification.
- **Introducing potential solutions:** *"However,"* signals a shift to a contrasting point or topic, in this case, shifting from the implications of acidification to what is needed to address it.

5.3.3.2 Ensuring Logical Consistency

A key characteristic of sound arguments in scientific articles is the consistency of logical Reasoning throughout and across arguments. One essential consideration is the usage of *consistent and concise terminology* to ensure clarity and avoid confusion. For example, in studies on climate change, researchers should consistently use terms such as "global warming" or "climate variability" rather than switching between these terms and similar phrases, such as "environmental heating" or "atmospheric instability." Consistent terminology should be used throughout the title, statement of the problem, and research questions, creating a cohesive narrative throughout the research. Newman and Covrig (2013) emphasize that consistency enhances the quality of research plans and reports by promoting coherence and minimizing ambiguity in interpretation.

A more challenging element of consistent Reasoning in preserving the integrity of scientific arguments is the avoidance of **logical fallacies**. Faulkner (1941) argue that recognizing and addressing logical inconsistencies, such as straw man fallacies, false dilemmas, or ad hominem arguments, enhances the credibility and reliability of a scientific article. For instance, a researcher studying biodiversity loss should avoid the straw man fallacy of claiming that critics believe "deforestation has no impact on biodiversity." In reality, critics may argue that the impact of deforestation is simply overstated in specific regions or contexts. Misrepresenting nuanced arguments oversimplifies opposing views and weakens the researcher's credibility by addressing an exaggerated or fabricated Claim rather than the actual argument. Instead, researchers should address opposing views fairly and with evidence-based Reasoning. Let's consider two main types: a false dilemma, and an ad hominem argument.

A **false dilemma** (also known as a false dichotomy) occurs when only two options are presented as possible solutions, thereby ignoring other viable alternatives. For example, in debates about energy production, a commonly used false dilemma is exemplified in this paraphrase:

> We must either rely entirely on fossil fuels to meet our energy demands or face economic collapse, as renewable energy sources are currently unreliable.

This statement ignores other options, such as combining renewable energy with advancements in energy storage, nuclear power, or energy efficiency improvements.

An ***ad hominem* argument** attacks a person's character or credentials rather than addressing the argument itself. For example, during early climate change debates, some skeptics dismissed climate scientists' findings by expressing this notion:

> The lead authors on that article cannot be trusted because they are environmental activists, not objective scientists.

Attacking the scientists' character avoids engaging with the actual evidence of climate change. Logical fallacies can detract from rational, evidence-based discussions on critical scientific issues, especially in echo chambers on social media.

In contrast, there are many ways to improve logical consistency. **Robust supporting studies** improve logical consistency. An argument often refers to other studies and sources that either support the main reasoning chain (e.g., standard methodologies and field-specific arguments) or provide validating Context. For example, an analysis of the effects of greenhouse gases on global temperatures may rely on peer-reviewed sources, such as data from the Intergovernmental Panel on Climate Change (IPCC), to substantiate its Claim. By using robust, peer-reviewed sources, researchers enhance the credibility and persuasiveness of their arguments. Varpio (2018) has found that appeals to credible sources strengthen an argument and build trust with the audience.

Critical thinking involves questioning assumptions and evaluating Evidence to strengthen arguments. For example, in research on biodiversity loss, a scientist might critically examine whether data on species decline accounts for regional variability, long-term oscillations, and potential confounding factors such as climate change or habitat fragmentation. Macagno and Rapanta (2019) argues that critical thinking enhances scientific arguments by exposing logical gaps and fostering more nuanced conclusions. We dive deeper into this complex skill set in Part II.

Precision and accuracy are also fundamental to a scientific study's logical consistency, particularly in the Reasoning that supports the quality of the Evidence: its uncertainty and limitations, and the robustness of the study design. ***Precision*** reflects the consistency or repeatability of the Evidence. In experimental research, precision refers to obtaining measurements or observations that are consistent and close to one another across multiple trials. ***Accuracy*** refers to the degree to which Evidence aligns with objective facts or reality, ensuring that an argument is based on solid ground.

It is important to note that precision and accuracy in Evidence are independent of each other and can lead to different interpretations of a Claim, as illustrated in Figure 5.5. When our measurements are far from the expected outcome and widely dispersed, then the Evidence is "inconclusive" and does not support the Claim. When the measurements are far from the expected outcome but tightly dispersed, then the Evidence may be "biased" or "skewed," making the Claim suspect. When the measurements are consistent with the expected outcome but widely dispersed, then the Evidence may be "indicative" of the Claim but may not rule out other interpretations. The best result is when the measurements are consistent with the expected outcome and tightly clustered making both the Evidence and the Claim "robust."

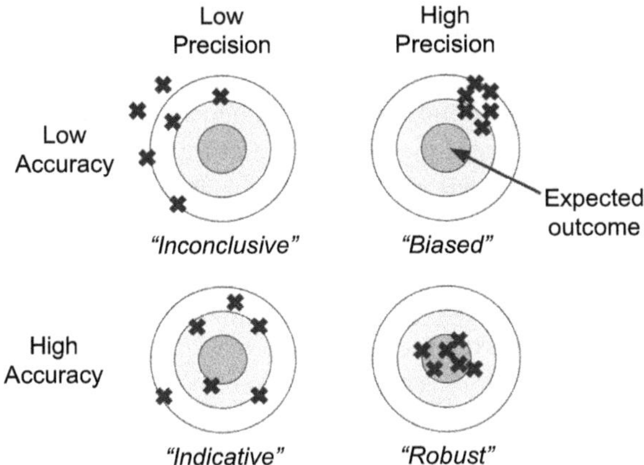

Figure 5.5 Precision and accuracy in evidence are independent and lead to different interpretations.

In a scientific research study, the accuracy of the Evidence is often argued using control cases where the outcome is known, or calibration procedures that are standard for the field. Accuracy is also reflected in the scope of the Evidence, such as the sample size or range over which measurements were made, which in turn defines the limitations of the Evidence and ultimately the Claim. Finally, precision and accuracy address different forms of uncertainty, ***statistical uncertainty*** (repeatability) and ***systematic uncertainty*** (correctness), respectively, which are often represented separately in experimental studies.

Precision and accuracy can be inferred from the language of a scientific article. Consider the following statement of a *hypothetical* study's main finding:

> The use of drug Z reduced systolic blood pressure by 10%.

A more precise and accurate description of this hypothetical finding would provide further details of the study design, uncertainties, and limitations of the Evidence, such as:

> A randomized controlled trial involving 514 men aged 45–65 found that drug Z reduced systolic blood pressure by 10% (±4% statistical and ±2% systematic) when 20 mg was taken once daily for six months.

We must also be confident that these values are accurate. If the actual rate of improvement was only 8% but the authors decided to round up to 10%, or if the investigators cherry-picked their sample to highlight the best outcomes, then the Claim is inaccurate and therefore not trustworthy.

5.3.4 Flexibility in Reasoning

One of the trickiest aspects of conducting research is balancing high precision and accuracy with the flexibility to follow a study to wherever the data leads. Many aspects of a study can change after it is underway, including the methodology, sampling data

Box 5.1 Better Practice: Spotting Pseudoscience Using Reasoning

Sound scientific argumentation sets out competing explanations, gathers Evidence that can rule them in or out, quantifies uncertainty, and engages with counterarguments. Pseudoscience often copies the appearance of this structure while bypassing the hard analytic work beneath it. Mental alarms should ring when an author offers only one explanation and never tests alternatives.

Clowe et al. (2006) illustrate the gold standard. They pitted the theory of dark matter against that of modified gravity, designed observations that could discriminate between the two, and reported their results with explicit error estimates (e.g., "detected at 12 σ").

Pseudoscientific writing, in contrast, tends to:

- cherry-pick or massage data,
- dismiss or ignore rival explanations,
- equate correlation with causation, and
- gloss over uncertainty or limitations.

A telltale sign is language that leans on authority – phrases such as "scientists have proven" or "it's obvious" replace the careful disclosure of methods, assumptions, and statistical margins of error that genuine research provides.

When the argument feels slick but thin, pause and ask four quick questions:

1. What Claims are being compared, and by what criteria?
2. How were the data collected and analyzed?
3. Which assumptions underlie the Reasoning?
4. What Evidence, limitations, or alternative explanations remain unaddressed?

Takeaway: A study that answers these questions openly invites confidence. One that skirts them may be performing science in name only.

analysis procedures, calibration procedures, comparative studies and samples, lines of Reasoning, and even the Claim itself. If a researcher starts a study with a particular hypothesis, but finds that the data either contradicts that hypothesis or leads to a new one, the best practice from a Reasoning perspective is to follow the data to the actual outcome. Forcing data to fit a preconceived idea is a fallacy in Reasoning called ***cherry-picking*** *or* ***pseudoscience***.

In addition to being mindful of where the Evidence leads us, characteristics of a research study, such as its scale, nature, and motivation, can influence the degree of possible flexibility in Reasoning. For instance, fields addressing large-scale, complex phenomena such as urban social environments or stellar populations across the Milky Way galaxy permit more adaptability in Reasoning, particularly in the interpretation of results. On the other hand, studies focused on small-scale, highly controlled systems, like genetic knockout experiments in Wistar rats, require more exacting standards, narrowing the scope of valid and reliable outcomes.

Climate science modeling highlights the need for flexibility in Reasoning to account for the numerous factors that influence the climate. Quantities such as temperature, greenhouse gas concentrations, glacier melt rates, and feedback mechanisms vary on spatial and temporal timescales that are often too fine-grained to be computationally modeled. Rigid adherence to the more simplified conceptual models could lead to contradictory results. Conversely, studies of quantum systems, such as Bose–Einstein condensates, demand strict exactness because measurements at the atomic scale leave little room for variability. Similarly, in ecology field studies, researchers observing animal behavior in the wild may need to adjust their methodologies and Reasoning to incorporate real-time environmental or population changes. In contrast, microbiology lab experiments studying bacterial growth must adhere to strict experimental protocols, such as consistent nutrient media and temperature controls, to ensure reliable results.

Let's consider two additional examples of flexibility in Reasoning, one that is appropriate for the experimental design and one that is inappropriate.

Example 1. Appropriate Use of Flexibility in Reasoning
Research project: Interpreting Arctic tern migration data in response to shifting climate zones.

- **Flexibility in Reasoning:** Researchers initially hypothesize that terns migrate based solely on seasonal changes in daylight. However, upon observing anomalies in their flight patterns, they adjust their Reasoning to include factors like oceanic temperature shifts and wind currents. This adaptation allows them to refine their explanatory framework while staying aligned with the broader goal of understanding migration behavior.
- **Why this is appropriate:** Adjusting Reasoning to incorporate unforeseen influencing factors like weather enables researchers to make sense of dynamic phenomena in a way that reflects evolving observations.

Example 2. Inappropriate Use of Flexibility in Reasoning
Research project: Interpreting the results of a clinical trial for a new cancer drug.

- **Variations in reasoning:** Midway through the trial, researchers notice significant variability in the drug's efficacy across patient subgroups. Instead of adhering to the original framework of testing uniform efficacy, they shift their focus to testing the drug's effectiveness for specific subgroups without reevaluating their sample design.
- **Why this is inappropriate:** This adjustment undermines the study's validity, as it introduces bias (shift to a subgroup) without sufficient Evidence (testing the full population), potentially leading to misleading conclusions about the drug's efficacy.

In summary, rigorous research requires nimble thinking but only when that flexibility is anchored to transparent methods, explicit assumptions, and field-based standards of Evidence.

5.4 Reasoning Compared to Other Elements

Of all the CERIC elements, Reasoning is often the hardest to pin down – especially when it overlaps with Evidence. While Evidence consists of observations, data, and measurements, Reasoning explains how those observations support a Claim through logic, justification, or comparison. Understanding how these two elements interact helps readers trace the logic of an article's argument and determine whether the study's conclusions are well-supported.

5.4.1 Reasoning and Evidence

When novice researchers first approach the CERIC elements in an article, the most common confusion is how to distinguish Reasoning from Evidence. This confusion is not surprising, given the strong connection between these logical elements and the fact that some forms of Evidence (e.g., models) can also play a role in Reasoning (e.g., comparing models to data). It is helpful to consider how Evidence and Reasoning support each other when deciding how to classify elements of a scientific study.

Reasoning as the framework for Evidence: Reasoning fundamentally serves as the logical structure that connects Evidence to Claims, ensuring that the former is meaningfully and accurately interpreted and analyzed. The chain of Reasoning that makes up an argument involves applying logical principles to make sense of the Evidence (e.g., its uncertainty and its limitations) and to derive coherent conclusions (Ford, 2012).

Evidence as the basis for Reasoning: In turn, Evidence provides the empirical foundation upon which Reasoning is built. Evidence consists of observations and data that can either support or challenge a hypothesis, but it's the logical argumentation of Reasoning that determines which of these outcomes emerge (Allen & Baker, 2016).

Let's look at how these distinctions emerge in three examples that are based on our main reasoning types:

1. **Deductive Reasoning and Evidence**: In physics, the theory of gravity predicts that objects fall at a constant acceleration near the Earth's surface, assuming negligible air resistance. A classroom experiment measuring the position of a falling object as a function of time provides empirical *Evidence*. We can then compare these measurements to the predicted motion, assuming acceleration is constant, which reflects a deductive *Reasoning* framework.
2. **Inductive Reasoning and Evidence**: In ecology, observations of the distribution of plant species across various habitats serve as *Evidence*. The researchers then search for patterns among these distributions concerning the habitats' environmental conditions, such as temperature or rainfall, to determine what factors influence biodiversity, using an inductive *Reasoning* framework.
3. **Abductive Reasoning and Evidence:** In medicine, researchers record a patient's symptoms and collect test results as *Evidence*, then consider these within the context of the typical response and prevalence of various diseases to determine the most likely diagnosis, illustrating an abductive *Reasoning* framework.

Box 5.2 Common Confusion: Is Data Reduction Reasoning or Evidence?

A common confusion for novice researchers is distinguishing between Evidence and Reasoning when it comes to data reduction or data cleaning, the process of transforming "raw data" (e.g., an image acquired when observing an astrophysical source) into "science data" (a calibrated measurement of the source's brightness). This confusion is common when one is less familiar with the field-standard approaches to data analysis, or when standard approaches are modified.

For example, an article may simply state that "standard reduction techniques" were used. It is tempting to lump this analysis into the Evidence, since the science data seems to be the starting point of our argument. However, what if the standard techniques are not appropriate for this particular source? Do we need to then consider those techniques as an aspect of Reasoning? Alternatively, a very detailed description of the data reduction methods could signal that the methods are unique and potentially impact the interpretation of the data. In this case, the data reduction appears to be relevant to justifying the uncertainties and limitations of the Evidence, part of our Reasoning ecosystem.

So how do we know if data reduction is Evidence or Reasoning? When confronted with this quandary, I often ask my students, "Does the Claim depend critically on how the data are reduced?," or "If the data were reduced in a different (but still correct) way, would that significantly change the Claim?" If the answer to either question is "yes," then the reduction methods should be considered as part of the Reasoning. On the other hand, if the data reduction procedures are straightforward and use field-standard approaches (e.g., using mean and standard deviation, or comparing significance through a standard t-test), then they are just part of the underlying Evidence. Delving into this ambiguity often leads to important discussions on how measurements are made, reduced, analyzed, reported, and challenged in a given field.

Box 5.3 Knowledge Check: Question 5.1 Which statement best distinguishes Reasoning from Evidence in a scientific article?

A. Reasoning is the data collected during the study, and Evidence is the logical interpretation that connects the data to the claim.
B. Evidence encompasses everything written in the methods, while Reasoning always appears only in the discussion.
C. Evidence is the relevant data used to support a Claim, whereas Reasoning is the logical framework or chain of argument connecting that data to the Claim.
D. Evidence and Reasoning are functionally interchangeable because both contain numbers and statistics.

(Check your understanding using the Knowledge Check Key at the end of the book.)

5.4.2 Reasoning and Context

Another common confusion is the distinction between Reasoning and Context (Chapter 7), particularly when it comes to the motivation of a particular study. The Context of a research article helps to establish the motivation and framework of a study by articulating its background (what work has been done before), rationale (why new work is needed), and objectives (what this work aims to achieve). These elements provide reasons to conduct the study, so are they Reasoning? For research studies, the answer is no, because each of these elements is independent of the study itself; we can have background, rationale, and objectives for a study without doing any work. More importantly, these elements can exist without any Evidence (i.e., no data need be collected) or a Claim (i.e., none of these establishes a declarative statement of fact). As such, the reasons presented in an article that aims to motivate a study are elements of Context and are not part of the study's internal Reasoning.

This ambiguity becomes even more confusing when we consider another form of scientific work: the ***proposal***. Scientific proposals are used to secure resources (telescope time, lab materials, computing hours, and funding) to pursue a research objective. Proposals generally focus on elements of Context and Evidence, including background, motivation, objectives, study design, and expected outcomes, to convince an organization to provide resources. In essence, the Claim of a proposal boils down to "you should provide resources for this study," and these elements form part of the logical chain of argument, the Reasoning, that justify that Claim. That being said, this "proposal Claim" is not a declarative statement of fact, but rather a request for resources to pursue a study aimed at establishing a valid scientific Claim.

If this is confusing, it is important to remember that proposals are distinct from the primary literature. The former motivates a study to test a scientific hypothesis; the latter establishes a Claim based on the results of a study. If a proposal is on your upcoming to-do list, we take a deeper dive into presentations in Chapter 11 and literature reviews in Chapter 12.

5.4.3 Reasoning and Claim

In some experimental studies, Reasoning can take the form of intermediate results that appear to be claims but are ultimately used to justify the primary Claim (see Chapter 2). For example, a study may need to establish the degree of precision or sensitivity achieved in a measurement to establish the existence of a detection as the primary Claim.

For example, in the article by Abbott et al. (2016) that reports the first detection of gravitational waves from a merging binary black hole (see Chapter 3), the authors justify their detection as a real merger through multiple statements of fact. For example:

> On September 14, 2015, at 09:50:45 UTC, the LIGO Hanford, WA, and Livingston, LA, observatories detected the coincident signal GW150914…. (p. 060112–2)

> … matched-filter analyses that use relativistic models of compact binary waveforms [44] recovered GW150914 as the most significant event from each detector for the observations reported here. (p. 060112–3)

Both sentences could be interpreted as "statements of new knowledge." However, both ultimately serve to support the primary Claim, the detection of gravitational waves, and are hence statements of Reasoning. Indeed, key adjectives such as "coincident" and "significant" are important Reasoning descriptors that signal establishing the reliability of the detection.

5.5 Finding Reasoning

While Reasoning serves the critical role in establishing an article's logical scientific argument, it is rarely labeled outright. To recognize Reasoning in primary literature, we must learn to identify its various forms, their corresponding locations within an article, and the associated language.

5.5.1 Where Forms of Reasoning Are Found in Research Articles

Reasoning's various forms, from shaping and supporting a logical argument that links Evidence to a Claim, to supporting and contextualizing the Evidence and Claims directly, can be found throughout a research article. Here, we focus on three main categories of Reasoning and where to look for them outside the abstract using the IMRaD headers.

Reasoning that develops the main logical argument: The primary logical chain connecting Evidence to Claim is generally found in the **methods** and **discussion** sections of an article. These elements of Reasoning are central to the study and will often be highlighted through section headings. In addition, a key figure or table in an article can serve to illustrate the main throughline of Reasoning, such as a plot comparing observations to a model or a table that justifies the efficacy of a drug by comparing it to a placebo treatment.

Reasoning that supports the study design: Reasoning that motivates the study design or justifies the uncertainties and limitations of the Evidence is typically found in the **methods** section and sometimes in the **introduction**. These elements of Reasoning can provide a detailed rationale for the experimental design, data collection, and analytical approaches used, or justify why particular methodologies, tools, or techniques were chosen. For example, a study evaluating drug efficacy employs a double-blind, placebo-controlled trial, a choice aimed at minimizing bias and enhancing the credibility of the results.

Reasoning that supports the Claim: As illustrated in Figure 5.4, elements of Reasoning can further support a Claim by highlighting confirmation studies, rebutting counterfactuals, or contextualizing a Claim's uncertainty and limitations. These elements of Reasoning are generally found in the **discussion** section. For example, a study that claims a new organic molecule synthesis method is highly effective may support this claim by noting that other synthesis methods are typically more effective than traditional distillation. Similarly, the discussion section of this article may note that the result only applies under specific temperature and pressure conditions with a catalyst, and that its effectiveness would need to be tested in other conditions, a form of Reasoning that constrains the scope of the Claim.

5.5.2 Language that Signifies Reasoning

The sentence builders that signal Reasoning depend on what form of Reasoning is being applied. Here are a few examples of the language that signals common Reasoning statements:

1. **Logical development**: These elements of Reasoning are scaffolded by classic logic constructions such as: "if X then Y," "X implies Y," or "X rules out Y."
   Hypothetical example: **If** *no electrons are detected*, **then** *we can confirm the quantum nature of the photoelectric effect.*
2. **Supporting reasoning**: These elements of Reasoning aim to support or clarify the main logical thread and are thus presented in terms like "is justified by," "is based on," or "can be assumed because."
   Hypothetical example: *Oxygen isotope ratios can be converted to temperature using the following conversion*, **which has been calibrated** *against present-day measurements.*
3. **Compare and contrast**: These elements of Reasoning are used when a comparison between data and model, or between the predictions of two hypotheses, is used to support a Claim. Common phrasing includes "the data agree with the model," "X is smaller than predicted," or "the two measurements disagree significantly."
   Hypothetical example: *The experimental group exhibits* **a significantly higher response** *to the treatment than the control group, indicating that the intervention is efficacious.*
4. **Establishing significance**: These elements of Reasoning support compare and contrast statements by quantifying or qualifying the degree of agreement or disagreement, or to signal clear outliers. Common phrasing includes "is statistically significant," "differs by 5-sigma," or "has a p-value < 0.01."
   Hypothetical example: *The signal is found to be* **5.3 times above the background noise**, *indicating a* **significant detection**.
5. **Qualifying or quantifying uncertainty or bias:** These elements of Reasoning explain how measurement uncertainties are determined or how biases in the measurements or the sample are mitigated, often through statistical tests or control samples. You will usually find this kind of Reasoning in tandem with establishing significance. Common phrasing includes "we determined our uncertainties by," "we controlled for variance in X by," or "we compared our measurements to."
   Hypothetical example: **To assess our detection thresholds**, *we implanted simulated stars in our image*, **defining the detection region** *as those combinations of separation and brightness in which 90% of simulated sources are correctly recovered.*
6. **Qualifying or quantifying design choices:** These elements of Reasoning modify the Evidence by justifying the choices of the study design. Common phrasing includes "was done to," "is justified by," or "was chosen because."

Hypothetical example: *The control sample* **was selected** *to have the same age, gender, and socioeconomic distribution, thereby* **removing** *biases associated with these characteristics.*

7. **Qualifying or quantifying limitations:** These elements of Reasoning modify either the Evidence or the Claim by defining the scope of the work or acknowledging where the assumptions of the analysis may break down. Common phrasing includes "measurements were constrained to," "this result is applicable for," or "this result should only be used for."
   Hypothetical example: *The trend between income and health* **is only valid** *between half and twice the median income, and* **only applies** *to the demographic groups examined in the study.*
8. **Supporting confirmation from other studies**: These elements of Reasoning provide further support for a Claim by bringing in the results of outside studies with similar or parallel findings. Common phrasing includes "study X has also shown," "these results are comparable to," or "our findings are consistent with."
   Hypothetical example: **Previously reported** *measurements of the color magnitude diagrams of the globular clusters NGC 7652 and Tucanae* **show similar** *splits in the Main Sequence, indicating that multiple populations are common in these systems.*
9. **Evaluation of counterfactuals**: These elements of Reasoning also support a Claim by examining alternative scenarios that test the causality of a relationship; whether conditions are sufficient or necessary to lead to our outcome. Common phrasing includes "if we consider the opposite case," "outcomes for the control group," or "the contrary scenario also supports these findings."
   Hypothetical example: *Our study finds that smoking increases the likelihood of lung cancer 10 times compared to those who never smoke. However, the rate of lung cancer was still 3%* **among the control group**.

In summary, identifying these linguistic cues can reveal Reasoning in every section of an article, from Methods to Discussion, allowing you to follow how Evidence connects to and supports each Claim.

Box 5.4 Knowledge Check: Question 5.2 Which phrase most clearly signals that a statement is part of a research article's Reasoning (rather than just stating data)?

A. We measured 3.2 micrograms of the substance in each sample.
B. Control and treatment groups had the same initial size (n = 50) to reduce sampling bias.
C. We found a significant *p*-value of <0.01, indicating we reject the null hypothesis.
D. Figures 2 and 3 summarize the spectral data over a 24-hour period.

(Check your understanding using the Knowledge Check Key at the end of the book.)

5.5.3 Is Reasoning Missing?

Because Reasoning is so central to establishing the validity of a Claim, it should be the primary focus of a review or critique of an article. Here we describe several issues to look for when assessing whether an article has sufficient Reasoning to justify its Claim.

Missing or incomplete uncertainties: Many lines of Reasoning rely on statistical significance to distinguish between two hypotheses or to indicate a detection, and this, in turn, depends on accurate measures of uncertainty. For experimental measurements, this includes accounting for both statistical and systematic uncertainties. For sampling, uncertainty must factor in the sample size and inherent biases. If the experimental uncertainties are missing, incomplete, or poorly justified (e.g., "we uniformly adopt an error of 0.3 m/s"), then the Reasoning may not be sufficient to support the Claim.

Missing or weak support for Reasoning: The individual logical steps that are used to connect Evidence to Claim often require their supporting Reasoning to justify them. For example, consider the following dubious, hypothetical Claim:

> Our measurements show that drug T is effective at reducing dry skin. Since dry skin and psoriasis both affect the skin, drug T is likely effective at resolving psoriasis.

While this statement grammatically parses as a logical chain, it should be clear that the phrase "Since dry skin and psoriasis both affect the skin" is a logical fallacy, attempting to equate two completely different conditions simply because they both affect the skin. The researchers would have to demonstrate that dry skin and psoriasis are fundamentally or causally related (they are not) to make this Claim.

Failure to account for other factors: This often happens when studies confuse correlation with causation. For example, Goldfarb et al. (2014) conducted a meta-study that indicated that sitting down to eat family meals was correlated with a decrease in various teenage risk behaviors (i.e., alcohol, smoking, drug use, and school delinquency, etc.). This result was widely interpreted to mean that eating as a family would reduce teenagers' risky behaviors – a causative rather than correlative Claim. This leap in interpretation overlooks other underlying factors. Indeed, the discussion section of Goldfarb et al. (2014) notes that the observed relationship could be due to socioeconomic factors (e.g., more affluent families are more likely to have the time and resources to eat meals together) or depend strongly on past and inherited risk behaviors, rather than being a causal connection.

Failure to account for alternative interpretations: A common source of incomplete Reasoning is the failure to account for alternative interpretations to explain a result. This oversight is a common manifestation of confirmation bias, a cognitive bias in which one tends to focus on Evidence and Reasoning that supports a presupposed Claim. A historical scientific example of this oversight is the persistence of the geocentric model of the Universe, in which the Earth (no doubt, an important place for humans) was assumed to be at the center of the Universe. Even when the motions of planets like Mars were found to be inconsistent with this picture, early astronomers found it easier to simply modify their model by adding epicycles rather than accept the alternative interpretation that the Earth is not at the center of the Universe.

Unwarranted generalizations: Inductive Reasoning develops a generalized Claim based on patterns in Evidence. However, generalizations must be carefully constrained to match the scope of the Evidence; otherwise, we risk overgeneralization. Population studies are often susceptible to unwarranted generalizations. For example, suppose a study shows that applying finely ground basalt to a single temperate-forest test plot in Oregon accelerates silicate weathering and boosts carbon sequestration. It would be an unwarranted generalization to assume the same treatment will work equally well in tropical rainforest soils, arid grasslands, or agricultural fields worldwide. These cases would need to be studied separately to establish their validity. In effect, the Reasoning has failed to consider the limitations on the Claim.

Ignoring counterfactuals: As discussed in Sections 5.2.4.4 and 5.3.1, counterfactuals are a powerful means of examining the causality between a set of conditions and associated outcomes, thereby testing the validity of a Claim of causality. An illustrative example in Physics is the photoelectric effect, where light shining "(photo-)" on a metal can cause electrons to be emitted ("-electric"). This phenomenon enables solar panels to generate electricity from sunlight. In 1905, Einstein proposed that this process occurs because light is grouped into bundles of energy, now known as photons, with the energy of each bundle being proportional to the frequency of light. This quantization of light was one of the first steps in the development of the theory of Quantum Mechanics. When the energy, and hence frequency, of light was large enough, electrons were released by a metal surface. This theoretical model had two predictions (1) the light had to be above a minimum frequency to have enough energy to kick out an electron and (2) above this frequency, more intense light (i.e., more photons striking the surface of the metal) would release more electrons. Both predictions were experimentally verified.

This model also predicted a counterfactual: If light with a frequency below the minimum threshold illuminated a metal surface, no electrons should be emitted, regardless of the light's intensity. This counterfactual was also confirmed experimentally, showing that light frequency is a *necessary condition* for the photoelectric effect to occur. This groundbreaking work earned Einstein the 1921 Nobel Prize in Physics.

The importance of peer review: Because there are many ways to reason an argument, and many of the Reasoning steps tend to be field-specific, peer review remains the standard way to determine if an article's Claim is justified. By drawing on the experience and expertise of practitioners in the field, who may have different perspectives, insights, and biases than the authors, peer review enables the evaluation of Reasoning from various perspectives. As a result, journals typically require two or three peer reviewers, and some of the highest-profile journals, such as *Science* and *Nature*, require multiple peer reviewers to review, critique, and ultimately accept or reject a study. Articles that have not undergone peer review (and even some that have; see Chapter 10) should be considered with a high level of skepticism.

Box 5.5 Question 5.3 Which scenario best exemplifies a missing or incomplete Reasoning step that undermines the study's conclusions?

A. A research article reports a drug's efficacy with a thorough explanation of the control group and statistical analysis.
B. An article observes a correlation between reading speed and test performance and then explains potential cognitive links and confounding factors.
C. An astronomer takes raw images, applies field-standard calibrations, and provides robust uncertainty estimates before concluding a planet discovery.
D. A team claims a new vaccine prevents disease but never clarifies how they handled biased sampling or alternative explanations for improved patient outcomes.

(*Check your understanding using the Knowledge Check Key at the end of the book.*)

5.6 Worked Example

Let's return to the Long et al. (2020) biology article we examined in Chapter 4 for Evidence, and see how the article's Reasoning is articulated. As a reminder, Long et al. (2020) investigated why some individuals exposed to SARS-CoV-2 develop severe symptoms – such as respiratory distress and fever – while others remain asymptomatic or mildly affected. Here, we will focus on the Reasoning that connects their immunological Evidence to the Claim that asymptomatic individuals mount a weaker immune response. As we will see, **Long et al. (2020) employ inductive Reasoning** by building a general Claim from patterns observed in the data.

- Title: "Clinical and immunological assessment of asymptomatic SARS-CoV-2 infections"
- Citation: Long, Q. X., Tang, X. J., Shi, Q. L., … & Huang, A. L. (2020). *Nature Medicine, 26*(8), 1200–1204. https://doi.org/10.1038/s41591-020-0965-6
- Abstract:

> The clinical features and immune responses of asymptomatic individuals infected with severe acute respiratory syndrome coronavirus 2 (SARS-CoV-2) have not been well described. We studied 37 asymptomatic individuals in the Wanzhou District who were diagnosed with RT–PCR-confirmed SARS-CoV-2 infections but without any relevant clinical symptoms in the preceding 14 days and during hospitalization. Asymptomatic individuals were admitted to the government-designated Wanzhou People's Hospital for centralized isolation in accordance with policy. The median duration of viral shedding in the asymptomatic group was 19 d (interquartile range (IQR), 15–26 d). The asymptomatic group had a significantly longer duration of viral shedding than the symptomatic group (log-rank P = 0.028). The virus-specific IgG levels in the asymptomatic group (median S/CO, 3.4; IQR, 1.6–10.7) were significantly lower (P = 0.005) relative to the symptomatic group (median S/CO, 20.5; IQR, 5.8–38.2) in the acute phase. Of asymptomatic individuals, 93.3% (28/30) and 81.1% (30/37) had reduction in IgG and neutralizing antibody levels, respectively, during the early convalescent phase, as compared to 96.8% (30/31) and 62.2% (23/37) of symptomatic patients. Forty percent of asymptomatic individuals became seronegative and 12.9% of the symptomatic group became

> negative for IgG in the early convalescent phase. In addition, asymptomatic individuals exhibited lower levels of 18 pro- and anti-inflammatory cytokines. These data suggest that **asymptomatic individuals had a weaker immune response to SARS-CoV-2 infection**. The reduction in IgG and neutralizing antibody levels in the early convalescent phase might have implications for immunity strategy and serological surveys.

The **Claim** is near the bottom:

> These data suggest that asymptomatic individuals had a weaker immune response to SARS-CoV-2 infection.

Motivation for the study: The authors begin by justifying the need for their research, citing work by Bai et al. (2020) and Hu et al. (2020) to emphasize the distinct role of asymptomatic carriers in viral spread:

> We studied 37 **asymptomatic individuals** in the Wanzhou District who were **diagnosed with RT–PCR-confirmed SARS-CoV-2 infections but without any relevant clinical symptoms** in the preceding 14 d and during hospitalization.

This section is formally part of the Context of the study, as it establishes the research gap: the immune profiles of these carriers remain poorly understood. To address this gap, the authors ensure meaningful comparisons for an **inductive argument** by organizing participants into three well-defined groups:

- **Positive control:** *Symptomatic* SARS-CoV-2-*positive* patients confirmed via RT-PCR;
- **Negative control:** *Asymptomatic* SARS-CoV-2-*negative* individuals confirmed via RT-PCR; and
- **Experimental group:** *Asymptomatic* SARS-CoV-2-*positive* patients confirmed via RT-PCR.

All participants were matched by sex, age distribution, and comorbidities within the Wanzhou District of China at the same time. This careful matching of sample groups is a key piece of **methodological reasoning**: It helps minimize potential confounding variables, improves the comparison across groups, and improves the study's internal validity.

The authors then implemented a range of standard medical diagnostic tools – RT-PCR tests, antibody assays, chest radiographs, and complete blood counts – to assess immune response. They analyzed this Evidence using statistical methods, such as box plots and Mann–Whitney tests, which are forms of **mathematical/statistical reasoning**. From the abstract:

> The median duration of viral shedding in the asymptomatic group was 19 d (interquartile range (IQR), 15–26 d). The asymptomatic group had a **significantly longer duration** of viral shedding than the symptomatic group (log-rank P = 0.028). The virus-specific IgG levels in the asymptomatic group (median S/CO, 3.4; IQR, 1.6–10.7) **were significantly lower** (P = 0.005) relative to the symptomatic group (median S/CO, 20.5; IQR, 5.8–38.2) in the acute phase. Of asymptomatic individuals, 93.3% (28/30) and 81.1% (30/37) had reduction in IgG and neutralizing antibody levels, respectively, during the early convalescent phase, **as compared to** 96.8% (30/31) and 62.2% (23/37) of symptomatic patients.

The bolded statements highlight the distinctions between the experimental group and the positive control group, with mathematical reasoning employed to establish statistical significance ($p < 0.005$). These quantified assessments of significance strengthen the main Claim that asymptomatic patients had weaker immune responses.

Long et al. (2020) also draw on **analogical reasoning** to support their Claim, comparing SARS-CoV-2 to biologically similar viruses like SARS-CoV and MERS-CoV. For example, in the main text:

> In SARS-CoV, viral RNA was detectable … () for as long as 4 weeks after disease onset. **In MERS-CoV infections**, viral shedding in respiratory secretions persisted for at least 3 weeks (p. 1203–04).

By drawing on past work that shows that other CoV viruses show prolonged RNA shedding, they argue that SARS-CoV-2 does as well. They extend this analogical reasoning by noting that prior MERS-CoV research has found that asymptomatic individuals shed the virus for a median of 19 days – longer than symptomatic individuals, suggesting a similar delay could occur for SARS-CoV-2 as well.

Next, the authors employ **causal reasoning** to support the interpretation that a longer duration of shedding in asymptomatic individuals leads to delayed immune clearance. For example, in the main text:

> IgG levels and neutralizing antibodies in a high proportion of individuals who recovered from SARS-CoV-2 infection start to decrease within 2–3 months after infection (p. 1204).

The timeframe for waning IgG immunity (i.e., humoral or antibody-based immunity) is significantly shorter than that for other CoV viruses. Reported IgG durability varies widely across coronaviruses: about 8–12 months for common cold strains, 2–3 years for SARS-CoV-1 and typical MERS-CoV cases (occasionally longer), and ≥ 1 year for most SARS-CoV-2 infections to date. However, the authors qualify this interpretation. They note the presence of viral RNA does not equate to infectiousness, and RT-PCR cannot detect whether virus particles are viable. These **limitations** are part of a reasoning framework that acknowledges uncertainty and constrains the Claim.

Long et al. (2020) use additional **methodological reasoning** to explain differences across studies. They note that definitions of shedding duration, severity classification, and testing frequency all influence findings. Further **epistemic reasoning** – acknowledging the limitations of experimental studies in the scope of an evolving viral outbreak – helps readers understand why studies vary and why generalizations require caution.

Finally, Long et al. (2020) employ **policy-based reasoning** to generate Implications – a somewhat uncommon approach in the research literature. If immunity wanes quickly (i.e., 2–3 months), as their IgG data suggest, policies, such as immunity passports, may be premature. This logic connects individual-level immunological data to societal-level decisions.

5.6.1 Critiquing Reasoning

A crucial benefit of clearly defining an article's Reasoning is that it enables us to assess whether the arguments presented by the authors – and hence the article's main Claim – are robust. We will delve into using CERIC for Critique in Chapter 10, but let's take a moment to evaluate the Reasoning in this article on COVID-19 immunity for some early practice.

Long et al. (2020) present a compelling inductive argument, building from carefully collected Evidence to a cautious but consequential Claim about the immune response. They acknowledge uncertainty and limitations and invite future studies to test the durability of their conclusions – a hallmark of strong scientific Reasoning. They also draw on other forms of Reasoning – methodological, mathematical/statistical, analogical, causal, epistemic, and policy-based – to strengthen their arguments.

Despite the extensive argumentation, the paper nevertheless contains some Reasoning gaps. A more straightforward explanation for the **causal mechanism** behind asymptomatic individuals clearing the virus more slowly could reinforce the stated causal claims; for example, the cause could be lower levels of initial immune response. More explicit **probabilistic reasoning** – such as quantifying the likelihood of reinfection – would sharpen their treatment of uncertainty. Their existing **epistemic reasoning** could be extended by comparing study design trade-offs, such as the benefits and limits of cross-sectional vs. longitudinal approaches. Some of these lines of Reasoning may appear in an online Supplement, and for an article with global health significance, it may be worth searching.

5.7 Chapter Key Takeaways

1. **Reasoning is the logical bridge that connects Evidence to scientific Claims**. Sound Reasoning links past knowledge, observations, models, or data to conclusions through a structured and logical process of argumentation. At its core, Reasoning seeks to make findings, conclusions, beliefs, assertions, or decisions understandable and acceptable to others by offering a "robust argument" that is clear, relevant, logical, and sufficient for the study.
2. **Reasoning builds on different types of logical arguments:** Deductive arguments test general hypotheses through specific experimentation. Inductive arguments draw general conclusions from specific experimental results. Abductive reasoning attempts to infer the best explanation from a set of observations or facts. These methods may appear singly in specific types of research or may be combined.
3. **Causal arguments require special attention** to establish necessary and sufficient conditions that distinguish them from correlation arguments. Counterfactuals can provide a useful approach to testing these conditions.
4. **Modeling and simulation**, including conceptual, physical, mathematical, statistical and computational models, allows researchers to explore complex datasets or systems, and to distinguish noise from expected signal in measurements.
5. **Regular versus critical arguments** highlight how researchers either operate within accepted frameworks or challenge them.
6. **Sound reasoning** depends on accurate and precise premises and transparent acknowledgment of limitations or uncertainties. Logical fallacies (e.g., "we must

use fossil fuels or society collapses"), flawed premises (e.g., "penguins are cold-blooded"), or unjustifiably broad generalizations (e.g., "all dissolving substances must be salt") undermine an argument's logic and cast doubt on an article's Claim. It is also important to follow the data to where it leads; forcing data to fit a preconceived idea leads to flawed Reasoning practices of cherry picking and pseudoscience.

6 Implications

The implications of our knowledge of the stars are far more profound than merely revealing the nature of distant worlds; they change our understanding of our place in the universe.

—Carl Sagan (*Pale Blue Dot*, 1994)

6.1 Overview

Chapter 6 focuses on the CERIC element of Implications, providing a framework for understanding their role as indicators of future work and guides for expanding scientific research. This chapter begins by defining Implications and how they manifest in different research areas. It explores the common types of Implications, indicating incompleteness in the present work, pointing toward future validation studies, and suggesting future research directions. This chapter discusses how today's implications often become tomorrow's research questions, underscoring their importance in driving the progression toward new knowledge. This chapter distinguishes Implications from Claims and Context, providing strategies for differentiation.

6.2 Defining Implications

Implications are forward-looking statements that project the significance of a study beyond its immediate results. While Claims rest on the article's Evidence and Reasoning, Implications sketch the logical next steps: follow-up experiments, practical applications, or policy uses that flow from the study's Claim, Evidence, and Reasoning (CER) core. Seeing how authors move from "what we found" to "what should happen next" helps readers trace the arc of scientific progress – today's discovery becomes tomorrow's research question.

Because authors usually place future-oriented comments where readers will remember them, Implications are among the easiest CERIC elements to locate. They appear most often in the closing paragraphs of the Discussion or Conclusion, where the text pivots from a results summary to broader meaning. Watch for cue phrases such as *this suggests*, *future studies should*, or *these results may inform*. Abstracts and introductions may also include a brief look ahead at possibilities that still require testing.

6.2.1 Implications in Research Articles

In this chapter, we define an Implication in research as a **future-looking narrative** about what comes after the present research study. Implications emerge logically from the CER core and address at least two questions:

1 How is the result significant beyond the immediate findings?
2 Where does the research go from here?

Let's pick apart this definition of an Implication.

1. How is the result significant beyond the immediate findings? An Implication is a forward-looking narrative, highlighting future research based on the current study's findings. A well-constructed Implication can convey in a few sentences what researchers could logically examine in subsequent studies. There are at least two common ways to break down the forward-looking time aspect of the narrative:

a. **The immediate significance of the result** directly suggests the logical next steps. For example, A pilot study reveals that high-resolution airborne LiDAR (a laser surveying method) detects recent surface ruptures on a previously unmapped fault beneath the outskirts of Santiago, Chile (Maldonado et al., 2021). The most direct next step is to run trenching, dating, and geophysical surveys along the entire fault trace to determine slip rates and recurrence intervals. Those data could then be folded into regional seismic-hazard models and municipal building-code updates.
b. **The broader significance of the result** indicates mid- or long-term impacts on larger systems, fields, or society. For example, researchers demonstrate a machine-learning data-assimilation scheme that ingests satellite radiances and ground observations into a mesoscale weather model, cutting 72-h precipitation-forecast error by about 30% (Sun et al., 2025). In the medium term, this improvement could extend flood-warning lead times, allowing emergency managers to stage resources more effectively. Over the long term, embedding the same assimilation architecture in coupled atmosphere–ocean–land models could raise the fidelity of decadal climate projections, providing policymakers with more reliable guidance for infrastructure planning and carbon-mitigation targets.

By distinguishing the immediate next experiment (i.e., fault-trenching) from the broader chain of benefits (i.e., better flood response and enhanced climate policy), we see how a well-crafted Implication maps today's result onto tomorrow's research agenda and, potentially, society-level outcomes.

2. Where does the research go from here? An Implication describes what needs to be done to build on the present research study. While the nature of knowledge in Implications is speculative, substantial implications are drawn from the CER core to address what logically comes next in the future research continuum. What comes next should never be wishful, but a direct extension, expansion, or application of the study's main argument. Let's consider a real-world example of how wishful Implications can go wrong.

6.2.1.1 Example: Real-World Harm from a Wishful Implication

In 1990, oceanographer John Martin published a short, provocative essay in *Paleoceanography* outlining what he called the "iron hypothesis" (Martin, 1990). Laboratory and bottle experiments had convinced him that phytoplankton growth in vast, nutrient-rich but iron-poor reaches of the Southern Ocean was stunted only by a lack of the micronutrient iron. Martin described this with a flourish in a 1988 lecture at Woods Hole Oceanographic Institution: "Give me half a tanker of iron and I'll give you an ice age" (Weier, 2001). The sentence quickly took on a life of its own. Readers leapt from the paper's modest Evidence – iron limitation of surface-water algae – to the alluring Implication that a few shiploads of powdered iron could draw down enough atmospheric CO_2 to cool the planet.

Over the next two decades, that wishful Implication attracted entrepreneurs who moved faster than the science. Many invoked John Martin's untested "iron hypothesis" to justify commercial ocean-fertilization schemes. Planktos Inc. first proposed dispersing iron filings over 10,000 km^2 of the Pacific, a plan quashed by international regulators yet widely publicized (Courtland, 2008). Soon after, the Haida Salmon Restoration Corporation – bankrolled by a C$2.5 million loan from the Indigenous community of Old Masset – dumped 100 tons of iron-sulfate west of the Pacific islands of Haida Gwaii, promising revived salmon runs and lucrative carbon credits; instead, the project triggered a federal investigation for breaching the 2008 London Dumping Convention's moratorium and coincided with toxin-producing Pseudo-nitzschia blooms (Tollefson, 2012). By then, 13 controlled iron-addition experiments had shown that most carbon captured by fertilized blooms fails to sink permanently, implying that large-scale dumps could offset only a few percent of annual emissions while courting ecological risk (Boyd et al., 2007). The accumulating evidence prompted an amendment to the London Convention in 2013 to mandate case-by-case permits for any future ocean-fertilization research, curbing further wishful ventures (London Protocol, 2013).

The iron-fertilization saga illustrates how an enticing but weakly supported Implication can escape the confines of scholarly debate and cause real-world harm – financial, regulatory, and ecological. It is a textbook warning to researchers everywhere: before broadcasting an Implication, make sure the science truly supports the leap.

Box 6.1 Common Confusion: Implications versus Conclusion

A common confusion point with implications is how they differ from the conclusion. While implications sketch out the logical next steps for future work, conclusions summarize what the present study has already established. Put another way, conclusions are anchored in now, whereas implications look toward next. Both are often colocated at the end of an article.

For example, Alvarez et al. (1980) concluded that an anomalously high global iridium layer at the Cretaceous–Tertiary boundary "is best interpreted as the result of a large extraterrestrial impact" (p. 1105). The Implication pointed forward: If an impact caused the mass extinction, researchers should "*search for* a suitably sized crater of approximately 100 km diameter and of latest Cretaceous age" (p. 1105). The first sentence reports a finding already supported by data – the Claim – whereas the second proposes a concrete line of inquiry for future studies – the Implication.

Takeaway: Distinguishing these two time horizons keeps us from mistaking what the article has found for what it recommends exploring next.

6.2.2 The Nature of Implications Is the Future and after the Present Work

As scientific research progresses, today's implications become tomorrow's research questions. The march of ideas – from Implications to new findings to more Implications – is a significant reason why a research career can be thrilling and rewarding over many decades. Tracing the Implications of one study to the next is a focused strategy to follow the narrative thread of an idea's evolution. By focusing on the Implications and the research questions they generate, it is possible to navigate the ocean of new studies without drowning.

6.2.2.1 Case Study: When Implications Mature into New Research Questions

One way to visualize how Implications become research questions is to consider a case study of how one topic developed over time, using the Implications from prior research as the basis for new research, repeating this approach over many years.

Cancer immunotherapy offers a clear illustration of the *Implication → Research Question → Implication* cycle. Each step below pairs (a) the key implication an article raised with (b) the question that the very next study set out to answer.

Case Glossary

First, let's define some key terms:

- **Anti-PD-1 antibody:** Lab-made antibody (e.g., nivolumab and pembrolizumab) that blocks the PD-1 checkpoint on T cells, preventing tumors from switching them off and enabling a more potent immune attack.
- **Cancer immunotherapy:** Treatment approach to cancer that harnesses the patient's immune system to locate and destroy cancer cells by lifting inhibitory signals, boosting stimulatory cues, or supplying tumor-specific cells or vaccines – often achieving longer-lasting control than standard therapies.
- **T cells:** Thymus-trained white blood cells that patrol for infected or malignant cells, directly killing threats or orchestrating broader immune defenses.
- **Tumor mutational burden (TMB)**: Count of DNA mutations in a tumor (i.e., somatic per megabase). A high TMB yields many new antigens and often predicts better responses to checkpoint immunotherapy.

Implication → Research Question → Implication Cycle

With our key terms defined, let's dive into the situation and cycle. Because patient responses to cancer immunotherapy vary widely, we need tools that can more accurately predict who will benefit.

1. **Mixed responses call for new predictors**

Topalian et al. (2012) demonstrated that the first anti-PD-1 antibody could control skin (melanoma), kidney, and lung cancers in some individuals. Still, it helped only a small fraction of patients who had already tried many other treatments (like chemo, surgery, and radiation).

a. Topalian et al. (2012) highlighted PD-L1 expression as a preliminary biomarker that **could be important for future trials**.
b. Research question that followed: *Which tumor features forecast benefit from PD-1 blockade?*

2. **Tumor mutational burden (TMB) emerges as a predictor**
Rizvi et al. (2015) found that non-small cell lung cancers (NSCLC) with many DNA mutations – creating lots of new tumor antigens – were far more likely to respond to treatment and remain stable for longer periods.
a. The Implication was practical: Use TMB to **refine future work** of patient selection and trial design.
b. Research questions now being pursued: *What is the optimal TMB cut-off, and can other genomic or immune markers improve its precision?* (e.g., Hellmann et al., 2018).

3. **Chemotherapy may boost the power of anti-PD-1 therapy**
Preclinical evidence showed that certain cytotoxic drugs trigger immunogenic cell death (ICD), a form of tumor cell demise that decorates dying cells with "eat-me" signals (calreticulin) and releases danger molecules (such as HMGB1).
a. These signals activate dendritic cells, which in turn, prime tumor-specific T-cells:
   - **Doxorubicin**: Doxorubicin-treated tumor cells release HMGB1; blocking the HMGB1–TLR4 interaction abolishes the tumor's protection in mice (Apetoh et al., 2007).
   - **Oxaliplatin:** Oxaliplatin, but not cisplatin, exposes calreticulin and releases HMGB1, inducing an antitumor response from T cells in mouse models (Tesniere et al., 2010).
b. Building on the ICD concept, Langer et al. (2016) randomized previously untreated, advanced NSCLC patients and asked, *what is the progression-free survival on triplet therapy (pembrolizumab + carboplatin + pemetexed) compared to chemotherapy alone?* The open-label phase-2 trial showed:
   - Objective response rate: 55% vs. 29%
   - Median progression-free survival: 13.0 months vs. 8.9 months

These encouraging data prompted the initiation of the double-blind phase-3 trial, which incorporated patient-reported quality-of-life measures. The chain of Evidence – from ICD in mice to improved outcomes with chemo-immunotherapy in patients – supports the hypothesis that chemotherapy can unveil additional tumor antigens and enhance the effectiveness of anti-PD-1 treatment.

4. **Confirming benefit and measuring quality of life**
KEYNOTE-189 provided the definitive clinical test of chemo-immunotherapy in advanced non-squamous NSCLC:
- **Efficacy (Gandhi et al., 2018):** A double-blind, placebo-controlled phase-3 trial (N = 616) showed that adding pembrolizumab to pemetrexed + cisplatin or carboplatin halved the risk of death (hazard ratio ≈ 0.49) and doubled median progression-free survival compared with chemotherapy alone.

- **Patient-reported outcomes (Garassino et al., 2020):** The cohort reported stable or slightly improved global health status and NSCLC-specific quality of life with the triplet regimen.

a. **These studies answered the key implication raised by Langer et al. (2016)**: triplet therapy (pembrolizumab + platinum + pemetrexed combined) extends survival and maintains patient-reported quality of life.
b. **Future work will now focus on longer follow-up**, biomarker-guided subgroup analyses, and the management of immune-related toxicities.

Takeaway: At every step, researchers treated the "next steps" proposed in one study as the working hypothesis for the next. In barely 10 years, that iterative hand-off achieved two parallel breakthroughs:

- **Uneven responses led to genomic biomarkers**: Early variability in anti-PD-1 benefit prompted a search for predictors, which led to the development of TMB and PD-L1 assays now built into clinical trials.
- **Hypothetical chemo-immunotherapy synergy led to first-line standard:** Preclinical hints of antigen release after chemotherapy inspired a phase-2 test, which in turn justified the phase-3 study and, ultimately, practice-changing guidelines that also document preserved quality of life.

Tracing research chains makes this principle clear: the Implications paragraph is not filler at the end of an article. It is the seed for the next experiment and, often, the next advance in research or patient care.

6.2.3 Implications in Proposals

Because Implications drive future research questions, they serve an essential purpose in research proposals. Proposals are all about the work you want to do next, and Implications organize the work in at least two crucial ways. First, the Implications of past research logically point to your proposed work. A strong proposal will synthesize major Implications from several lines of research into a new set of research questions. Second, the likely Implications of the proposed work, when completed, can make the proposal stand apart from competing proposals and be more compelling to funders. Let's consider a historical example from the field of epigenetics.

6.2.3.1 Example: Stress-Induced Epigenetic Modifications

Several lines of research in the early 2000s examined stress-induced epigenetic modifications. These are defined as changes in gene expression that occur without altering the DNA sequence, including the maternal influence on epigenetics (Szyf et al., 2005), paternal contributions to epigenetic inheritance beyond maternal influence (Dias & Ressler, 2014), and the dynamic interplay between environmental factors and epigenetic plasticity as reviewed by Feil and Fraga (2012). Integrating Implications from these lines of research allows for the generation of new research questions that could further explore epigenetic mechanisms. Two such possibilities are:

1. What role do early childhood environmental factors play in reinforcing or reversing stress-related epigenetic marks inherited from both parents?
2. Are there critical windows during prenatal and postnatal development when interventions (e.g., stress-reducing therapies) can reverse or mitigate inherited epigenetic modifications associated with stress?

Notice how we synthesized major Implications from several lines of research into a new set of research questions exploring possible interventions.

Next, we consider how, if successful, the proposed research on epigenetic modifications could have Implications for interventions and treatments, such as:

- New targeted epigenetic therapies,
- New public health strategies,
- Improved understanding of trauma transmission, and
- Improved models of gene–environment interactions.

To increase a proposal's chance of success, we want to focus on the Implications that specifically address the funders' mission or the call for research, making the proposal more compelling. For instance, if a call for proposals from the National Science Foundation (NSF) emphasizes interdisciplinary research, the Implications related to improved computational models of gene–environment interactions are likely to align best and stand out.

In summary, keep projected Implications proportionate and firmly anchored in the CER cores of past articles. Overly ambitious Implications, such as curing inherited cancers, though worthy, fall outside the scope of any study. The goal is to deliver a proposal that preserves scope and credibility for your career stage, aligns expectations with funders' missions, and ultimately, succeeds.

6.3 Types of Implications

Here, we review the main types of Implications that a research article can establish, depending on the study's findings, the strength of its CER core, and the connections to future work that logically emerge. This list is not exhaustive and represents typical types.

6.3.1 Implications that Improve upon the *Present Work*

While a fully developed scientific article provides the necessary evidence and reasoning to establish a Claim, it may nevertheless remain incomplete, leaving room for further study that may affirm or refute the Claim. Some common limitations to a study include a small sample size, simplifying assumptions made in the theoretical framework or analysis, marginal statistical significance of a result, or rejecting alternative explanations without sufficient proof. As it is not practical to address every possible limitation in a study, authors often identify limitations to motivate future work related to the study.

- **A current finding needs more work** when a study's results are preliminary or incomplete, requiring further investigation to validate or explore additional aspects. Often, this incompleteness stems from simplifying assumptions, study design limitations, or limited analysis. For example, a study reporting a revised rate of glacial melt might suggest more work is needed to better to understand the interplay of complex physics and environmental factors.
- **Replication of the current work** highlights the importance of repeating a study to verify its findings. For example, in hydrogen fuel development, suppose an article reports that a newly developed nickel–iron layered double hydroxide catalyst splits water at record-low overpotentials (a hot topic in energy conversion technologies like metal-air batteries and water-splitting in renewable fuels; Edao et al., 2024), releasing hydrogen for fuel. In that case, independent laboratories must reproduce the synthesis, electrode fabrication, and electrochemical testing with large sample sets to confirm the performance before the hydrogen fuel industry invests in pilot-scale electrolyzers.
- **Expansion of the current work** identifies ways to broaden a study by adding variables, increasing sample size, or testing new conditions. For instance, in quantum computing, if a prototype experiment shows that a pair of superconducting qubits (similar to bits in classical computing) achieves 99.9% gate fidelity at 20 mK, the logical expansion is to replicate the result across larger qubit arrays, multiple fabrication batches, and a wider temperature range to see whether the quantum architecture remains robust as the system scales.

6.3.2 Implications that Point to *New Work within a Field*

Implications are often used to motivate future work in the specific field of study. They can be couched as testable predictions that can further illuminate aspects of the stated Claim or present new methodologies or theoretical frameworks that can affirm or refute key results in the field. These Implications differ from the previous set of Implications in that they focus on the *future work that is possible* as a result of the Claim. Here are some examples:

1. **Future research directions** identify new questions or areas for further study that logically follow from the current findings. For instance, after discovering a correlation between PM 2.5 air pollution and respiratory diseases (Xing et al., 2016), future research could investigate specific pollutants' effects on special human populations (like children or people with compromised immune systems).
2. **Methodological Implications** suggest improvements or adjustments to research methods based on the study's findings or limitations. For example, a study on survey data collection revealed that participants' responses vary significantly depending on the time of day they are surveyed, suggesting the need for future studies to standardize survey timing or adjust for these variations to improve the accuracy of results.

Box 6.2 Knowledge Check: Question 6.1 Which of the following scenarios best exemplifies Implications that highlight future work related to the study?

A. Explaining the historical context of air pollution studies to show what was known before the study.
B. Proposing further tests using larger patient samples to confirm the efficacy of a new vaccine.
C. Citing well-known fundamental constants, such as the speed of light, that helped guide the current experiment.
D. Summarizing the current article's claim that identifies an Earth-sized exoplanet.

(Check your understanding using the Knowledge Check Key at the end of the book.)

3. **Strengthening or weakening of existing practices:** Suppose a study compares a machine-learning classifier against the widely used Box Least-Squares (BLS) pipeline for detecting exoplanet transit (e.g., Schanche et al., 2018). If the machine-learning classifier delivers higher sensitivity and fewer false positives, it could strengthen the case for adopting future AI-based search methods across upcoming astronomy missions, such as TESS and PLATO. Conversely, if the study uncovers systematic biases or missed detections in the output, it could weaken confidence in using machine-learning tools and reinforce reliance on traditional algorithms.
4. **Theoretical Implications** refer to how research findings contribute to, modify, or challenge existing theories within a field. For instance, precise measurements of neutrino oscillations were made by collaborations between two observatories: Super-Kamiokande in 1998 (Fukuda et al., 1998) and Sudbury Neutrino Observatory in 2002 (Ahmad et al., 2002). Each showed that neutrinos have mass. These findings contradicted the original Standard Model, which treated them as massless, and spurred the development of new theories, such as seesaw mechanisms, to explain neutrino mass generation. These discoveries contributed to the awarding of the 2015 Nobel Prize in Physics to Takaaki Kajita and Arthur McDonald, the leaders of the Super-Kamiokande and Sudbury Neutrino Observatory projects.

6.3.3 Implications across Fields or in Society

A study's findings may extend beyond the specific field in which it was focused, leading to impacts across disciplines or even changing society itself. Often, an original study may not recognize such transformative Implications, which are only recognized later. Here are some examples:

- **Ethical Implications** arise when research raises ethical concerns about its application, participant welfare, or broader societal impacts. For example, genetic research may raise ethical questions about privacy and consent, while animal testing in the pharmaceutical industry often sparks debates about humane treatment.

- **New research fields or topics** are less common but very important when they occur. Some examples include the detection of gravitational waves and the discovery of CRISPR-Cas9 as a gene-editing tool, both of which have opened new research fields and led to groundbreaking studies that were previously impossible.
- **Paradigm shifts** occur when research findings challenge or overturn established scientific beliefs, leading to a fundamental change in the understanding of a field. For instance, the shift from Newtonian mechanics to Einstein's theory of relativity redefined our understanding of space and time.
- **Policy Implications** show how research findings can inform or influence policy decisions. For example, seismologists found that when Oklahoma quadrupled the volume of salty wastewater pumped underground (2009–2013), the number of magnitude-3 or larger quakes jumped (Walsh & Zoback, 2015). In response, the Oklahoma Corporation Commission ordered lower injection rates and shut down some wells in 2016 (OCC, 2016).
- **Practical or applied Implications** focus on how research can solve real-world problems or enhance current practices. For example, in battery chemistry, demonstrating that lithium–iron–phosphate cathodes retain >80% capacity after roughly 2,000 charge–discharge cycles spurred their adoption in electric buses and grid-storage batteries, thereby boosting safety and lifetime relative to cobalt-based cells (Yan et al., 2024).

6.3.4 Speculative Implications

Implications can sometimes be framed as speculation on future outcomes that the present work cannot fully justify. These implications may include testable predictions that affirm or refute the speculation, thereby deepening the present study's impact. Let's consider some common scenarios:

- **Betting on marginal results:** Discoveries can initially appear as an anomaly in the data or a marginal discrepancy with the underlying theory, which is later shown to be statistically robust. For example, in 1992, physicists at Fermilab reported the first detections of the top quark, one of the six types of these fundamental particles. However, it was not until three years later that enough events had been detected to warrant a true "discovery" of this particle (Abachi et al., 1995; Abe et al., 1995; Liss & Tipton, 1997). While many marginal Claims will turn out to be false, speculative discussion about the potential Implications of a marginal Claim can motivate further work to affirm or refute the measurement.
- **Extending trends:** Going beyond the range of a study's experimental data is a standard method for making predictions about the underlying theory. For example, a trend in rising ocean temperatures over the past decade could lead to speculation about what ocean temperatures will be at the end of this century. Implications about long-term trends are usually based on simple mathematical models and can

motivate future studies to test whether the trend and, perhaps, the underlying theory are accurate.

- **Projecting small samples onto larger samples:** This is a common Implication for initial "pilot studies" that may focus on a limited sample with known or unknown biases. A preliminary study of a new drug may indicate it is an effective treatment for prostate cancer, but more clinical trials are required to substantiate the Claim.
- **Suggesting novel interpretations:** Occasionally, researchers observe a pattern they cannot yet explain. In 1978, Stark, Kole, Bowman, and Altman purified Escherichia coli ribonuclease P (RNase P) and showed it contained essential RNA; and nuclease digestion destroyed activity (Stark et al., 1978). Four years later, Kruger and colleagues reported that an intervening sequence from Tetrahymena thermophila rRNA excised and ligated itself in vitro without any protein catalyst (Kruger et al., 1982). Additionally, Guerrier-Takada et al. (1983) demonstrated that the isolated RNA moiety of RNase P was, by itself, a true enzyme. These Implications sparked a decade of work on "ribozymes" and, by 2000, X-ray crystal structures of the 50S (larger) ribosomal subunit confirmed that peptide-bond formation is catalyzed solely by ribosomal RNA (Ban et al., 2000; Nissen et al., 2000). Collectively, these studies rewrote biochemistry textbooks, extending catalysis from proteins-only to proteins-and-RNA, and they underpin contemporary research on prebiotic chemistry and RNA-based therapeutics. The foundational contributions of Sidney Altman and Thomas Cech were recognized with the 1989 Nobel Prize in Chemistry "for the discovery of catalytic properties of RNA" (Nobel Prize Outreach, 2025).

Speculative Implications can be among the most controversial aspects of published studies. Scientists must balance the possible real-world consequences of a Claim – should it prove correct – against the duty to avoid overhyping preliminary results, especially when those consequences could steer policy, public health, or significant investments.

Box 6.3 Knowledge Check: Question 6.2 Which statement most clearly signals a speculative Implication rather than a Claim or Context?

A. According to prior studies, gamma-ray bursts originate from merging neutron stars.
B. Our measurements prove that patients carrying gene X produce fewer immune cells.
C. Based on these pilot results, we predict that incorporating catalyst Y might dramatically increase reaction yields.
D. This study's sample included 100 low-mass stars, selected based on their temperature–pressure regime and proximity to our sun.

(Check your understanding using the Knowledge Check Key at the end of the book.)

Box 6.4 Common Confusion: When Do Implications Show Up?

We often hear two linked questions about Implications: How far can the influence of a single study travel? How long does that influence take to surface in later articles, patents, or products? The guideposts below offer a reality check (for a worked example, see the Case Study in Section 6.2.1.2).

- **Follow the citation hand-offs**: Implications in one article often reappear as explicit research questions in the next. Trace the chain with Web of Science, Scopus, or patent searches to watch each new project treat a prior "next step" as its starting hypothesis. Systematic tracking helps calibrate your sense of scholarly momentum.
- **Allow for long time horizons:** Brenner and Lerner (1992) introduced DNA-encoded combinatorial libraries to accelerate drug discovery. Thirty years later, the same idea powers high-throughput screens and precision-medicine pipelines. Transformative ripples of this scale commonly unfold over decades.
- **Scan for methods that migrate into new settings**: Companies such as X-Chem now run DNA-encoded libraries at industrial scale (X-Chem, 2025), far removed from the academic bench where the idea was born. When a technique solves problems outside its original context, its spread can outpace the formal literature. Observing such migrations keeps your view of impact realistic.

Takeaway: Before closing an interesting article, jot down one concrete experiment, tool, or policy that would logically follow if its findings held. Revisit the literature 6, 12, and 24 months later (set a calendar reminder). The growth curve will help you adjust your expectations about how quickly Implications evolve.

6.4 Implications Compared to Other Elements

Implications comprise a distinct category among the CERIC elements, commonly mistaken for Claims or Context. This confusion can arise for several reasons, such as the logical ordering of information in an article or a lack of understanding of the established work in a research area. Here, we discuss some ways to distinguish Implications from Claims and Context.

6.4.1 Implications versus Claims

Implications and Claims are often written as declarative statements, typically found toward the end of an abstract or article. Let's dig into three ways to tell them apart.

6.4.1.1 Clue 1: Time

First, time is the main way to distinguish a Claim from an Implication:

- Claims are about the **present** research.
- Implications are about **future** research.

6.4.1.2 Clue 2: Evidence and Reasoning

If the timeframe is unclear, a second distinguishing feature of a Claim is that it must have supporting Evidence and Reasoning in the study. Implications do not require these, but when written well, they extend the Claim into future research using their speculative lines of Evidence or Reasoning. Let's consider, as an example, an article reporting clinical trial results from a novel blood pressure medication by Chow et al. (2021):

- **Claim**: A phase-3, double-blind trial of 591 adults showed that an ultra-low-dose pill containing four medicines ("quadpill" = irbesartan 37.5 mg + amlodipine 1.25 mg + indapamide 0. + bisoprolol 2.5 mg) lowered unattended office systolic blood pressure by 6.9 mm Hg more than standard-dose irbesartan monotherapy after 12 weeks ($p < 0.0001$).
- **Evidence**: The measured blood-pressure differences between treated and control groups.
- **Reasoning**: Statistical analysis indicates that the difference is significant and unlikely to be due to chance.
- **Implication** (extends beyond the data): Because even small reductions in systolic pressure can significantly reduce cardiovascular risk, the authors suggest that starting treatment with an ultra-low-dose combination could lead to a decrease in heart-attack or stroke rates. They add:

> Implementation research **is now required** to understand how ultra-low dose combinations can best be integrated into current treatment algorithms globally … () Description of this strategy **should be incorporated** into hypertension guidelines, though currently implementation **is limited by** the availability of suitable product (p. 10).

In summary, the Claim rests on data collected and analyzed in the trial. In contrast, the Implication projects those findings into future research and clinical practice – without yet having direct evidence for how to best accomplish its implementation.

6.4.1.3 Clue 3: Squishy Language

A third way to distinguish Implications from Claims is language. Claim statements are often directly declarative: "The Sun <u>is</u> a star." Implication statements are often couched in softer, more qualified language, such as:

- From the example in Clue 2, "Implementation research **is now required**" and "this strategy **should be incorporated**."
- Catalyst L **may be** scalable for increasing yield.
- **If confirmed**, the top quark mass is far higher than predicted.
- Based on the evidence, we **speculate** that APOE ε4 is a key genetic marker for Alzheimer's disease.

In summary, Claims tend to use firm, declarative phrasing about the present study supported by Evidence and Reasoning. At the same time, Implications often appear with qualifiers like *may*, *if confirmed*, or *we speculate*, signaling future ideas that require further testing.

6.4.2 Implications versus Context

Like Claims and Implications, Context can also take the form of declarative statements, and distinguishing one from the other can be confusing. Again, **time** is our main clue:

- Context is about **past** research
- Implications are about **future** research.

As discussed more in Chapter 7, Context is the backward-looking narrative that places the current study within the specific field and provides the rationale for conducting the study. If a statement is based on prior research that existed before this study, then it is part of the Context; if it directly follows from this study, it is an Implication. Consider the following hypothetical examples:

- Atmosphere models **have suggested** that phosphine **should be** common in the atmospheres of gas giant exoplanets.
- Our research suggests that prior detections of very high-redshift galaxies **may be incorrect** due to inherent degeneracies in photometric redshift determination.

The first statement has the qualified language of an Implication ("have suggested. should be") but is based entirely on *past work*; there is no mention of the current study here. This statement is, therefore, part of the Context and potentially the motivation of the present study. The second statement, which also uses qualified language ("may be incorrect") and refers to past work, explicitly refers to the present work ("our research"). It is, therefore, an Implication.

In some cases, the authors may state an Implication that is based on a combination of the present study and past work. For example, the 2020 Lancet Commission report on dementia (Livingston et al., 2020) combines its own analyses with earlier epidemiological work to extend the field's Implications. Building on the nine modifiable risk factors identified in the 2017 Commission – limited early-life education, hypertension, hearing loss, smoking, mid-life obesity, depression, physical inactivity, diabetes, and low social contact (Livingston et al., 2017) – the 2020 update adds three more that newer cohort studies have linked to dementia incidence. These include excessive alcohol consumption, traumatic brain injury, and chronic exposure to fine-particulate air pollution.

The authors of the 2020 report integrated these 12 factors into a population attributable fraction model and conclude, "Modifying these risk factors might prevent or delay up to 40 % of dementias worldwide" (p. 414). Then, they immediately temper that projection:

> We cannot determine from current data whether multiple risk factors act additively or synergistically. Association does not by itself prove causation; however, falling dementia incidence in several high-income countries suggests that at least some of these factors exert a causal influence on clinical dementia expression (p. 429).

This example illustrates how researchers in a multifaceted domain, such as dementia research, synthesize past findings with fresh analyses to state an Implication rather

Box 6.5 Knowledge Check: Question 6.3 Which statement best characterizes the difference between Implications and Context?

A. Implications always summarize the key findings of the current study, while Context identifies future research directions.
B. Context looks to future studies in other fields, while Implications focus on immediate improvements within a study.
C. Context focuses on the past research findings, whereas Implications focus on what logically comes next in future work.
D. Implications can replace Context when strong data are unavailable.

(Check your understanding using the Knowledge Check Key at the end of the book.)

than a new Claim: The 40% figure is a forward-looking projection that rests on combined evidence but still requires future intervention trials to confirm causality and interaction among risk factors.

6.5 Finding Missing Implications

Let's consider a puzzling situation where authors provide no Implications. The landmark article by Penzias and Wilson (1965) that reported the first detection of what would become known as the cosmic microwave background (CMB) did not explicitly state the Implications of its findings, yet it is a stunning work that has had far-reaching Implications. We can uncover the Implications by following some fact-finding steps:

1. **What's the article's main result?** First, we carefully read the study results to gain a deeper understanding. Penzias and Wilson (1965) detected an excess antenna temperature in this case, which known sources could not explain.
2. **What do other authors say about it?** Textbooks, review articles, and later articles that cite this work discuss its significance. For instance, a quick search in Google Scholar reveals the articles that cite the original finding, many of which have thousands of citations each. These later articles indicate that the discovery of excess radiation is widely recognized as evidence of the CMB, which in turn supports the Big Bang theory.
3. **What was going on at the time?** Next, we need to understand a bit of the scientific context of the era. In 1965, cosmology was exploring the origins of the universe. The "Big Bang" idea was a hot, new topic, and many scientists were looking for evidence of the CMB.
4. **What do your instructors say about it**? Lecture notes or class discussions might illuminate why this discovery was pivotal. Professors or experts in astrophysics can provide insight into its broader meaning.
5. **What does theory add?** The Implications may become more apparent when we connect the findings to broader theoretical frameworks. In this case, the excess

antenna temperature aligns with predictions of relic radiation from the Big Bang, a foundational piece of evidence for modern cosmology.

These steps can help us deduce that Penzias and Wilson's (1965) accidental discovery of the CMB provided strong evidence for the Big Bang model, despite the authors not explicitly stating this (or anything like it) in their article. It also earned Arno Penzias and Robert Wilson the 1978 Nobel Prize in Physics.

6.6 Worked Example

In biology, James D. Watson and Francis H. C. Crick's twin *Nature* notes of 1953 (Watson & Crick, 1953a, 1953b) became classics because of the Implications they articulated. Their model – a double helix of two antiparallel DNA strands held together by specific base pairing – explained how genetic information is stored, copied, and transmitted. This breakthrough, built on the fundamental X-ray crystallography work of Rosalind Franklin and Raymond Gosling, helped explain how genetic information is stored, replicated, and passed on from generation to generation.

1 The First Watson & Crick Nature Note (25 April 1953)

1. Title: "Molecular structure of nucleic acids: A structure for deoxyribose nucleic acid"
2. Citation: Watson, J. D., & Crick, F. H. C. (1953a). *Nature*, 171(4356), 737–738. https://doi.org/10.1038/171737a0

Watson and Crick's structural proposal depended on companion X-ray fiber-diffraction articles in the same issue: Franklin and Gosling (1953) and Wilkins, Stokes, & Wilson (1953). Rosalind Franklin's celebrated Photo 51 and related patterns convinced Watson and Crick of the helical arrangement and base complementarity.

They highlighted two immediate implications (p. 737):

1. "This structure has novel features which **are of considerable biological interest**."
2. "It has not escaped our notice that the specific pairing we have postulated **immediately suggests a possible copying mechanism** for the genetic material."

2 The second Watson & Crick *Nature* note (30 May 1953)

- Title: "*Genetical Implications of the Structure of Deoxyribonucleic Acid.*"
- Citation: Watson, J. D., & Crick, F. H. C. (1953b). *Nature*, 171(4361), 964–967. https://doi.org/10.1038/171964b0

While the structural model itself was unchanged, Watson and Crick explored its consequences:

> Despite these uncertainties, we feel that our proposed structure … **may help to solve one of the fundamental biological problems** – the molecular basis of the template needed for genetic replication. (p. 966)

Then, they posed some pivotal questions as next steps for the research (p. 966):

- What precursors build DNA?
- How do the strands unwind and separate?
- What roles do proteins and chromosomal organization play?

This pair of articles is a heavyweight of Implications. From structure to theory, five years later, Crick (1958) distilled these insights into biology's central dogma, describing the directional flow of sequence information from DNA to RNA to protein. That concept unified genetics and biochemistry within a single explanatory framework.

The lasting Implications are profound. The double-helix model clarified DNA replication, transcription, and translation, and it underpinned technologies such as recombinant-DNA methods, gene therapy, high-throughput sequencing, and the Human Genome Project. It also laid the essential groundwork for *genomics* – a term coined in 1986 – while reshaping structural biology and the then-emerging field of molecular biology.

6.7 Chapter Key Takeaways

1. **Definition and importance:** An Implication in research articles is a forward-looking narrative about the logical next steps beyond the current study. It should be concise and clearly indicated by the Claim, Evidence, and Reasoning. It is distinct from opinions, assumptions, or a wish list of applications. Implications answer at least two questions:
   - How is the result significant beyond the immediate findings?
   - Where does the research go from here?
2. **Implication types vary:** Implications can be about new topics, future directions, ethics, policy, or even challenging existing ideas or methods. Strong Implications are always logical extensions of the Claim.
3. **Articles can have more than one Implication:** Research articles can have several or many Implications. The relative importance of the Implications can usually be inferred by considering their relative impact on the next set of possible research questions. More direct effects that drive expansion or create new lines of research are likely more important.
4. **Implications drive the subsequent research questions:** Today's Implications are tomorrow's research questions. What was once a significant Implication in the past might become an approach or method in the present, demonstrating how research progresses.

7 Context

Scientific statements are always in need of further testing and are always framed within certain assumptions, many of which are understood only within the research context of the time.

—Karl Popper (1945) The Open Society and Its Enemies

7.1 Overview

Chapter 7 focuses on the final CERIC element, the Context, which defines a research study's relevance, scope, and motivation. Context is generally found in the introduction section of a research article but may also appear in various locations to motivate aspects of Evidence, Reasoning, and Implications. This chapter discusses how Context functions in research articles and its types. This chapter considers how to use Context to identify the driving research question or problem (i.e., "the gap" or rationale) and illustrates how this shapes the current study. This chapter illustrates examples of well-contextualized research, discusses the challenge of identifying missing Context, and explores how to deal with the jargon maze common to Context. This chapter also explores how to distinguish Context from Evidence and Reasoning.

7.2 Defining Context

Context is the **backward-looking narrative** that does at least three things: locates a study inside a broader research framework, specifies its disciplinary scope, and explains why the work is needed. Its core job is to summarize what has already been learned and to point to the unanswered question – the **research gap** – that motivates the present study. When crafted well, that gap naturally and persuasively leads to the article's Claim.

In most articles, the Context section appears at the top of the manuscript, typically labeled as the Introduction, Background, or Literature Review. These sections precede the Methods section. Regardless of the heading, the rhetorical purpose remains the same: To show readers how prior research funnels to the specific hypothesis or objective that the authors will test.

You may notice, however, that CERIC places Context last, whereas the *IMRaD* format places it first. We do this deliberately. Beginning critical readers often must reread an article several times – annotating terminology, unpacking background concepts – before the contextual landscape comes into focus. It can be very *overwhelming*, as you may well know. Anchoring Context at the end of the CERIC sequence lowers the conceptual barrier to entry, allowing readers to concentrate first on the core CER argument. At the same time, the "three Ps" serve us well: *practice, pace*, and *patience*. As you gain fluency, the field's technical language and methods will begin to feel familiar, and you will see why an efficiently written Context section works so well for expert-to-expert communication.

7.2.1 Context in Research Articles

In this chapter, we define Context in research as a **backward-looking narrative** that explains what came before a study, where the study sits in its field, and why the work is needed. In practice, every good Context section delivers three elements:

1. **Historical narrative:** It surveys prior knowledge, key debates, and recent advances, creating a logical bridge from what is known to what is uncertain.
2. **Field and scope:** It situates the project inside a clearly bounded discipline or sub-discipline, flagging the specific unresolved issues that matter to that research community.
3. **Rationale:** By exposing limitations or contradictions in the literature, it identifies a research gap and shows how the present study's methods or insights aim to close it.

Seen from another angle, Context packages the Implications of past work – the unanswered questions and recommended next steps – and passes them to the current study as its motivating problem. We could update the cycle from Chapter 6 like this:

$$\text{Implications} \rightarrow \text{Context (Research Question)} \rightarrow \text{Implications}$$

A well-crafted Context section can do many things in a short span, including:

- **Orients readers:** It provides the background necessary to understand what problem is being addressed and why it is being addressed now.
- **Positions the argument:** By tying the study to ongoing debates or theoretical frameworks, it prepares readers to evaluate whether the new findings challenge, refine, or extend established knowledge.
- **Demonstrates expertise:** Thorough engagement with the literature signals credibility and guides interpretation of the Evidence and Reasoning that follows.
- **Shows relevance:** Linking the work to broader scientific or societal questions makes the Claim more compelling.

In summary, Context balances breadth and detail to frame the study's Claim and Reasoning within its broader intellectual landscape, setting the stage for the CER argument that follows.

7.2.2 Nature of Context Is the Past and before the Present Work

A research article opens by providing sufficient Context to justify the study and preview its main Claim. One helpful way to visualize that backward-looking narrative is the Introduction Funnel shown in Figure 7.1. The prose moves from a broad background to a narrow, precise research gap, then tips directly into the Claim. Although specifications vary by journal, the same three-stage logic appears across the natural sciences.

Box 7.1 Knowledge Check: Question 7.1 Which statement best captures the core purpose of Context in a research article?

A. Context ensures the authors' measurements are robust and properly calibrated.
B. Context defines the backward-looking narrative that justifies the study's motivation to fill that gap.
C. Context summarizes new Evidence and focuses on the future Implications of the findings.
D. Context is the forward-looking discussion of how the research will be applied in practice.

(Check your understanding using the Knowledge Check Key at the end of the book.)

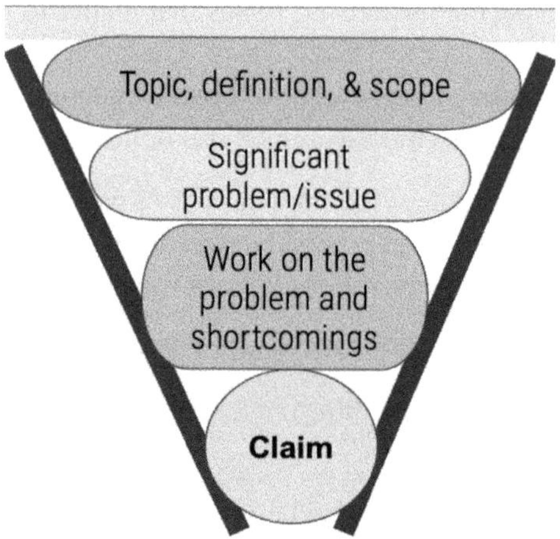

Figure 7.1 The introduction funnel.

7.2.2.1 Broad Field and Key Definitions

The funnel's wide mouth orients readers to the broader topic and clarifies essential terms. For example, consider a study that tests a new chemistry for rechargeable batteries. Its topic, definition, and scope would open with the global demand for efficient energy storage, the economic stakes for electric vehicles and grid backup, and definitions of core concepts, such as specific capacity and cycle life. This framing anchors the study in a crucial and active research arena.

7.2.2.2 Narrowing to a Subfield and Scope

The subsequent paragraphs zoom in on a specific sub-topic. For the battery example, the authors might examine shortcomings of current lithium-ion cathodes, such as limited cobalt supply, safety concerns, and lifespan degradation. Highlighting these unresolved issues situates the new work inside a well-defined niche and sharpens the study's motivation.

7.2.2.3 Identifying the Research Gap

Finally, a focused mini-review highlights what remains unknown or undone. Common lead-ins include:

- *Prior research has not ...*
- *There is as yet no study of ...*

These kinds of phrases signal that the area requires further investigation.

7.2.2.4 Claim Setup

In our hypothetical battery scenario, the concluding sentence of the Context could be:

> ***Because preliminary studies suggest that lithium-iron-phosphate doped with manganese could double cycle life without sacrificing capacity****, we report here a systematic evaluation of its electrochemical performance.*

That sentence marks the narrow end of the funnel: It states the **gap** and cues the article's principal Claim.

By guiding readers from a broad background through a clear gap to the study's objective, the Introduction Funnel makes the logical stakes of the work transparent and shows precisely how the new results will extend the field.

Box 7.2 Better Practice: Funnel Your Research Rationale

Draft an Intro Funnel for your current research project and then annotate the following:

1. Identify the topic, scope, problem, and why it's important.
2. Briefly summarize the existing research most relevant to the problem, explaining why each study has shortcomings. Identifying these is crucial because *shortcomings provide the logical basis for the next study.*
3. Collect those shortcomings into a short paragraph that identifies a gap in the existing research using phrases like *previous research has not* and *there is no study of.*
4. Add a final sentence that describes your research question, framed as a statement using a phrase like *this study will address that gap by investigating...*
5. When you have an advanced third draft, share it with peers or your advisor for some feedback.

Take a few more editing passes to revise this short paragraph into a single sentence that explains concisely what knowledge is missing from the field that your study will address.

If you struggle with literature reviews, you can work the Intro Funnel in reverse. Begin with your study's likely main Claim (at this point, your best guess about the outcome of your hypothesis) and work backward through the Intro Funnel, moving from a focused Claim to the gap to the broader framework.

7.2.3 Context in Proposals

Where Implications orient a proposal toward the future, Context anchors it in what is already known. Funders first ask, "Why does this problem matter, and why now?" These are key questions Context must answer clearly. A well-composed proposal, therefore, performs two tasks:

1. **Distills the state of the field**: Proposals must show how prior Evidence converges on an unresolved gap; and
2. **Justifies the proposed work:** Proposals must make clear the logical next step to close the research gap.

Below, we model how to weave those threads together using a worked example.

7.2.3.1 Historical Example: Subduction-Zone Earthquake Risk

In the late 1990s and early 2000s, three research streams converged to recognize the importance of the Cascadia subduction zone:

- Precise coastal-marsh stratigraphy revealed cycles of sudden subsidence at about 500-year intervals (Atwater & Hemphill-Haley, 1997);
- Offshore turbidite records corroborated synchronous great earthquakes along the margin (Goldfinger et al., 2003); and
- GPS networks detected centimeter-scale plate convergence locked beneath the forearc (Dragert et al., 2001).

Notice how the marsh, turbidite, and GPS studies function as Context, not Implications. They establish urgency and identify precisely *what* we do not yet understand in primary literature.

1. Synthesis of these Context Elements Clarifies a Critical Knowledge Gap

The gap is that the mechanical controls on rupture segmentation and tsunami generation remain uncertain. On that basis, a successful proposal could then pose research questions, such as:

1. How do along-strike variations in plate coupling govern the size and frequency of Cascadia megathrust earthquakes?
2. Can 3-D, physics-based rupture models assimilating geodetic and turbidite constraints reduce uncertainty in coastal tsunami forecasts?

There are many other possible research questions. This is where an individual researcher's (or team's) interest, creativity, and expertise come into play.

2. Tailor Context to the Funding Call

Just as Implications must suit a funder's mission, Context must resonate with the call's stated priorities. For example, if the NSF seeks proposals that integrate geohazard science with community resilience, foregrounding the Cascadia region's population exposure and emergency-planning needs will align the proposal with that mandate. Excessive background – for instance, a comprehensive history of plate tectonic theory – dilutes the focus away from the call's mandate. Stick to the Context that directly aligns with the funder's mandate and sharpens the research gap the project will address.

3. Enact Practical Guardrails
It is easy to lose focus and feel overwhelmed by what Context to include or exclude. Here are some practical guardrails:

- **Keep the background current and selective**: Cite seminal work and the newest key advances, not every paper ever published.
- **Quantify the gap where possible:** Saying that rupture models vary by "two orders of magnitude" in the predicted slip communicates the stakes more vividly than generic phrases like "poorly understood" or "higher risk."
- **Return to the Whys:** Ensure that every Context paragraph transitions smoothly to *why* your approach is uniquely positioned to solve the problem, and why now is the right time to do it.

By rooting a proposal in a gap-oriented Context with focused and precise language, researchers demonstrate mastery of the literature, frame a compelling rationale, and clear a path for reviewers to see both the necessity and the feasibility of the work ahead.

7.3 Types of Context

Since Context is past work that sets up the research article's motivation and central question, it is often found in the abstract and *before* the current study's methods or results. However, skillful authors may include essential aspects of Context throughout the article. This section reviews the types of Context and how to find them.

7.3.1 Context in the Abstract

Given the importance of Context, a well-written research article often places Context in the abstract, and again, in the Introduction or Literature Review sections (if allowed by the journal). Below, we examine how Context is found in an abstract, using a study by Shin et al. (2025) as an illustrative example. Notice the flow of Intro Funnel elements highlighted in bold.

- Title: "Skin autofluorescence is associated with glycemic variability in type 2 diabetes patients"
- Citation: Shin, E. S., Jung, H. J., Kim, J., Cho, M., & Kim, T. S. (2025). *Scientific Reports*, *15*(1), 24105. https://doi.org/10.1038/s41598-025-09517-7
- Abstract:

Type 2 diabetes mellitus (T2DM) is a chronic disorder characterized by insulin resistance and hyperglycemia. To prevent diabetic complications, blood glucose levels should be maintained within a target range. While glycated hemoglobin (HbA1c) is widely used to assess long-term glycemic control, it does not provide

> **information on glycemic variability (GV), which contributes to diabetes-related complications. Measuring skin autofluorescence (SAF) is a non-invasive method for measuring advanced glycation end-products (AGEs) in skin tissue and the accumulation of AGE is accelerated under hyperglycemic conditions**. Fifty T2DM patients and fifty non-diabetic controls were recruited. SAF was measured using an AGE reader, and GV was assessed through continuous glucose monitoring (CGM) for 14 days. CGM metrics, including Mean Amplitude of Glycemic Excursions (MAGE) and Coefficient of Variation (CV), were analyzed. Serum AGE levels were measured via ELISA. Bivariate and multivariate associations between SAF and GV were examined using Pearson correlation analysis and multiple linear regression. T2DM patients exhibited higher SAF and GV compared to controls. SAF was positively correlated with GV metrics, including CV and MAGE, in diabetic patients, even after adjusting for confounding factors. No significant correlation was found between GV and serum AGE levels both in normal and T2DM subjects. SAF correlates with GV in T2DM, suggesting its potential as a non-invasive biomarker for glycemic fluctuations and associated risk. Unlike serum AGEs, SAF reflects localized AGE accumulation, which may be a complementary biomarker to HbA1c in assessing glycemic variability related with diabetes complication. Further research, such as longitudinal study, is needed to confirm its clinical utility in predicting long-term outcomes.

This abstract is well-organized and transparent, making it straightforward to find the Context immediately. Notice how almost half of this abstract follows the classic Intro Funnel pattern. Let's dig into it.

1. **Broad Context:** The abstract begins broadly with the topic and importance of T2DM:

> Type 2 diabetes mellitus (T2DM) is a chronic disorder characterized by insulin resistance and hyperglycemia. To prevent diabetic complications, blood glucose levels should be maintained within a target range.

This statement serves to establish why the topic (T2DM) is important and introduces key terms (insulin resistance and hyperglycemia).

2. **Specific scope**: Next, the abstract sharpens to a measurement problem (GV not captured by HbA1c):

> While glycated hemoglobin (HbA1c) is widely used to assess long–term glycemic control, it does not provide information on glycemic variability (GV), which contributes to diabetes–related complications.

This narrows to a specific measurement issue (HbA1c misses GV), thereby signaling a shortcoming in current practice.

3. **Research gap:** The research gap is created by the juxtaposition of two earlier sentences. It proposes a potential fix using skin autofluorescence (SAF):

> Measuring skin autofluorescence (SAF) is a non-invasive method for measuring advanced glycation end-products (AGEs) in skin tissue, and the accumulation of AGE is accelerated under hyperglycemic conditions.

This sentence introduces SAF as a plausible tool, linking it mechanistically to hyperglycemia and further focusing the narrative. Next, it highlights the implicit gap, the unknown link between SAF and GV:

> While glycated hemoglobin (HbA1c) is widely used to assess long-term glycemic control, it does not provide information on glycemic variability (GV), which contributes to diabetes-related complications.

This sentence highlights the limitation of the current gold-standard metric, the HbA1c test. Because the abstract never states whether SAF tracks GV, that uncertainty forms the implicit research gap. Together, these sentences invite, but do not answer, the essential and implied research question that we deduce to be: *Can SAF – already linked mechanistically to hyperglycemia – serve as a practical proxy for the glycemic variability that HbA1c misses?* A better practice is to write the gap in the abstract using ultra clear and explicit language.

4. **Claim setup:** Finally, the abstract pivots to the current study's design:

> Fifty T2DM patients and fifty non-diabetic controls were recruited … (method sentences that follow) … Bivariate and multivariate associations between SAF and GV were examined using Pearson correlation analysis and multiple linear regression.

This sets up the Claim that SAF correlates with glycemic variability, and marks the narrow end of the funnel by explaining how the authors will test the gap. Based on this logical setup of the study's rationale, we can predict what CER core they might develop in a study.

7.3.2 Context in the Introduction

As we have seen Chapters 3–6, incomplete CERIC elements are common when focusing exclusively on the abstract of an article; hence, it is essential to strategically seek out the main text. Fortunately, Context is almost always contained in the Introduction section of an article (the "I" in *IMRaD*), and this is where we can find further detail to bolster the Context elements of the Shin et al. (2025) study.

Specifically, concerning the motivating research gap, the Introduction notes that there are studies examining SAF and glycemic control, and they describe two:

> There are studies that suggest a potential link between SAF and glycemic control in diabetes patients. In patients with cystic fibrosis-related diabetes, AGE levels were correlated with average glucose levels, standard deviation of glucose levels and coefficient of variation measured using a continuous glucose monitoring system (CGM)[10]. In prediabetes patients, the level of glyceraldehyde-derived AGE was elevated in groups with the higher mean amplitude of glycemic excursions. (MAGE)[11]

Next, they note an explicit gap as a lack of correlation between SAF and glycemic variability (GV):

> However, the relationship between SAF and GV in T2DM remains unexplored. Understanding this correlation could provide valuable insights into the utility of SAF as a marker for GV and its potential role in predicting diabetes-related complications.

Finally, they dive straight into the rationale:

> This study aims to investigate the correlation between skin autofluorescence and glycemic variability in patients with type 2 diabetes. By examining this relationship, we will

elucidate whether SAF can serve as a non-invasive biomarker for monitoring GV and assessing the risk of complications in T2DM patients.

These statements could be easily rearranged into a research question, for instance: *Does SAF serve as a noninvasive biomarker for monitoring GV and assessing the risk of complications in patients with T2DM?*

In summary, by reading deeper into the Introduction, we find an explicit discussion of the gap and rationale that were only implied in the abstract.

7.3.3 Context throughout the Article

Although most readers first meet Context in an article's opening paragraphs, skilled authors weave background information throughout the paper to help audiences see how and why each section matters. Here are two common situations:

- **Methods as Context:** A well-written Methods section situates the study by describing the setting, sample, and procedures that shaped the investigation. For instance, Daivadanam et al. (2019) designed a chronic-disease intervention only after mapping five nested "layers" of Context – family, community, health-care setting, district, and national policies – so that each choice in recruitment and data collection logically followed from local realities. In this sense, the Context in the methods explains why these procedures are applied with the participants, which functions as an extension of the study's rationale.
- **Discussion as Context:** When authors interpret their results, they join the scholarly conversation. Robertson et al. (2016), for example, evaluated a school-based physical-activity program by comparing their outcomes with earlier trials, thereby using prior literature as Reasoning to judge the program's added value and guide future interventions. Thus, the Context in the Discussion functions as Reasoning by framing what the findings mean in relation to what was already known.

7.3.4 Identifying and Remediating Missing Context

Journal word limits and format rules sometimes limit the inclusion of background information. If essential Context feels thin, readers can recover it with a compact "5 W" audit:

- **Who** conducted the seminal work in this area?
- **What** questions remain unanswered, and which ones does this article address?
- **When** was the key prior research completed? Decades ago or last year?
- **Where** were those earlier studies carried out (geographic site, model system, or instrument type)?
- **Why** is the present study timely or necessary now?

If Context is missing, try to answer these questions either from memory, a quick database search, or the article's citations. This information fills in essential background

Box 7.3 Knowledge Check: Question 7.2 Which of the following scenarios suggests that Context is missing or insufficient in a research article?

A. The introduction lays out a concise rationale by citing recent studies and highlighting a clear gap in knowledge.
B. Several references in the introduction guide the reader through prior work leading to the study's question.
C. The article launches straight into methods and results with no mention of prior research or problems they intend to solve.
D. The conclusion reiterates major findings and suggests future directions, building upon the existing literature.

(Check your understanding using the Knowledge Check Key at the end of the book.)

and helps clarify the authors' motivation. The main takeaway is that, even if Context is missing, we still need to critically evaluate how the authors set up the article's central argument. This kind of detective work is time- and effort-intensive, so we reserve only for articles that are essential for a proposal or literature review.

7.4 Context as the Main Research Question

The foremost job of Context is to justify – and clearly deliver – the study's primary research question phrased as a statement of how to fill the gap in the literature. We, therefore, need a standard for gauging whether that research question is well-built. This section offers guidance and works through some examples of research questions.

7.4.1 What Makes a Good Research Question?

A first-rate research question meets a Goldilocks test: It is neither so broad that it becomes unanswerable nor so narrow that it yields trivial insight. Specifically, a strong question:

- **Addresses a gap:** It targets knowledge that the literature has not yet resolved.
- **Fits the project**: Its scope aligns with the available data, time, and methodological tools.
- **Guides action:** It is clear, focused, and written in a way that allows appropriate methods to follow naturally (Creswell & Plano-Clark, 2017).
- **Invites exploration:** It is open-ended enough to allow deep analysis while remaining specific enough to steer the study.
- **Matters to the field:** Answering it promises new insights, practical implications, or both.

When Context sets up a question with these qualities, the study begins on firm logical ground and positions itself to make a meaningful contribution.

7.4.1.1 Examples of Research Questions

Let's consider a range of research questions – from bad to better – using three different topics.

1. Conifers and rising CO_2
   - **Bad:** How do conifers interact with CO_2?
     - **Why it does not work:** No gap (a well-established topic); vague and lacks any specificity to be actionable; unclear significance.
   - **Okay:** How does the atmospheric concentration of CO_2 affect the photosynthesis rate in conifers?
     - **What needs improvement:** The gap is vague, forest interactions are substantial, and likely exceed the scope of a single study. It lacks specifics (species, methods, measurements) and is overly broad (encompassing all species and any CO_2 level).
   - **Better:** Within a controlled greenhouse experiment, how does elevating CO_2 from 415 ppm (ambient) to 800 ppm alter net photosynthetic rate and stomatal conductance in 2-year-old Douglas-fir (*Pseudotsuga menziesii*) seedlings over a 30-day exposure?
     - **Why it works:** This question specifies the gap (response of a commercially important conifer under future CO_2 scenarios), sets measurable variables (photosynthesis, conductance), and keeps the scope feasible (single species, controlled duration).
2. Soil pH and tomato growth
   - **Bad:** Is soil important for tomato plant growth?
     - **Why it does not work:** Overly broad and vague, lacks any specificity to be actionable.
   - **Okay:** What is the impact of different soil pH levels on the growth rate of tomato plants?
     - **What needs improvement:** Generic (feasible if narrowed to cultivar and pH range); "different" is undefined; "growth rate" could mean height, biomass, yield; needs tighter bounds to avoid an unmanageable factorial design; moderate relevance increases if linked to real-world soil management.
   - **Better:** How does adjusting soil pH from 5.0 to 7.0 in 0.5-unit increments affect weekly dry-matter accumulation and final fruit yield of 'Roma' tomato plants under controlled growth-chamber conditions?
     - **Why it works:** This question focuses on an economically important cultivar, defines a realistic pH gradient, and names specific response variables, yielding a clear and actionable protocol.
3. Microplastics and migratory fish
   - **Bad:** Do migratory fish eat microplastics?
     - **Why it does not work:** Yes/no question does not invite investigation, and it probably can be answered by a Google search; no gap; no scope; unclear significance.

- **Okay:** How does the presence of microplastics in marine environments influence the reproductive function of migratory fish species?
    - **What needs improvement:** The single species and concentration range need to be specified to remain feasible; "presence" and "migratory fish species" are vague; reproductive metrics are unspecified; currently too broad, covering all migratory species and all plastics.
- **Better:** To what extent does chronic exposure to 0.1–10 μg L^{-1} polyethylene microplastic particles affect gonadosomatic index, plasma sex-steroid levels, and egg viability in Atlantic herring (*Clupea harengus*) during a complete spawning cycle?
    - **Why it works:** Narrows scope to one ecologically and commercially important fish, sets environmentally realistic concentrations, and states concrete reproductive endpoints – making the study doable and the results interpretable for resource managers.

7.4.2 Hunting for Research Questions

If you enjoy "thrifting" or shopping in second-hand stores for great items at super low prices, you might also enjoy hunting for the main research question in a published study. Consider searching a thrift store for an excellent quality, gently used storage bin. It might be found in various places, including home decor, kitchenware, furniture, clothing, and sports equipment. It might be empty on the floor, stacked, or filled with other goods. Hunting like this is part of the fun of thrifting.

Similar to hunting for a storage bin in a thrift store, we often must hunt for a study's main research question. It can be found in various sections of the article, including the abstract, Introduction, Literature Review, Methodology, Discussion, and Conclusion. Indeed, the research question may not even be written as a question, but rather as a statement buried in the middle of a long paragraph within the article. In the worst case, we may need to deduce the main research question from the Context clues found in the article.

7.4.2.1 Example from a Clinical Trial

Let's look at an example from Teong et al. (2023). They reported a clinical trial comparing early time-restricted eating (iTRE) plus exercise to calorie restriction (CR) and standard care in adults at risk of developing type 2 diabetes. The authors do an excellent job of stating the main research question throughout the article. Here are the parts of the article where the main research question appears:

- Title: "Intermittent fasting plus early time-restricted eating versus calorie restriction and standard care in adults at risk of type 2 diabetes."
- Citation: Teong, X. T., Liu, K., Vincent, A. D., … O'Connell, A. (2023). *Nature Medicine, 29*(4), 963–972. https://doi.org/10.1038/s41591-023-02287-7
- Abstract:

> Intermittent fasting appears an equivalent alternative to calorie restriction (CR) to improve health in humans. However, few trials have considered applying meal timing during the 'fasting' day, which may be a limitation. We developed a novel intermittent fasting plus

> early time-restricted eating (iTRE) approach. Adults (N = 209, 58 ± 10 years, 34.8 ± 4.7 kg m^{-2}) at increased risk of developing type 2 diabetes were randomized to one of three groups (2:2:1): iTRE (30% energy requirements between 0800 and 1,200 hours and followed by a 20-h fasting period on three nonconsecutive days per week, and ad libitum eating on other days); CR (70% of energy requirements daily, without time prescription); or standard care (weight loss booklet). This open-label, parallel group, three-arm randomized controlled trial provided nutritional support to participants in the iTRE and CR arms for 6 months, with an additional 12-month follow-up. The primary outcome was change in glucose area under the curve in response to a mixed-meal tolerance test at month 6 in iTRE versus CR. Glucose tolerance was improved to a greater extent in iTRE compared with CR (−10.10 (95% confidence interval −14.08, −6.11) versus −3.57 (95% confidence interval −7.72, 0.57) mg dl^{-1} min^{-1}; P = 0.03) at month 6, but these differences were lost at month 18. Adverse events were transient and generally mild. Reports of fatigue were higher in iTRE versus CR and standard care, whereas reports of constipation and headache were higher in iTRE and CR versus standard care. In conclusion, incorporating advice for meal timing with prolonged fasting led to greater improvements in postprandial glucose metabolism in adults at increased risk of developing type 2 diabetes. ClinicalTrials.gov identifier NCT03689608.

The article's title suggests the main research question. While this might initially appear to be a Claim, it only describes the intervention and population. There is no statement of new knowledge, such as whether it works. Therefore, we can infer that this situation is what the study tests, serving as the primary research question. This idea, which we suspect is the Claim, appears again:

> We developed a novel intermittent fasting plus early time-restricted eating (iTRE) approach. Adults (N = 209, 58 ± 10 years, 34.8 ± 4.7 kg m^{-2}) at increased risk of developing type 2 diabetes …

Let's rephrase this to an actual question:

> What is the effect of a novel intermittent fasting plus early time-restricted eating (iTRE) approach in adults (N = 209, 58 ± 10 years, 34.8 ± 4.7 kg m^{-2}) at increased risk of developing type 2 diabetes?

This statement rephrased as a question is straightforward, makes sense, and is actionable, supporting the idea that it is the main research question. We might even guess what methods they will use to test it, perhaps by comparing their novel intervention to standard approaches.

Methods: They embed the study's main question in the Methods section by stating the study's primary objective is to:

> () … assess differences in glucose tolerance in response to a mixed-meal in iTRE versus CR at 6 months.

A telling phrase is "primary objective," which implies the main question by identifying the primary endpoint. In addition, their methods explicitly test the novel approach we highlighted in the title and abstract using standard techniques, further supporting our hypothesis regarding the central question driving this line of research.

Discussion: The authors underscore that the novel approach showed a

> () … *modest benefit for postprandial glycemia in response to mixed-meal tolerance test* compared to daily calorie restriction.

This comparison reflects the study's central aim of examining glucose tolerance as a primary outcome to compare iTRE with calorie restriction and standard care. Again, we see the question and methods pulled through the discussion, confirming our prediction of the main research question.

7.4.3 Context as a Jargon Maze

Context sections of most research articles are loaded with jargon, offering up a jargon maze for novice researchers. Jargon is "the technical terminology or characteristic idiom of a special activity or group" (Rakedzon et al., 2017). One benefit of successfully navigating the jargon maze is that the Context can help novices learn important concepts because mastering a field involves learning its specific jargon and methods. On the other side, novice researchers can become hopelessly lost in confusing and overwhelming passages that can burn through precious hours of research time without ever getting to the article's main argument – a trap for novice critical readers, as illustrated in Figure 7.2.

While Context is beneficial to get oriented to a new field, the jargon poses real barriers for newcomers. Studies consistently show that highly technical wording feels harder to process (Bullock et al., 2019; Kreiger & Gallois, 2017; Petty & Cacioppo, 1986; Schulman et al., 2020). Because people tend to avoid unnecessary effort, most tend to ignore or dismiss material that requires too much mental work (Schulman & Sweitzer, 2018). Researchers refer to this phenomenon as *processing fluency* – the subjective ease with which information is understood (Schwarz, 2011). Low processing fluency and the resulting cognitive overload is why we place Context last in the CERIC sequence for novice critical readers. If your fluency is higher, by all means, dive into Context first.

If reading research articles seems painfully difficult, you are not imagining it; they are! Articles are written for experts, not novices. The good news is that the same critical-reading habits you develop throughout this book will also build your fluency with technical terms. Create personal glossaries, maintain systematic notes, and revisit key concepts. Chapter 8 demonstrates how to integrate these strategies in a comprehensive CERIC review. With steady practice, once-daunting jargon will become part of your scientific vocabulary.

Figure 7.2 Context is a Jargon Maze.

Box 7.4 Knowledge Check: Question 7.3 Which phrase in an article's Context would most likely indicate a potential "jargon maze" for novice critical readers?

A. Prior research did not address this question, which motivates the present study.
B. The sample population included 50 adult participants, ages 18–35, enrolled at a local college.
C. Traditional HPC methodologies incorporate advanced QNN-embedding transformations for data preconditioning.
D. To measure reliability, we performed a Cronbach's alpha test of internal consistency.

(Check your understanding using the Knowledge Check Key at the end of the book.)

7.5 Context Compared to Other Elements

Context is distinct from other CERIC elements because it frames or sets up the study's argument, whereas the other elements are the argument and its logical Implications. Context is also related to information that existed before the current research study and stands whether this study exists or not; hence, past-tense phrasing is often a distinguishing feature of Context.

7.5.1 Distinguishing Context from Evidence

Context looks backward, reminding the reader what is already known and why a new study is necessary. Evidence looks to the present: It is the brand-new data the authors have gathered. Even when Context and Evidence rub shoulders in the same section, we can separate them by asking, "Did this information exist before the study began?" If yes, it is Context; if no, it is Evidence. A good rule for distinguishing Context from Evidence is critique:

- Critique of Context asks, *Was this study necessary and well-motivated?*
- Critique of Evidence asks, *Are the data and analyses valid, reliable, and trustworthy?*

Here are some additional examples to practice distinguishing between Context and Evidence:

- **Astronomy** (Riess et al., 2016): The authors begin by reviewing decades of disputed Hubble constant measurements (*Context*). Only later do they unveil fresh Hubble Space Telescope data on Cepheids and supernovae (*Evidence*).
- **Biology** (Dias & Ressler, 2014): Standard mouse details align the work with earlier epigenetic studies (*Context*). Novel odor-conditioning routines and DNA methylation assays create the dataset that supports transgenerational inheritance (*Evidence*).

- **Chemistry** (Zhu et al., 2020a): Earlier catalysts (cobalt-porphyrin and Ni–N–C) are cited to define performance expectations (*Context*). The authors' own electrochemical measurements of a composite catalyst demonstrate that it outperforms the field (*Evidence*).
- **Earth Science** (Keranen et al., 2014): Historical earthquake records and wastewater-injection data suggest a possible link (*Context*). The team's own relocated earthquake catalog and well-log analysis supply the new test of that link (*Evidence*).
- **Physics** (Abi et al., 2021): The introduction recalls the 2006 Brookhaven E821 result for the muon's anomalous magnetic moment, which already hinted at a possible tension with the Standard-Model value. That historical number frames why a fresh, higher-precision test is needed (*Context*). The article's data on Fermilab Run-1 measurements of the muon precession frequency yield a determination with 0.46 ppm precision. This newly recorded value, not the earlier Brookhaven figure, is the empirical basis for evaluating agreement or mismatch with the Standard-Model prediction (*Evidence*).

Keeping these distinctions in mind will make it easier to tell where the background ends and the argument truly begins.

7.5.2 Distinguishing Context from Reasoning

Context and Reasoning can also form a confusing pair. Context reviews the past and what is already known and pinpoints the research gap. Reasoning takes the new Evidence in the present study, aligns it with (or against) that background, and shows why the study's Claim follows. When past studies motivate the experiment ("Because no spectra exist, we measured …"), they are Context. Only the logical bridge that links the new data to the Claim counts as Reasoning. A quick way to tell them apart in critique is to consider how each functions in critique:

- (Again) Critique of Context asks, *Was the study necessary and well-motivated?*
- Critique of Reasoning asks, *Does the logic connect the Evidence to the Claim?*

Here are some additional examples to practice distinguishing between Context and Reasoning.

- **Astronomy** (Riess et al., 2016): Past Hubble-constant disagreements justify gathering new data (*Context*). Statistical analysis of those data explains why the authors believe their higher value is correct (*Reasoning*).
- **Biology** (Dias & Ressler, 2014): Previous reports of epigenetic inheritance mark a gap (*Context*). Comparing odor-conditioning results across generations provides the logical steps that close the gap (*Reasoning*).
- **Chemistry** (Zhu et al., 2020a): The introduction notes that most covalent organic frameworks (COFs) used in drug delivery are two-dimensional and that synthetic routes to dynamic three-dimensional frameworks with high-connectivity nodes are scarce, noting a gap (*Context*). After synthesizing a 3D cage COF, the authors

use single-crystal X-ray diffraction, nitrogen-sorption isotherms, and structural modelling to argue that the material is truly 3D, highly porous, and mechanically flexible – thereby closing the stated gap.
- **Earth Science** (Keranen et al., 2014): A reviewer might ask, "Did earthquakes really rise only after disposal began?". Spatial correlation alone cannot prove causation.
- **Physics** (Abi et al., 2021): The discussion cites a 2006 muon g-2 value to frame expectations. The crucial reasoning step is the match (or mismatch) between Fermilab's fresh precession data and the Standard-Model prediction, not the historical number itself.

In summary, we use a quick checklist for telling the difference between Context, Evidence, and Reasoning:

- If a statement answers **why** the study needed to be done, it is Context.
- If a statement explains **how or why** the new Evidence supports the Claim, it is Reasoning.
- If a statement reports **what** or which data were collected by the authors, it is Evidence.

With these distinctions in mind, it is easier to discern one CERIC element from another, especially when the writing is unclear or an element is implied. In Part II, we will apply these distinctions to improve our presentations and literature reviews.

7.6 Worked Example

For this chapter's worked example, we consider an unusual Physics article published by Procureur et al. (2023) that used cosmic rays to study the interior of an ancient pyramid. The abstract of this article closely follows an Intro Funnel, demonstrating how to set up a clear Context.

- Title: "Precise characterization of a corridor-shaped structure in Khufu's Pyramid by observation of cosmic-ray muons"
- Citation: Procureur, S., Morishima, K., Kuno, M., … & Elkarmoty, M. (2023). *Nature Communications 14*, 1144. https://doi.org/10.1038/s41467-023-36351-0
- Abstract:

> Khufu's Pyramid is one of the largest archaeological monuments all over the world, which still holds many mysteries. In 2016 and 2017, the ScanPyramids team reported on several discoveries of previously unknown voids by cosmic-ray muon radiography that is a non-destructive technique ideal for the investigation of large-scale structures. Among these discoveries, a corridor-shaped structure has been observed behind the so-called Chevron zone on the North face, with a length of at least 5 meters. A dedicated study of this structure was thus necessary to better understand its function in relation to the enigmatic architectural role of this Chevron. Here we report on new measurements of excellent sensitivity obtained with nuclear emulsion films from Nagoya University and gaseous detectors from CEA, revealing a structure of about 9 m length with a transverse section of about 2.0 m by 2.0 m.

Let's break down this abstract using the Intro Funnel model.

1. **Broad Context**

   Khufu's Pyramid is one of the largest archaeological monument all over the world, which still holds many mysteries.

The first sentence sets up the broad Context of this work, notably in archaeology, although physical science techniques (e.g., cosmic-ray muon radiography) are deployed to explore its "many mysteries."

2. **Specific Scope**

   In 2016 and 2017, the ScanPyramids team reported on several discoveries of previously unknown voids by cosmic-ray muon radiography that is a non-destructive technique ideal for the investigation of large-scale structures. Among these discoveries, a corridor-shaped structure has been observed behind the so-called Chevron zone on the North face, with a length of at least 5 meters.

Here we find the specific area of study and its connection to physics, namely that prior work has established that "cosmic-ray muons" (a class of fundamental particles that arrive to Earth from outer space) can be used to study voids in pyramids, and that prior work had already identified a structure in Khufu's Pyramid.

3. **Research Gap**

   A dedicated study of this structure was thus necessary to better understand its function in relation with the enigmatic architectural role of this Chevron.

This sentence highlights the research gap: Although a structure has been identified, its relationship to the Chevron architectural component of the pyramid remains unclear, necessitating further data. We are thus set up for the primary Claim of the study that addresses this research question by providing a more precise measurement of the structure.

4. **Claim Setup**

   Here we report on new measurements ... revealing a structure of about 9 m length with a transverse section of about 2.0 m by 2.0 m.

The first paragraphs of the main text provide further detail on these elements of the Context, including more detail on the history and structure of Khufu's Pyramid, the importance of the Chevron architectural feature, and the technique of using cosmic-ray muons to study the insides of massive structures such as pyramids, volcanoes, and nuclear reactors.

7.7 Chapter Key Takeaways

1. **Definition and purpose of Context:** Context is the backward-looking narrative that situates the current study within its broader field, identifies a research gap, and sets up the research question in the current study.
2. **Types of Context and where found**: Context establishes background and scope by reviewing past research and highlighting key developments; identifies unresolved questions or gaps in knowledge, providing a rationale for the study's purpose; and

demonstrates relevance within societal, theoretical, or practical frameworks, while enhancing the study's credibility by showcasing familiarity with the field. Context is found primarily in the *Abstract*, *Introduction/Background*, and *Literature Review* sections. Additional elements of context may appear in the *Methods* (e.g., describing research settings) and *Discussion* (e.g., interpreting findings in relation to existing literature, where the comparative usage of Context functions as a line Reasoning).

3. **Context as the main research question:** A successful Context section zooms in from the broad topic and scope and identifies the research gap and the rationale (or motivation) to do a study. This sets up the main research question, which is typically written as a declarative statement, not as a question. These Context items should directly align with the article's main Claim.
4. **Distinction from Evidence and Reasoning:** Context establishes the prior research leading to the rationale of a study. Evidence presents findings and data that address the research question, which can include prior or archival research in that body of data. Reasoning may include past Context to justify a study's Claim, but does not motivate the study itself. Critique of Evidence or Reasoning focuses on the strength of the article, whereas critique of Context explores whether an article is needed or not.
5. **Common challenges and practical tips:** Missing or unclear Context can obscure a study's rationale, while excessive jargon can hinder comprehension, especially for novice researchers. Use the Introductory Funnel approach to frame the main Context elements: broad topic and its importance, narrow scope, and research gap motivating the main research question.

8 Putting It Together in a CERIC Review

The whole is greater than the part.

—Euclid, *Elements*, Book I, Common Notion 5 (quoted from Euclid & Heath, 1956)

8.1 Overview

Chapter 8 provides a framework for conducting CERIC reviews of research articles, synthesizing Chapters 1–7's approaches to identifying and analyzing Claims, Evidence, Reasoning, Implications, and Context. It discusses the purpose and scope of the CERIC Review and presents a basic review template to scaffold the organization and evaluation of these elements, with suggestions on how to capture information efficiently through ethically paraphrased notes. This chapter then describes ways of improving CERIC reviews, including identifying CERIC elements, dealing with articles with confusing terminology or structure, tailoring reviews to specific article types, and knowing when reviews are sufficient. This chapter also describes how to build on individual CERIC reviews to generate narrative summaries of individual articles and construct personal libraries of CERIC reviews for reference.

8.2 Creating a CERIC Review

Chapters 3–7 provided detailed examples of how the five CERIC elements emerge in various research studies, but ultimately, we need to put these elements together to create a holistic picture of a single article. This is where **CERIC Reviews** come in, defined as a structured approach to organizing the Claims, Evidence, Reasoning, Implications, and Context of an article, which can be subsequently used for other research tasks, such as peer review (Chapter 10), journal presentations (Chapter 11), and literature reviews (Chapter 12). We begin by describing the purpose and scope of a CERIC review, followed by strategies for constructing a comprehensive and concise review document.

8.2.1 The Purpose and Structure of a CERIC Review

The purpose of the CERIC review is to provide a concise but reasonably complete summary of the main CERIC elements of a single article. It serves as both a summary of the article and as a foundation for building more advanced research reviews. The

CERIC review bridges the typical set of informal notes or annotations created when reading an article from start to finish, and the more structured formats of annotated bibliographies and literature reviews. A well-structured CERIC review captures the primary Claim(s), the supporting Evidence and Reasoning, and the broader Context and Implications in such a way that it becomes the primary reference point for utilizing the article in other research tasks.

In this chapter, we introduce a ***CERIC Review Basic Template,*** which scaffolds the article's bibliographic information, the main CERIC elements, and synthesis and learning elements particularly useful to beginning readers, as shown in Figure 8.1 (see Supplemental Material for this chapter). We also provide a worked example and suggestions for modifying this template in Section 8.5.

The first half of the CERIC Review Basic Template focuses on the primary information contained in the article as follows:

- **Article bibliographic information:** This section provides a condensed version of an article's primary identifying information, including title, authors, publication year, journal, volume, page, and DOI or URL, which enables efficient access to the article for subsequent reading. Sufficient citation information is essential for building up the network of research articles, and it is helpful to maintain consistent citation formatting following the style established by your field, institution, or publisher (see Chapter 1). The template also includes a section for keywords, allowing readers to organize collections of articles for their personal libraries.
- **CERIC elements**: The next main section provides space for readers to compile their notes for each of the CERIC elements, including where the information is found in the article. The latter is particularly useful when it is necessary to verify information or draw direct quotes for analysis. Each heading includes guiding questions on how to identify the CERIC element and where the element is likely to be found in the text. The notes compiled here can be a combination of quoted and paraphrased text, written description of tables and figures, and the reader's commentary, depending on the reader's style and goals. We discuss the particulars of ethical note-taking further.

The second half of the CERIC Review Basic Template focuses on analysis of the article's content, and is particularly useful for assessing the quality and increasing your understanding of the argument:

- **CER analysis:** A primary goal of the CERIC method is to help readers develop an understanding and critique of an article's core research argument: How the Evidence and Reasoning presented support a Claim. This section encourages readers to critically reflect on the article's arguments by addressing three critical questions:
  - o Does the Evidence and Reasoning support the Claim, and if so, how exactly?
  - o Are there weaknesses in the article's CERIC elements, and if so, how can they be addressed?
  - o Evaluate the article's quality and significance, including argument and impact.

These questions scaffold a critical analysis of primary literature through critique, a topic that is the focus of Chapter 10.

- **Article summary:** Here, the reader can transform their CERIC notes into a narrative summary of the article's primary findings, supporting arguments, and an assessment of the article's quality.
  - The emphasis here is on a **plain language** summary, written in the reviewer's own words and using language that is appropriate to their level of understanding.
  - This section provides a practice opportunity to develop a reader's ability to accurately and succinctly summarize an article's key points in a straightforward narrative.

We return to this component in the Worked Example at the end of this chapter in section 8.5.

- **Additional notes:** The last page of the review provides the reader space to compile additional information about an article's context and implications.
  - The first part provides the opportunity to define jargon, expert terminology, and advanced concepts in the article that may be new to the beginning reader or researcher. It is especially useful when approaching a new topic as a place to park questions.
  - The second part enables readers to identify other primary sources that are highly relevant to this article, such as those that are prominently or frequently cited in the main text or are found in the article's citation chain, which may serve as a basis for a follow-up review.

Any of these sections can serve as the basis for a follow-up review or an advanced comparison.

8.2.2 Paraphrased Notes versus Direct Quotes

Balancing brevity with accuracy in note-taking is crucial for crafting a high-quality CERIC review that is both informative and useful for future reference. Three methods for citing evidence are **quoting, summarizing**, and **paraphrasing** (Harvard College, 2025). Which method you use may depend on your intended use of the note-taking. For example, when summarizing a study on the effects of exercise on mental health, direct quotations can capture the author's technical point without losing any nuance or meaning, and summaries can distill the author's quotation into its key points. Likewise if you are a student in a German-speaking university, you will likely not be allowed to paraphrase, whereas in an American university, it is expected. Always check with your instructors.

In comparison, paraphrasing is an essential scholarly practice that makes your writing more concise while highlighting the elements of the article most relevant to your purpose. Van der Sande et al. (2023) have emphasized that paraphrasing allows readers to focus on the vital aspects of the research, making their notes both manageable and reliable for later use in analysis or review. Shaw University (2024) further highlights that paraphrasing avoids plagiarism and fosters a more cohesive understanding of the content. This practice is especially relevant for reading scientific research articles, as it ensures that notes succinctly and accurately reflect the original ideas without being bogged down by lengthy direct quotes.

Finally, incorporating personal reflections or questions into your CERIC review notes can deepen engagement with the content over time, prompting critical thinking and aiding in the retention of complex concepts. Self-reflection is distinct from the key CERIC elements of an article, requiring more careful consideration of the article's supporting arguments and impact in the field. By integrating both paraphrasing and reflective writing, readers can create concise, accurate, and insightful notes that are valuable resources for future reference and avoid common ethical problems, a topic we turn to next.

8.2.3 Acceptable versus Unacceptable Reuse of Text

Let's examine a few scenarios that illustrate the balance between when it is acceptable and unacceptable to reuse text in scientific research.

8.2.3.1 Scenario 1: An Acceptable Scenario of Sharing Protocols in Collaborative Research

Two collaborating laboratories are studying marine ecosystems and share a standardized experimental protocol across publications to ensure reproducibility. Here, reusing the protocol text in reports or articles is acceptable because:

- The shared protocol is attributed to the collaborative effort.
- The shared protocol avoids unnecessary rewriting of established procedural descriptions, maintaining consistency and transparency.

8.2.3.2 Scenario 2: An Unacceptable Scenario of Self-Plagiarism in a Results Section

A physicist publishes a set of experimental results in a journal article and then includes the exact text and figures in a conference proceedings article, in the latter case without disclosing that the results had been previously published elsewhere. Here, the reproduction of material is unethical because:

- The duplicated material misleads readers and journal editors by implying the results are novel in both articles.
- The duplicated material can unfairly inflate the researcher's publication record, potentially skewing academic or professional evaluations.

The author can address this ethical issue by: (1) explicitly citing the original journal article in the text and the figure caption, (2) paraphrasing the relevant text in the conference proceedings (while still including a citation), and (3) obtaining permission from the journal to reproduce the article text in the conference proceedings.

8.2.3.3 Scenario 3: Acceptable and Unacceptable Reuse of Text in a Grant Application

A biochemist applies for multiple grants and reuses text and figures from a previously submitted grant proposal to describe the significance of their research. This scenario could be either ethically acceptable or unacceptable, depending on how the text is used:

Acceptable if:

- The reused text is their original work from a prior submission.
- The grant agencies do not prohibit the resubmission of similar content or proposals.
- The biochemist tailors the application to meet the specific goals or priorities of the new funding agency.

Unacceptable if:

- A collaborator or another party initially wrote the reused text, and proper attribution is not provided.
- The grant agency explicitly requires original text for each application or prohibits resubmissions with reused content.

In this situation, the author should verify the specific guidelines and transparency requirements of the funding agency and ensure that the same coauthors are involved in both grants; otherwise, they should modify the text and figures significantly.

Ethical considerations in duplicating text are crucial for maintaining the integrity and originality of scientific communication. While there are scenarios where reusing text is acceptable, such as with proper attribution or shared methodologies, other situations, such as plagiarism or undisclosed self-reuse, cross-ethical boundaries. Understanding the nuanced differences between acceptable and unacceptable duplication practices ensures that researchers uphold transparency, credibility, and respect for academic standards.

Box 8.1 Common Confusion: Why Can't I Just Copy-Paste?

Copy-paste is second nature in the modern world, yet research communication follows stricter rules. Most universities, journals, and professional societies define plagiarism as presenting someone else's words or ideas as your own, and they prohibit copy-paste outright. Less obvious, but equally important, is self-plagiarism, the unacknowledged reuse of your own previously published text or figures in a new article. Both practices undermine scientific integrity and can lead to severe academic or career penalties.

If you grew up copying snippets from websites – or now rely on AI tools (see Chapter 14) – these restrictions may feel puzzling. Remember, though, that transparent, original reporting is the bedrock of research credibility. Building good habits early protects you when the stakes rise for a dissertation, grant proposal, or journal submission.

When working with passages from a research article, choose one of two ethical options:

1. Quote the exact wording and give a page-specific citation.
2. Paraphrase the idea in your own words and cite the source (author, year).

What you must *never* do is drop the original text into your draft without quotation marks and treat it as your own prose. Start the quoting-or-paraphrasing habit now during notetaking. It is far easier to establish a sound practice than to unlearn a risky one later, or worse, have an article rejected because plagiarism checkers are now standard and getting better by the day.

8.2.4 Summary of Better Practices for Using Text from Research Articles

To protect the integrity of your work and uphold academic and scientific communication standards, only use language from original research articles following **ethical procedures**. Proper attribution is essential for ethical writing; always cite the original source, whether citing text, facts, or ideas. If the exact phrasing is necessary, use **direct quotations** and ensure the reference is clear and precise. Whenever possible, **paraphrasing** is preferred because it demonstrates comprehension and enables you to offer a unique perspective.

Self-plagiarism is another essential consideration. Reusing previously published text is permitted as long as it is fully cited. Continuously review and follow the individual regulations of journals or publishers regarding material reuse. Transparency is essential in these situations. Acknowledge when methods, results, or text are adapted from previous work, and ensure that your collaborators are aware of and approve of any reused shared information.

When reusing text, only use what is most relevant and adds logical value to the new work. Avoid including quotations only for convenience or to pad the manuscript. Furthermore, reused materials must be tailored to meet the goals and requirements of the new document or submission. Adherence to the guidelines established by publishers, journals, or funding authorities is equally vital. When in doubt, consult an editor or compliance officer

Finally, **obtaining permission** is essential when considerable text or figure reuse is required, particularly from third-party sources or collaborative works. Always document permissions and agreements for future reference. Finally, respect your audience by eliminating redundancy in publications aimed at the same readers. Reused material should improve the clarity and focus of your work, not detract from its originality or value. Researchers who follow these standards can ensure their writing is ethical, trustworthy, and professional.

8.2.5 Capturing Information from Figures, Tables, and Equations

Figures, tables, and equations are often the key sources for an article's Claim, Evidence, and Reasoning, capturing essential information in a concise framework. Extracting accurate interpretations and meaningful insights from these elements requires different strategies from those used in the main text.

Research emphasizes the importance of examining the hierarchy and organization within complex visuals, such as flowcharts or layered tables, to understand logical progressions and relationships (e.g., Guo et al., 2020). For instance, in ecological studies, flowcharts depicting energy transfer across ecosystems require careful attention to arrows, labels, and formatting to fully grasp the depicted processes. Additionally, visual cues such as color, shading, or gradients can be used to distinguish categories or highlight significant changes in the data. In pharmaceutical research, a heatmap illustrating gene expression levels might employ gradient colors to convey variations in activity across a tissue sample. Multilayered visual cues require careful interpretation of legends and scales to avoid misunderstanding.

Box 8.2 Knowledge Check: Question 8.1 Which of the following scenarios aligns with the acceptable forms of text reuse by a researcher?

A. Reproduces an unchanged methods section from a prior article and cites the study.
B. Copies paragraphs from a review article into a new introduction without citing it.
C. Submits identical manuscripts and results to two journals to see which one will publish the result first.
D. Lifts a collaborator's prose for a new proposal without attribution to save time.

(Check your understanding using the Knowledge Check Key at the end of the book.)

Figures are often used to communicate key mathematical relationships between variables that are essential for establishing Reasoning and justifying Claims. Buckley and Nerantzi (2020) emphasize the importance of identifying trends and deviations in visual data to assess whether the evidence aligns with the study's hypotheses. For example, a time-series graph tracking the economic impact of public health policy changes may reveal sudden deviations that warrant further investigation into external factors influencing the results. Equally important is scrutinizing the methodology behind creating figures and tables to evaluate their reliability and validity. For instance, when analyzing survey results presented in a table, paying attention to the sample size and statistical methods enables us to determine whether the results reflect a global trend or a statistical fluke.

Equations can be the most difficult elements of an article to interpret, particularly if the variables are poorly defined or the mathematical notation is complex (e.g., matrix operations, specialized operators, or functions). Fundamentally, equations describe the relationships between quantitative variables, allowing us to deduce the relative importance of variables in the measured or predicted quantity.

Equations can also be used to articulate the inherent assumptions in a model (e.g., simplifications to the Navier–Stokes equation in fluid dynamics for incompressible fluids) or to provide a framework for interpreting the measurements of a system (e.g., a logistic growth model for population dynamics). In these cases, equations can be considered either Evidence describing the input models used to interpret empirical data or Reasoning describing the relationship between variables. In cases where equations serve one of these key roles, they should be articulated in a CERIC review.

8.3 Effective CERIC Reviews

Effective CERIC reviews are more than just identifying parts of a research article; they demand strategies for analyzing how those parts function within the argument that supports the Claim. This section provides targeted methods for recognizing, interpreting, and evaluating each CERIC element, supported by examples across disciplines, to help you deepen your critical reading and develop transferable academic skills.

8.3.1 Ensuring CERIC Elements are Complete

Let's run through each of the five elements and highlight ways to determine whether our capture of the element is complete.

Finding the Claim(s): As discussed in Chapter 3, the fundamental conclusion of an article is its Claim (or multiple Claims); hence, every CERIC review should be centered around the central Claim. However, as Rawlings (2019) noted, readers may struggle to distinguish between Claims and secondary assertions. A proper Claim is one supported by Evidence and Reasoning that is presented in the article. This contrasts with statements of fact based on previous findings (Context) or assertions that do not have sufficient Evidence or Reasoning to support them (Implications). Rawlings (2019) suggests carefully comparing assertions made in the abstract with those restated or elaborated upon in the Discussion section. Rawlings also suggests readers assess how authors position their Claims within the context of existing literature, as filling a current "gap" through novel contributions or as confirmation/refutation of previous work. Finally, a reader should verify that a Claim is supported by specific Evidence and Reasoning in the article. If these elements are lacking, then either the Claim is insufficiently supported (a weak finding), or it is not a Claim of this article.

Finding and Evaluating Evidence: As one of the supporting elements of an article's Claim, Evidence must be identified and evaluated for a robust review of the article. Finding Evidence means locating the key data, models, citations, and experimental results that directly support the Claim. Evaluating Evidence means critically assessing its credibility and relevance to the study's motivating research question. For example, when examining a study of climate change factors, one might consider temperature records, ice core data, and peer-reviewed articles supporting the methods used to generate that Evidence. Readers should verify that the measures used to gather data are correctly validated, possibly comparing multiple data sources to ensure the results are consistent with the stated methodologies and not unduly influenced by isolated findings (van der Sande et al., 2023). These steps often require some level of expert understanding of how a particular field generates Evidence and what counts as valid Evidence.

Understanding Reasoning Patterns: Reasoning traces the logical progression of an argument that directly connects Evidence to a Claim. As discussed in Chapter 5, logical patterns can take on many different forms (e.g., deductive, inductive, and abductive), and more than one approach may be deployed to establish a Claim. Identifying and evaluating these patterns is crucial for a thorough CERIC review. Pinninti (2024) proposes strategies such as examining transitional phrases and language cues that signal shifts in Reasoning (e.g., "therefore," "thus," "for instance"), as well as visually mapping out argument structures to help readers follow the logical

Box 8.3 Knowledge Check: Question 8.2. Which strategy helps readers distinguish a research article's primary Claim from secondary assertions?

A. Compare the Claim stated in the abstract with the version restated in the discussion to see how the authors position their findings in the literature.
B. Track transitional words such as therefore and thus to map the flow of inductive or deductive reasoning through the argument.
C. Verify that the study's data-collection measures have been validated in previous peer-reviewed work before accepting any findings as evidence.
D. Examine the policy or practice implications the authors draw to decide which statement they regard as the main conclusion.

(Check your understanding using the Knowledge Check Key at the end of the book.)

flow and identify the core Reasoning steps. As a reader gains experience, it becomes easier to see how specific patterns of logical argument are needed to connect certain types of Evidence and Claims; for example, establishing a high level of significance to conclude the existence of a weak signal in noisy data. Understanding these Reasoning patterns then allows readers to assess whether an article's Reasoning is sufficient to support a Claim.

Identifying Implications: Implications connect a study's findings to a broader research context, such as policy, theoretical frameworks, or practical applications. For example, a study by Kobos et al. (2006) reported declining costs of renewable energy sources, implying that governments should invest more in solar and wind power. This Implication may, in turn, influence environmental policy (e.g., supporting a transition away from fossil fuels) and economic investments in renewable power infrastructure (e.g., shifting from hydroelectric to wind generation). This example illustrates how Implications could go well beyond the immediate next steps of research, connecting findings to social, economic, or governmental frameworks.

Considering Context: Context examines what came before the immediate motivating factors of a research study; it encompasses cultural, historical, or disciplinary factors that influence the interpretation and significance of the findings. For instance, in a study on educational outcomes, understanding students' socioeconomic background and prevailing academic policies is essential for interpreting results meaningfully. As Pilcher and Cortazzi (2024) emphasize, these kinds of contextual details help to ground an article's research interpretations in the scope of the field of study. The authors recommend strategies like analyzing how disciplinary norms shape methodological choices and identifying historical trends that influence research focus or outcomes.

8.3.2 Dealing with Confusing or Ambiguous Information

Handling confusing or ambiguous information in research articles requires deliberate strategies to ensure active engagement and comprehension. Techniques such as rereading and cross-referencing sections can help readers piece together unclear ideas, while consulting external sources or glossaries provides additional context and clarification. For example, encountering complex terminology in a quantum mechanics article might require annotating the relevant sections and referring to external resources or a glossary to demystify the content. Connor et al. (2014) emphasize that active approaches are effective for overcoming comprehension barriers while engaging with challenging material.

Additional active reading strategies can also be used to build on these foundational approaches. Chang and Hu (2018) highlight various techniques, such as examining word parts, consulting outside resources, and using context to determine word meanings. Pijeira-Díaz et al. (2013) further advocate the use of metacognitive strategies to enhance comprehension, including planning, monitoring, and evaluating your learning as it progresses. These approaches encourage us to engage with the text actively, continually assessing understanding and refining interpretation.

8.3.2.1 Prereading Techniques

Prereading strategies are essential for efficiently navigating and understanding scientific research articles, enabling readers to identify key objectives, structure, and findings before delving into the details. Skimming abstracts, introductions, and conclusions helps pinpoint the main goals and results, allowing readers to develop focused questions about the study. Additional pre-reading techniques, such as scanning headings and subheadings, identifying keywords, and reviewing figures and tables, build a structural and visual summary of the research (van der Sande et al., 2023).

Other empirical studies refine our understanding of effective prereading. Clarifying a reading purpose, previewing headings and figures, and forming predictions channel our attention toward the article's goals and architecture (Iwai, 2016). Briefly skimming key sections promotes the "mental scaffolding" required to follow complex scientific arguments (De-la-Peña & Luque-Rojas, 2021).

Keshav's (2007) three-pass technique supplies a practical sequence: pass one delivers a bird's-eye overview, pass two secures basic comprehension, and pass three uncovers fine-grained details. Active reading moves complement this structure: reading moves – annotating margins, taking concise notes, summarizing sections, and posing six guiding questions (Carey et al., 2020) that map neatly onto the CERIC elements. Combined, these pre-reading tactics conserve time, deepen understanding, and prepare readers to locate the CERIC elements with efficiency and confidence.

8.3.2.2 Active Reading and Annotation

Active reading and annotation are pivotal strategies for deeply engaging with scientific research articles, enabling readers to highlight key terms, claims, and data points while taking structured notes aligned with the CERIC framework. For example, when reading a study on diet and heart health, annotating terms such as "cholesterol," "blood pressure," and "dietary fiber," along with their contextual roles, facilitates an active effort to understand the findings. Active practices enhance comprehension and retention, allowing readers to analyze and integrate critical information systematically (Pinninti, 2024). In Chapter 13, we examine a specific form of annotation, known as social collaborative annotation (SCA), which utilizes online platforms for real-time peer collaboration to annotate research articles. Finally, in our scan of recent reading research, we identified several key developments:

- **Recent studies provide practical nuance to active-reading and annotation practices:** Marking complex passages – highlighting key phrases, segmenting dense ideas, and pausing to reflect – enhances engagement and comprehension (Lloyd et al., 2022). In digital settings, finding reciprocal online annotations enhances collaborative learning (Yeh et al., 2017, while concrete moves such as underlining, vertical margin bars, and in-text questions keep readers engaged in dialogue with the text (Kalir & Garcia, 2021). Social annotation broadens perspectives and cultivates shared critical thinking by allowing peers to confirm, extend, or challenge one another's interpretations (Hodgson, Kalir & Andres, 2023).
- **Social-collaborative annotation (SCA) pushes these gains further:** Real-time peer comments on research articles raise comprehension and confidence, prompt critical questioning, and correct misunderstandings, especially when instructors embed prompts that direct attention to Claims, Evidence, and Reasoning (Bjorn, 2023; more on this in Chapter 13). Preview strategies complement these tools: the Question–Predict/Confirm routine encourages readers to forecast experimental outcomes after skimming abstracts and introductions (Sun, 2020), thereby priming deeper analysis when results are presented. Teaching the material to someone else reinforces understanding and exposes gaps (McGraw Center, 2025).

By combining predictive, collaborative, and reflective strategies with traditional annotation techniques, readers can effectively engage with scientific research articles. These approaches enhance comprehension, encourage critical analysis, foster deeper understanding, and promote active participation in scientific discourse.

8.3.3 Considerations for Different Types of Research Articles

The CERIC Review Basic Template is intentionally adaptable to various research and learning goals of readers and instructors, as well as the practices of different scientific fields. Here, we describe some ways the Basic Template can be modified to suit these needs.

Conceptual or theoretical articles: In studies that propose new models, frameworks, or hypotheses as their Claim without supporting empirical data, the support of the Claim extends more broadly across the other four CERIC elements. Evidence may focus on the framework or assumptions that underpin the model or hypothesis. At the same time, the Reasoning addresses the soundness of the proposed ideas, often comparing them to previously reported empirical data. Context and Implications may be essential for supporting the Claim, demonstrating the alignment of a model with prior measurements and established theories, and the potential to validate the model with future research.

For example, in a theoretical physics article that proposes a new model of dark matter particle interactions, the Evidence might focus on the model's logical structure and underlying assumptions, the Reasoning on the model's alignment with the behavior of normal matter and consistency with existing empirical limits, and the Implications on the types of future experiments for which the model could be affirmed or refuted. For this type of article, Implications implicitly support the Claim, as a model that cannot be validated experimentally is not an appropriate advance for the field.

Experimental validation articles: Articles that focus on empirical data to validate or refute theoretical ideas or models often emphasize Evidence and Reasoning to support their Claim, reducing discussion on Context or Implications. Authors may allocate considerable space in their Evidence section to describe experimental protocols and measurements, and in their Reasoning section to validate their experimental sample, calibrate measurement biases, determine statistical uncertainties, or exclude alternative hypotheses. For example, a medical study examining disability-adjusted life years (DALYs) in public health interventions may discuss in detail their specific samples and data collection procedures, evaluating whether correlations could be caused by secondary factors such as smoking, levels of stress, or history of heritable disease.

Literature review articles: Articles that synthesize the existing literature may posit many new Claims, particularly around the past or future development of a field, but Evidence and Reasoning are based mainly on existing research studies. As such, Context and Implications take primary roles in establishing the scope of existing and needed investigations. For example, an article addressing the current understanding of climate change impacts on coral reef ecosystems may contain an extensive section on current research (Context), using this to identify the present research gaps (multiple Claims) and the next steps forward (Implications).

New experimental methods and techniques, or engineering articles: For articles that describe novel methodologies, tools, or procedures, Implications become more crucial, highlighting potential applications across various research scenarios. For example, a biomedical engineering research article that introduces a novel prosthetic limb prototype may discuss specific target populations that would benefit most from the prosthetic to justify further investment and development of the prototype.

The most important takeaway here is that the Basic Review Template is merely a starting point. It can and should be adapted as needed.

8.3.4 When a CERIC Review Is Complete

Knowing when a CERIC review is complete is a common challenge for new readers, who may be unsure what details are essential, whether an argument is complete, or what information may be missing either in their review or in the article itself. Because the CERIC method is meant to be an efficient approach to the primary literature, it is essential to find a balance between thoroughness and time. Here, we highlight some methods for handling incomplete or missing elements, offering solutions for identifying gaps and evaluating their significance. Here is some guidance on balance:

- **Make sure all the CERIC elements are present:** Any complete research article must have each of the CERIC elements represented; missing elements suggest a review is not complete. If a reader focuses exclusively on the abstract for their review, they are likely to miss one or more of these elements. In this case, the reader should focus on the section(s) most likely to contain that missing element, such as the Introduction for Context or the Discussion for Reasoning.
- **It is not necessary to detail all the Evidence:** In experimental articles, the description of samples, research protocols, analysis, calibration, and other methodological details, may be extensive, and a reader may be unsure what is essential among these elements. It is helpful to focus on the aspects of the experiment that most directly support the Claim, as well as those aspects of data collection or analysis that are unique to the study. For example, a study on the efficacy of an educational intervention in an introductory Physics course may apply an AVONA analysis to standard survey protocols, such as the Force Concept Inventory (Halloun & Hestenes, 1985), in a typical college course curriculum. None of these aspects requires detailed examination, as they are common approaches in the field. Instead, the actual intervention, its implementation, and the sample size and composition will be unique and likely relevant to the outcome and should therefore be detailed in the CERIC review.
- **Make sure the CER argument is complete:** One reason the CERIC review Basic Template includes CER-based questions is to ensure the reader has captured the entirety of the argument supporting a Claim. Once a review is complete, it is essential to take a step back and map out the key pieces of Evidence and Reasoning that support the main Claim, ensuring they are sufficient to establish the Claim. If not, the reader should consult the relevant sections (Methods, Discussion, Tables, Figures) to determine if any information was overlooked. It may turn out that key elements of Evidence and Reasoning are missing, which provides a starting point for critiquing the article.
- **Context should focus on the motivating problem or gap**. The introductory sections of articles span a wide range, from comprehensive reviews of a subject to terse summaries of the problem at hand. For a CERIC review, the key contextual information to capture should be the motivating problem or gap, as well as any past work that directly motivates that question or supports the primary Claim. For example, a CERIC review of an article assessing the efficacy of a drug treatment protocol for a particular disease should describe the manifestation and impact of the disease, the current "gold standard" treatments and their efficacies (if they exist), and the rationale for the need for a new intervention. Discussion of related diseases

or the full scope of all secondary treatments is generally not necessary unless they directly inform the new treatment protocol.

- **Stick to the article's stated Implications:** When reading a research article, it is common to form our own interpretation of the impact and future work stemming from a research study, particularly if the authors do not provide much detail on these topics. When analyzing the CERIC content of an article, be sure to note what Implications are discussed in the article itself, separating these from your interpretations of the article. If the article lacks clear Implications, this should be included as part of the critique of the article.
- **Time management:** Setting defined time limits for a CERIC review can help protect you from getting bogged down in a particularly challenging article. As novice reviewers gain proficiency in identifying the elements, speed and accuracy will improve. Iterative reviews are also a successful approach to keeping CERIC reviews efficient: a first pass can analyze the overall structure. In contrast, subsequent passes can explore specific portions further, addressing any nuances or holes.
- **Finally, ask yourself, what is missing?** Once you have completed a review, consider whether there are questions you have about the justification of the Claim or the quality of the article that aren't answered in your notes. For instance, if Reasoning is provided but supports a different Claim than the article makes, what does that mean for the integrity of the core CER argument? It often helps to talk through inconsistencies and logical gaps with a study group or journal club, as others may give you new ideas and help you develop a more complete analysis of an article if you still have unanswered questions.

As you gain experience with CERIC reviews, many of these suggestions will become second nature. Until then, it may be helpful to create a meta-cognitive-style checklist to run through with each review.

8.4 Going beyond the Review

As we will see in Chapters 9–14, the CERIC review serves as a foundation for more advanced critical reading and thinking activities. Here, we will begin with the first derivative product, a **CERIC Narrative Summary**, which enables us to capture all the main points of an article.

8.4.1 Writing a Narrative Summary

A concise narrative summary is essential for effectively communicating an article's key claims, arguments, and quality, as illustrated in Figure 8.1. Like a news article or a book review, Narrative Summaries should proceed in a logical, time-ordered structure, while ensuring the most important element – the Claim – is

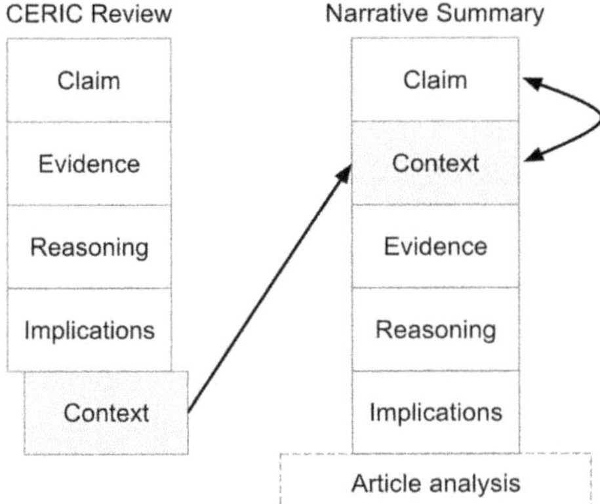

Figure 8.1 Organizing a CERIC review into a narrative summary.
A storyboard shows how CERIC notes shift in presentation order to become a narrative summary.

easily accessible. While the abstract can capture some of the elements of a Narrative Summary, the brevity of abstracts means that key details are often omitted, and there is typically no self-assessment of an article's quality. We rarely get the peer reviewers' or editors' perspectives. By creating our own narrative, we can capture all the CERIC elements and conduct our own evaluation of the impact and quality of a research study.

A coherent narrative follows a logical time order, so that Context should precede the main elements of the article (Evidence and Reasoning) and the Implications. Depending on style, you may also choose to swap the order of the Claim and Context. An advanced Narrative Summary can also include a critical analysis of the article at the end.

Fortunately, the CERIC review framework provides a straightforward way of transforming our notes into a narrative. Here's a sequence of steps that readers can follow to create a Narrative Summary from their review:

1. Start by copying your CERIC notes and rearranging the order to CCERI (Context, Claim, Evidence, Reasoning, and Implications). The first sequence is preferred as it places the most crucial information, the Claim, at the top of the narrative. The latter sequence, CCERI, mirrors the *IMRaD* format.
2. Make sure you've properly paraphrased any quoted text, possibly bringing in definitions from your Additional Notes.
3. Now edit your notes into a coherent paragraph, using complete sentences and plain language. This may take a few passes to ensure the text is clear and concise.

If desired, you can add a critical assessment of the article based on the CER argument analysis. Again, start by copying the notes from your analysis, ensuring they are paraphrased. Then, condense them into a few sentences to capture the article's strengths, weaknesses, and overall quality. If this seems a little unclear at this point, we provide a worked example of a narrative summary drawn from a CERIC review of Jao et al., (2018) in Section 8.5.

8.4.2 Identifying and Responding to Missing or Incomplete CERIC Elements

A systematic approach is crucial for identifying and addressing gaps in a CERIC review. Use the five elements as a checklist – Claim, Evidence, Reasoning, Implications, and Context – to sweep for omissions methodically. This structured approach ensures that gaps can be addressed precisely, and no element is overlooked.

When missing components are identified, specify precisely where the omission occurs and explain why it matters to the overall argument. For example, if the Reasoning is unclear or missing, the review should note how this hinders the connection between the Evidence and the Claim. Missing Reasoning weakens the central Claim, whereas missing Implications or run-on sentences in the Context do not. Clarifying any gaps helps reviewers understand the weaknesses in the article and their impact on its persuasiveness.

Adaptability is key when evaluating whether a missing element is critical to the article's argument. Some omissions may be significant, while others can be reasonably inferred. For instance, a detailed reporting of intermediate experimental steps might be absent in an article describing the synthesis of a new drug. However, if the final product's characterization data (e.g., purity and yield) are robust and consistent, reviewers can infer the integrity of the intermediate steps based on these outcomes. Journal page limits are real limits to reporting.

CERIC reviews can also address gaps by posing constructive solutions. You might include additional citations you've identified in the article or other articles. You can also detail the weaknesses in the argument of an article using sentence frames such as

- The authors could strengthen this argument by ______, or
- The information that would complete this argument is ______.

Sentence frames reduce cognitive load by giving us the language while also supporting critical thinking by not giving us the answers. This proactive approach goes beyond identifying issues and helps build toward formal critique through peer review, the subject of Chapter 10.

8.4.3 Capturing Citations

One final recommendation is to capture citations correctly. Appropriate citation is a foundational skill in academic and research writing, as it ensures proper attribution of past work and increases the reputation of your research – citation is how discourse continues. Citation management software, such as Zotero, EndNote, or Mendeley, can help you organize your references, generate citations in various formats, and interact easily with writing software like Google Docs or Microsoft Word. To avoid losing important information, always capture full bibliographic details while reading, including authors, publication year, title, publisher, and DOI or URL (i.e., Section 1 of our CERIC Review Basic Template). You can also add annotations or notes to each reference to highlight its relevance and summarize key points for future usage. Building a well-organized and correct reference database saves time during the writing process while also maintaining the integrity and rigor of your research.

8.5 Worked Example

Here, we provide a fully worked CERIC review using the Basic Template on an astronomy research article by Jao et al., (2018). In the remainder of this section, we step through each of the main elements of the review template.

8.5.1 Research Article Information

Table 8.1 Key bibliographic and content information for Jao et al. (2018), reporting the discovery of a gap in the lower main sequence using Gaia DR2 data.

Key Elements	Content Information
Title	A Gap in the Lower Main Sequence Revealed by Gaia Data Release 2
Lead Author (Year)	Jao et al. (2018)
Journal, Volume, Page, URL or DOI	*The Astrophysical Journal Letters, 861*(1), L11. https://doi.org/10.3847/2041-8213/aacdf6
Abstract	"We present the discovery of a gap near $M_G \approx 10$ in the main sequence on the Hertzsprung-Russell Diagram (HRD) based on measurements presented in Gaia Data Release 2 (DR2). Using an observational form of the HRD with M G representing luminosity and G_{BP}–G_{RP} representing temperature, the gap presents a diagonal feature that dips toward lower luminosities at redder colors. The gap is seen in samples extracted from DR2 with various distances, and is not unique to the Gaia photometry – it also appears when using near-IR (NIR) photometry (J – K_s versus M_{Ks}). The gap is very narrow (~0.05 mag) and is near the luminosity-temperature regime where M dwarf stars transition from partially to fully convective, that is, near spectral type M3.0V. This gap provides a new feature in the HRD that hints at an underlying astrophysical cause, and we propose that it is linked to the onset of full convection in M dwarfs."
Keywords	Hertzsprung–Russell and C–M diagrams – stars: distances – stars: late-type *cool stars, stellar main sequence, Gaia, stellar evolution*

Note: The keywords listed here include those from the original paper (at the top) and those of our own construction (at the *bottom*). Many articles include keywords after the abstract, and some fields have curated a set of predefined subject keywords, such as the Unified Astronomy Thesaurus.

8.5.2 CERIC Elements

1. Claim

A plain language, declarative answer to a scientific question that asserts new knowledge. *What is the central Claim or hypothesis this paper states? Are there additional secondary Claims being made?* Usually found in the abstract and title, as shown in in Table 8.2.

2. Evidence

What are the data, measurements, observations, models, computations, and other analytical components that the authors use to support their Claim? Usually found in the abstract, the Results section, and in tables and figures, as shown in Table 8.3.

3. Reasoning

What is the logical basis for this argument? What connects the Evidence and the Claim? Do the authors conduct analysis? This could refer to a specific analysis methodology or how authors were able to rule out alternative interpretations. *Do the*

Table 8.2 Identification of the main and subclaims in Jao et al. (2018): the primary Claim describes the discovery of a narrow gap near MG ≈ 10 on the main sequence, while the secondary Claim links the gap to the transition from partial to full convection in M dwarfs.

Notes	Page/Section
Main Claim: There is a narrow gap on the main sequence *"near $M_G \approx 10$"* (absolute G magnitude)	Title, Abstract, Section 2, Figure 1
Subclaim: The gap occurs at the *"transition from partially to fully convective"* in M dwarfs (convection is heat-driven circulation of material inside star)	Abstract, Section 4, Figure 7

Note: We differentiate between the main Claim, the discovery of the gap, and a sub-claim that the gap is caused by the partial-to-full convection transition. Both are Claims since both are based on Evidence and Reasoning in the article. We also indicate text directly from the article with quotation marks and italic font to ensure we paraphrase later, providing some additional text to explain the jargon in the quote for future paraphrasing.

Table 8.3 Evidence used in Jao et al. (2018) to support the Claim of a main sequence gap, including Gaia DR2 stellar samples, 2MASS photometry, extended Gaia data to 130 pc, and evolutionary models (PARSEC and YAPsI).

Notes	Page/Section
242,585 stars within 100 pc, from Gaia DR2 with M_G and G_{BP}–G_{RP} measurements, cleaned of *"low-quality sources"* (contaminants)	Abstract, Section 2, Figures 1 & 2
J & K_s magnitudes from 2MASS	Section 3 & Figure 3
Gaia data for more distant stars out to 130 pc	Abstract, Section 3 & Figure 4
PARSEC and YAPsI evolutionary models	Section 4, Figures 7 & 8

Note: We list the sample size as it helps to indicate statistical power in the results. However, specifics regarding the existing catalogs, Gaia and 2MASS, and models such as PARSEC and YAPsI are not necessary because we use these as standard sources of data.

authors use theory or prior research to link the Evidence and Claim? Usually found in the Analysis section (Table 8.4).

4. Implications

How is this result significant beyond the immediate findings? What has been learned? What further work is needed? Usually found in the Discussion section and possibly in the abstract (Table 8.5).

Table 8.4 Reasoning in Jao et al. (2018) linking evidence to the Claim of a main sequence gap, including direct observation, replication across photometric systems and distances, elimination of alternative explanations, and consistency with stellar evolution models predicting the partial-to-full convection transition.

Notes	Page/Section
Direct observation: we see a gap! In the plot of M_G vs G_{BP}–G_{RP}	Abstract, Section 2, Figure 1
Ruling out artifacts: Gap in seen in both M_G vs G_{BP}-G_{RP} and M_{Ks} vs J–K_s; therefore, this is a real feature and not a Gaia artifact	Abstract, Section 3, Figure 3
Ruling out artifacts: Gap in seen in sources out to 130 pc, so not an artifact of local sources	Abstract, Section 3, Figure 4
Why first discovery: Seen for the first time because the Gaia photometry and parallax data are better than prior measurements	Section 4, Figure 4
Ruling out alternatives: Not due to known biases in the Gaia data which only occur for brighter stars	Section 4
Ruling out alternatives: Not due to multiple main sequences like those in globular clusters	Section 4
Ruling out alternatives: Is not related to previously identified "Wielen dip" or "Kroupa dip" features which occur at brighter magnitudes	Section 4 & Figure 5
Comparison to models: Mass at which the gap occurs is at the same spectral type (M3.0V) and mass (≈0.35 $M_\odot$) that evolutionary models and prior research indicates stars go from partially to fully convective	Abstract, Section 4, Figures 7 & 8

Note: We have indicated in the table the types of logical arguments used to connect Evidence to the Claim, which is useful for assessing the nature and quality of the Claim.

Table 8.5 Implications identified in Jao et al. (2018), emphasizing directions for further research, including studies of unresolved multiples, measurement of stellar parameters, rotation, and photometric variability to probe the physical causes of the main sequence gap.

Notes	Page/Section
Further research: Study the stars inside the gap to determine if they are *"unresolved multiples"* (binaries)	Section 5
Further research: Measure the *"masses, radii, and metallicities for stars on both sides"* and in the gap (fundamental physical parameters)	Section 5
Further research: Measure *"the rotational periods for stars in and around the gap to see if there are any trends"* (rotation may be relevant to convection transition)	Section 5
Further research: Measure "photometric variability" to assess magnetic activity/"topology" (shape) which might *"provide clues as to the cause of the gap"*	Section 5

Note: We indicate the nature of the implications in each case; note that all propose further research on the physical phenomenon as opposed to work needed to validate the findings or the broader scientific/social implications of the article.

Table 8.6 Context for Jao et al. (2018), situating the discovery of a main sequence gap within the broader use of the Hertzsprung–Russell Diagram, the importance of parallax distances, the precision of Gaia measurements, and prior discoveries of stellar subpopulations, leading to the rationale for examining M dwarfs.

Notes	Page/ Section
"Hertzsprung–Russell Diagram (HRD)" important for determining stellar properties – *"size, metallicity, evolutionary state"*	Section 1
Parallax distances are necessary for this, but *"most red dwarfs within 100 pc did not have distance measurements."*	Section 1
New satellite mission Gaia provides unprecedented precision down to faint magnitudes, *"initiated a new era of astrometry measurements"*	Section 1
Prior Gaia measurements revealed new few features in HRD, including two populations of stars due to metallicity differences, and multiple populations of white dwarfs	Section 1
Main Rationale: Does Gaia show a new HRD feature for M dwarfs?	Section 1

Note: The primary rationale is explicitly stated here, although it was not expressly stated in the text. We can infer it from the lead-up Context that this is a novel (and unexpected) discovery rather than one that was predicted ahead of time.

5. Context
What is the scientific context of this work? What broader questions or problems are relevant to this work? What is the rationale for conducting this study? Usually found in the Introduction or Context sections (Table 8.6).

8.5.3 CER Analysis

1. Does the Evidence and Reasoning Support the Claim, and If So, How Exactly?
Yes, the Evidence and Reasoning support the Claim in several ways.

1. Direct observation of the **gap** in the data
   - The primary evidence comes from Gaia data, which shows a clear gap in the main sequence at $M_G \approx 10$ (Figure 1). This gap is consistent and measurable, appearing as a narrow feature of approximately 0.05 magnitudes.
   - The gap is also evident in multiple Gaia datasets, underscoring its significance as a genuine feature in the data.
2. Supplementary evidence
   - 2MASS J and K_s magnitudes (Figure 3) provide independent verification of the gap‘s presence, confirming the robustness of the observation across different datasets.
3. Alignment with theoretical models
   - The PARSEC and YAPsI stellar evolution models (Figures 7 and 8) support the claim by providing a physical explanation for the gap.
   - These models indicate that the gap occurs at a mass that corresponds to the transition from partial to full convection in M dwarfs

4. Rule-out analyses
   - The authors demonstrate that the gap is not an artifact of biases, errors, or data inconsistencies.
   - They also rule out alternative explanations, such as the presence of multiple main sequences.
5. Why is it seen now?
   - The authors argue that the unprecedented sensitivity and large set of data allow us to discern the gap for the first time.

2. Are There Weaknesses of the Article in Its CERIC Elements, and How Can They Be Addressed?

Yes, there are a few minor weaknesses that could be addressed as follows:

- **Context:** Although this study is based on discovery enabled by Gaia data, there is no discussion on whether such a gap might have been predicted based on theoretical models, that is, it is unclear whether this is a novel or unexpected result.
- **Evidence:** The study describes their data sample as "clean" based on "cuts discussed in Lindegren et al. (2018)," but these are not clearly specified in the text.
- **Reasoning:** The study rules out multiple main sequences but notes that the clusters these would be seen in are either too sparse or too distant to make similarly precise measurements. While the study argues that the gap occurs at the convective transition, its use of models does not show as clear a gap in synthetic HRD as the observations.
- **Implications**: The authors discuss future studies on the gap, but do not propose any theoretical work that would clearly demonstrate the gap is due to convection and how convection would produce a gap.

3. Based on This Analysis, Is This a "Strong" or "Weak" Primary Source? Why?

This is a "strong" primary source for several reasons:

- **Credibility:** The study is based on data from the Gaia mission, a highly reliable and widely respected source in the field of astronomy. The claim is supported by figures that directly demonstrate the gap, and the authors construct several arguments to rule out biases, errors, or data inconsistencies.
- **Peer review:** The publication appears in a reputable, peer-reviewed journal, ensuring that experts have critically evaluated the methods, evidence, and conclusions.
- **Novelty:** The study introduces a new finding – a gap in the main sequence for mid-M dwarfs – which contributes to the scientific understanding of stellar evolution.
- **Methodological rigor:** The authors systematically analyze the data, employ established methods for validation, and address potential biases and alternative explanations.
- **Contextualization:** The study situates the discovery within the broader context of stellar evolution and compares it to known features, such as the "Wielen dip" and the "Kroupa dip."

8.5.4 Article Summary

Using the CERIC elements and CER analysis, write a plain language, narrative summary of the article and your assessment of its quality in terms of establishing a robust claim.

Note. As discussed in Section 8.4, the narrative summary paragraph was generated by first copying the notes from our CERIC elements in "CCERI" order (Claim, Context, Evidence, Reasoning, Implications) and the CER analysis, then reformatting that text into a coherent paragraph. It reads like a paraphrased abstract of the article, but with an explicit analysis of the scientific argument. Our narrative summary, below, is a more informative and comprehensive summary and assessment of the article.

Jao et al. (2018) report the discovery of a gap in the stellar main sequence in the Gaia Hertzsprung–Russell diagram, at a mass at which stars transition between partially and fully convective interiors. The discovery is made possible by the unprecedented photometric and astrometric precision of the Gaia satellite, which, unlike previous surveys, can reach most red dwarfs within a distance of 100 parsecs. The study drew 242,585 high-quality stars within 100 pc from Gaia Data Release 2. Their M_G and G_{BP}–G_{RP} measurements clearly show a narrow gap ($\approx$0.05 mag) at $M_G \approx 10$ mag, consistent with an M3 dwarf. They validate the presence of this gap by identifying it in Gaia samples extending to 130 pc, finding that the gap is present in the near-infrared based on photometry from 2MASS. The authors rule out known biases in the Gaia data, as well as the presence of multiple main sequences and alignment with the Wielen and Kroupa dips. Instead, the authors claim the feature aligns with a mass of 0.35 solar masses where stars transition between partially and fully convective interiors, based on comparison to the PARSEC and YAPsI evolutionary models. The authors propose further investigation of stars in the gap region to directly measure their masses, radii, and metallicities; assess the fraction of binary systems; and probe their magnetic properties through trends in rotation. Overall, this article presents a novel discovery that connects to broader issues in stellar evolution and provides clear evidence and reasoning to support their claim, including ruling out data bias and other effects. The study could have provided more theoretical context as to whether such a gap was previously predicted and how the transition between partial and full convection would manifest as a gap in the Hertzsprung–Russell diagram.

8.5.5 Additional Notes

1. Unclear Jargon and Terms

Use this space to list any unclear or unknown jargon or terminology and their definitions, as shown in Table 8.7.

2. Relevant References

Are there other articles cited in or by this article that may be useful for further investigation? Indeed, there are, as shown in Table 8.8.

Table 8.7 Glossary of unclear jargon and terms from Jao et al. (2018), including definitions of stellar processes, observational features, and astrophysical concepts such as convection, isochrones, luminosity functions, and magnetic topology.

Jargon/Term	Page/ Section	Definition and Source
Stellar Convection	Section 1	A process by which stars convey energy from core to surface by the mixing of hot material upwards; this is in contrast to radiation where photons carry the energy. Stellar interiors are divided into convective and radiative layers. https://en.wikipedia.org/wiki/Convection_zone
Isochrone	Section 3	A curve on the HRD that shows where stars of the same age and composition are located. https://en.wikipedia.org/wiki/Stellar_isochrone
Luminosity function	Section 4	Distribution of stars as a function of luminosity (radiant power), typically expressed as the number or density within a given volume. Source: Wikipedia: Luminosity Function (astronomy)
Wielen dip	Section 4	A break in the stellar luminosity function at $M_V \approx 7$ (mid-K type stars) caused by the onset of H^- opacity in stellar atmospheres (Kroupa et al. 1990: https://ui.adsabs.harvard.edu/abs/1990MNRAS.244...76K/abstract)
Kroupa dip	Section 4	Dip in the stellar luminosity function at $M_V \approx 9$ (Kroupa et al. 1990: https://ui.adsabs.harvard.edu/abs/1990MNRAS.244...76K/abstract)
Stellar (or initial) mass function	Section 4	Distribution of stars as a function of mass, typically expressed as the number or density within a given volume https://en.wikipedia.org/wiki/Initial_mass_function
Photometric variability	Section 5	Variation in a star's brightness as a function of time, can indicate a star's rotation or episodic events (e.g., stellar flares) https://en.wikipedia.org/wiki/Variable_star
Magnetic topology	Section 5	The structure or shape of the magnetic field around a star https://en.wikipedia.org/wiki/Stellar_magnetic_field

Table 8.8 Relevant references cited in Jao et al. (2018), highlighting additional parallax studies of nearby stars and the Gaia DR2 release as key sources for understanding main sequence structure and red dwarf populations.

Abbreviated Citation (Author, Year, Journal)	Notes on relevance to this article
Finch, Zacharias, & Jao (2018, AJ)	Reports parallaxes for stars within 100 pc, derived from ground-based observations, providing additional insights into nearby main sequence stars.
Gaia Collaboration, Babusiaux, van Leeuwen et al. (2018, A&A)	Discusses Gaia Data Release 2 (DR2), which includes parallaxes for stars within 100 pc, enabling detailed studies of main sequence stars and their distribution.
Henry, Jao, Winters, et al. (2018, AJ)	Addresses parallaxes for nearby stars, with a particular focus on red dwarfs within 100 pc. Complements Gaia data by highlighting specific stellar populations.

The worked example in this section demonstrates the utility of the CERIC review template in systematically dissecting and evaluating a primary research article, offering a structured approach to assess Claims, Evidence, and Reasoning. This example assesses the rigor of Jao et al.'s (2018) discovery of a gap in the lower main sequence, while also highlighting the interplay between observational data, theoretical models, and methodical validation. Extending this approach across articles could foster deeper insights, which is the topic of the next chapter. Also, more worked examples in other scientific fields using the Basic Template are available in the Supplement.

8.6 Chapter Key Takeaways

1. **A CERIC review organizes critical reading**: The CERIC review provides a concise and complete summary of the key CERIC elements in an article, serving as a useful reference for building to more advanced research reviews. A well-structured CERIC review captures the primary Claim(s), the supporting Evidence and Reasoning, and the broader Context and Implications in a way that makes it easy to assess and utilize the article in other research tasks.
2. **Readers are encouraged to use structured templates**: The Basic Template demonstrated in Section 8.5 provides a framework for dissecting the key elements of an article and beginning the task of synthesis and critique. We emphasize importance of approaching these reviews ethically through accurate citation and paraphrasing; capturing information from figures, tables, and equations; and ensuring CERIC elements are accounted for, and the review is complete. We also discuss how to modify CERIC reviews for different types of research areas of methodologies.
3. **The CERIC review is a jumping-off point for more complex literature tasks**: One can quickly transform CERIC review notes into a Narrative Summary as well as compile multiple reviews into our own research libraries. In the next few Chapters, we will see additional ways of building off the CERIC review for research synthesis, critique, presentations, and literature reviews.

Part II

Critical Thinking

9 Using CERIC to Compare Articles

[Kepler] had to realize clearly that logical-mathematical theoretizing, no matter how lucid, could not guarantee truth by itself; that the most beautiful logical theory means nothing in natural science without comparison with the exactest experience. Without this philosophic attitude, his work would not have been possible.

—Albert Einstein, in Kepler & Baumgardt (1951). *Johannes Kepler: Life and Letters*. p. 13.

9.1 Overview

Chapter 9 focuses on using the CERIC method to compare two research articles. The chapter contextualizes this practice as an essential activity in scientific research. It then discusses how one finds similar articles for comparison, including how articles are related to each other; for example, how references, citations, and "similar article" searches can be used to find related articles, and which criteria are useful for selecting the most suitable articles for comparison. This chapter then delves into using the CERIC elements to conduct a detailed comparative analysis across texts, describing methods of visually comparing CERIC elements to aid in interpreting commonalities and differences. Finally, this chapter guides readers through turning CERIC reviews into a comparative narrative summary through a worked example.

9.2 Comparing Research Articles

Comparing research articles is a cornerstone of the scientific process, enabling researchers to evaluate the validity of a research study in the context of similar studies and to understand its contribution to a field's advancement. The comparison practice starts at the level of identifying differences and similarities between individual related articles, which when repeated across many studies, forms the foundation of a cohesive understanding of the research landscape. The CERIC elements offer a structured approach for making these comparisons, helping researchers analyze key components of the articles systematically. By mastering the skill of article comparison, readers cultivate the critical thinking needed to move beyond description of a single article's argument into a more analytical perspective, an important transition to research critique.

If you have ever heard an advisor or professor discussing key aspects of multiple articles, this person is likely having a mental conversation with themselves about each article's findings and how they collectively advanced the field. Let's imagine a conversation between this professor and a student; our hypothetical example might go something like this:

Scene: Department coffee lounge. *Max*, a first-year graduate student, approaches *Prof. Carter*, who is absorbed in a tablet full of research articles.

MAX: "Professor Carter, could we talk about this carbon-capture membrane paper? I'm getting lost in the terminology."

PROF. CARTER: "Of course! Do you mean Lu et al. (2024)? That's a great study. They extend the Huang–Zhang composite-interface model from Huang et al. (2020), and borrow the GO-PEG grafting protocol from Kim et al. (2022). Their big result is that they push CO_2/N_2 selectivity to 72–*72!* – which finally matches the dual-mode adsorption coefficients predicted by Crespo's (2016) Monte-Carlo work.

MAX [pulling out a notebook and pen]: "Right … Lu et al. (2024)."

PROF. CARTER: "Now keep in mind, dual-mode just blends the Langmuir and Henry regimes. Lu's group fit isotherms with the same β parameter Kratochvîl et al. (2017) reported for PIM-1 particles, then they corrected free-volume fraction with the Voronoi tessellation method from Ghosh et al. (2022). Plot log D versus 1/T and it simply collapses onto the Arrhenius line as predicted by Berneli et al. (2018). So all the "terminology" you might be getting lost in is just shorthand for that cascade."

MAX[furiously writing]: "Okay … Arrhenius line … Voronoi tessellation … Um, I think I need Cliff's notes or something. Is there a review article I can start with?"

PROF. CARTER: "Well, reviews in the field are already hopelessly outdated. I suggest you start with Huang et al. (2020) for the mixed-matrix paradigm shift; then read Zhang et al. (2021) on interface compatibilization. After that, go to Karan et al. (2019) for the Angstrom-scale sieving argument – after all, Lu's free-volume discussion is basically a rerun of Karan's slit-pore model. Finally, read Park et al. (2023); they show those selectivity gains vanish in humid flue gas. It's all interconnected."

MAX [staring at his notes]: "That's … a lot of papers. Could you maybe just break down Figure 3 of Lu's article?"

PROF. CARTER [bringing up the article on his tablet]: "Let's see … ah yes! Figure 3 is very straightforward – look, it's a Robeson upper-bound plot. The novelty here is the deviation angle θ they compute with the Kline–Ghosh fit, exactly as Li et al. (2014) applied to perfluorinated polymers. Overlay Li's data, and Lu's membranes fall right on the extrapolated 90°C trend. That reveals their segmental-mobility constraints – confirmed, incidentally, by DMTA in Table S6 of the supplement.

MAX [clearly overwhelmed]: "I … guess I'll take a look at those other articles, then."

PROF. CARTER: "Ha, science is a conversation, Max! You'll be able to follow along once you've internalized the literature lattice. So, start tonight with Huang et al. (2020), Zhang et al. (2021), and Karan et al. (2019), plus maybe Li et al. (2014) – after that, Lu et al. (2024) will read just like a memo."

MAX: "Right, no doubt it'll read just like a memo."

The conceptual gulf between Professor Carter's rapid article cross-referencing and fluent technical jargon versus Max's bewilderment reminds us that scientific expertise is not built on familiarity alone. It rests on three intertwined abilities:

1. Knowing the essential articles in a field, their Claims and central arguments;
2. Understanding and critiquing the methods and findings of these articles, and
3. Weaving the insights from these articles into a coherent intellectual map.

To student researchers, this can feel overwhelming – part vocabulary drill, part detective work, part cartography, part epic memorization. The task becomes more manageable, however, when we frame it as joining an ongoing scholarly conversation rather than scaling a solitary peak.

A practical first step is a systematic comparison of articles. By setting two or more articles side-by-side – asking how their research questions, methods, Evidence, Reasoning, findings, and Claims align or diverge – we can start to spot patterns, expose gaps, and trace the evolution of key ideas. Comparison, in turn, lays the foundation for full critique, a high-level skill we will tackle in Chapter 10. The remainder of this chapter, therefore, concentrates on the nuts and bolts of comparison: How to line up studies on a common grid, extract their core CER arguments, and integrate their contributions into the growing mental framework every graduate researcher needs.

9.3 Identifying Related Articles for Comparison

Identifying related articles is a crucial first step toward building a comparative analysis. This task is made difficult by the sheer number of research articles available, with well over a million new articles added each year (global estimates vary by database; see trends in Bornmann & Leydesdorff, 2018; Hanson et al., 2024). There are several strategies and tools that can help researchers identify related articles that leverage references, citations, and article keywords to trace an investigative thread or explore a broader research ecosystem. This section explores a handful of practical tools and strategies for locating similar research articles efficiently.

9.3.1 How Articles Are Related

Before diving into the ocean of research articles, it is important to first reflect on how articles are related to each other. Even when focusing on just two articles, there are several ways these can interrelate based on the connection between their findings and the broader research ecosystem, as illustrated in Figure 9.1.

Development: A development sequence is where the findings from one article enable or motivate the research undertaken in a second article. This relationship

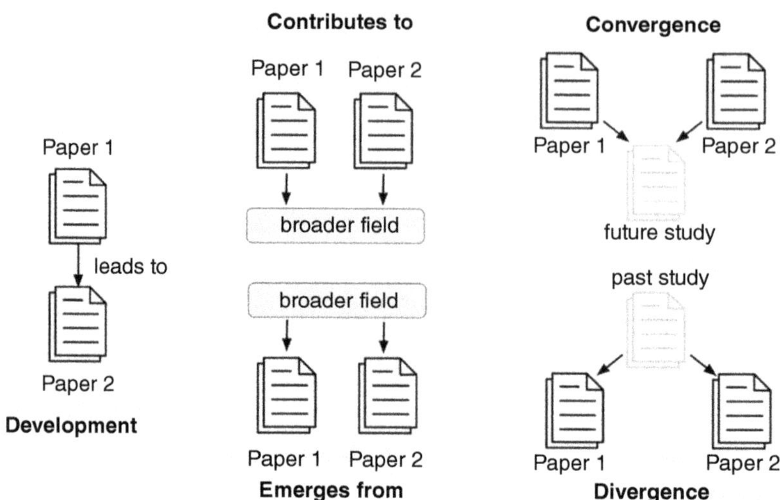

Figure 9.1 Illustrating the ways in which two research articles may be related to each other. The diagram illustrates two focal studies and their shared ecosystem links.

could be an iterative stepping stone in research, or it could be a response that supports or refutes a finding. An example of this relationship in medical science is the development of the CRISPR (Clustered Regularly Interspaced Short Palindromic Repeats) gene-editing technique, first reported in Jinek et al. (2012), which has enabled a wide range of peer-reviewed medical studies on new gene-based disease interventions.

Contributes to: This relationship involves two parallel articles that both contribute to a broader field of study or specific science topic. An example in astronomy is a series of articles published in the *Astrophysical Journal Letters* in 2017 on the detection of gravitational waves that resulted from the merger of two neutron stars. These articles include Abbott et al. (2017) which reports the detection of both gravitational waves and gamma ray radiation from the source, and Murguia-Berthier et al. (2017) which compares a model for the merger to the observations. Both articles contribute to the larger body of work on neutron star mergers that followed this first detection.

Emerges from: This relationship involves two parallel articles that have separately emerged from a common origin. An example in physics includes parallel studies on different but chemically related varieties of high-temperature superconductors, such as those based on sulfur hydride (H_3S; Drozdov et al., 2015) and lanthanum superhydride (LaH_{10}; Somayazulu et al., 2019).

Convergence: This relationship is a variant of "Contributes to" and occurs when two articles directly lead to a third article that merges their findings. This type of relationship can emerge when there are conflicting findings in the literature which a third article may resolve or when two findings are brought together to advance the field. An example in chemistry is the Nobel Prize-winning work of Yves Chauvin, Robert H. Grubbs, and Richard R. Schrock who collectively contributed to developments in olefin metathesis, an organic reaction that rearranges molecules containing carbon–carbon double bonds, common to industrial processes such as gas production. Hérisson & Chauvin (1971) first developed the Chauvin catalytic cycle, by which metal-carbon molecules could catalyze olefin metathesis. Later, Schrock et al. (1990) identified molybdenum- and tungsten-based catalysts for olefin metathesis that proved to be highly effective, but only in certain conditions. Grubbs and his collaborators built on these (and other) studies to create a new generation of ruthenium-based catalysts (Schwab et al., 1995) that were both effective and stable enough to be used in air, making it an optimal choice for industrial applications. Chauvin, Grubbs, and Schrock shared the Nobel Prize in Chemistry in 2005, partly for this series of convergence articles.

Divergence: This relationship is a variant of "Emerges from" and occurs when two articles emerge directly from an earlier third article. This type of relationship is common when we look at the citing literature of a landmark study. For example, the classic climate science result by Keeling (1960) demonstrating a steady increase of CO_2 concentration in the atmosphere has led to various independent research threads focused on identifying the primary sources of anthropogenic CO_2 emission (e.g., Hansen et al., 1998); and the physical, social, and economic impacts if we do not halt the continued climb in CO_2 concentration (e.g., Solomon et al., 2009).

9.3.2 Searching the Reference and Citation Network

To explore these relationship pairs, we first must find the articles. One of the most effective methods for finding peer-reviewed research on a common topic is through the references and citations of articles, as discussed in Chapter 1. This approach includes techniques such as **referential backtracking**, where researchers examine the reference list of an article to identify earlier foundational works; and **researcher checking**, which involves exploring the works of other authors frequently cited in the article or in the field more broadly. Additionally, **journal scouring** – reviewing the journals where cited articles are published – can reveal recurring themes and related studies (Alexander, 2020). These techniques allow researchers to trace the intellectual lineage of a topic, uncovering key works that provide both the background for and follow-up work from their research.

Another helpful strategy is to use citation tracking databases such as Clarivate's Web of Science (Clarivate, 2025b) and Elsevier's Scopus database (Elsevier, 2025; both subscription-based), or Google Scholar (Google, 2025; freely accessible). These platforms allow researchers to locate studies that cite a particular article (**forward citation tracking**) or are cited by it (**backward citation tracking**). By visualizing a "citation network," researchers can map the connections between articles and identify clusters of related research (Martín-Martín et al., 2018). The citation network method is especially helpful for pinpointing high-impact articles that have shaped the trajectory of a field. These approaches form a reliable framework for discovering relevant articles and constructing a network for a comprehensive literature review, as illustrated in Figure 9.2.

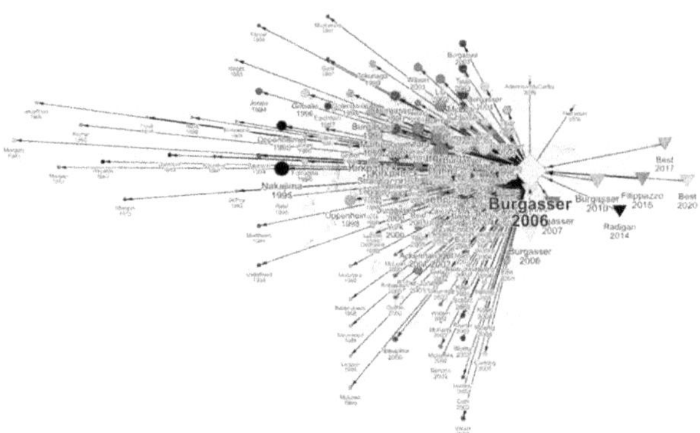

Figure 9.2 Reference and citation network.
The citation map depicts Burgasser 2006's influence within its local research network. Created by authors G. Bjorn and A. Burgasser using the reference and citation network constructed from Burgasser et al. (2006) using the Local Citation Network application developed by Tim Wölfle.)

Article-search platforms – Google Scholar, Web of Science, and Scopus – are indispensable for locating relevant studies, yet their citation coverage and accuracy differ markedly across disciplines. Let's compare them across a few main criteria.

Overall coverage: In a cross-field comparison of 252 subject categories, Google Scholar captured a median of ≈94% of all citations, far surpassing either Web of Science or Scopus (Martín-Martín et al., 2018). Much of Google Scholar's edge, however, comes from sources outside the traditional journal literature – dissertations, books, and conference papers – which are cited less often and may not be peer-reviewed.

Citations and accessibility: Both Google Scholar and Scopus usually retrieve more citations per article than Web of Science (Kulkarni et al., 2009). Scopus is especially strong at indexing non-English publications and review articles, whereas Google Scholar can miss citations for papers with group authorship or industry funding. However, Google Scholar is free and nearly universally available, a major advantage for accessibility.

Field- and time-sensitive strengths: Citation counts fluctuate by discipline and publication year (Bakkalbasi et al., 2006). Scopus excels in tracking current oncology research, while Web of Science performs better for older oncology papers and for niche areas such as superconductivity. Google Scholar captures more unique citations.

Field-specific databases complement these global search engines. For example, astronomy and astrophysics researchers rely on NASA's Astrophysics Data System (ADS; Smithsonian Astrophysical Observatory, 2025); health and medical scientists consult the National Institutes of Health's PubMed engine (NIH, 2025); and education scholars turn to the Education Resources Information Center (ERIC) database (Institute of Education Sciences, 2025). These subject-focused tools curate journals, conference proceedings, and gray literature (non-peer-reviewed studies) that global platforms sometimes miss.

Box 9.1 Knowledge Check: Question 9.1 Which statement best describes referential backtracking for finding research articles on a similar topic?

A. Using PubMed's "Similar Articles" function to find articles sharing overlapping MeSH terms.
B. Examining an article's reference list to trace earlier foundational studies on the topic.
C. Mapping citation networks in visualization software to locate highly cited clusters.
D. Limiting a Scopus search to non-English publications and review articles.
E. Comparing Journal Impact Factors within the same subject category to rank sources.

(Check your understanding using the Knowledge Check Key at the end of the book.)

Both global and specialized search engines also supply "related-article" functions that speed the discovery process. PubMed's Similar Articles algorithm draws on indexed terms, keywords, and MeSH (Medical Subject Headings) descriptors to surface papers that share themes, methods, or findings with a chosen study. Google Scholar offers a comparable Related Articles link, using citation patterns and textual similarity to suggest work that clusters around the original paper. This capability is especially valuable for interdisciplinary questions or for journals that fall outside PubMed's biomedical remit.

Because coverage depth, search filters, and cost vary widely, no single database is best in every situation. Researchers should select tools that match their disciplinary scope, budget, and need for supplementary features such as non-journal literature or multilingual indexing. Most university and some public libraries maintain their own collection search engines (including journals) and often provide online guides that demonstrate effective search strategies – an excellent first stop when deciding where to begin a literature search.

9.3.3 Criteria for Selecting Articles for Comparison

Selecting appropriate articles for comparison ensures that analyses are both meaningful and reliable. However, it can be difficult for novice critical readers to identify articles that are "comparable" due to the wide range of articles out there and the need to have a minimal level of understanding of a field to assess equivalence. Here, we outline several **study-level factors** and approaches for evaluating comparability between articles to facilitate effective synthesis.

Common analysis approaches: Data analysis and interpretation are fundamental for experimental studies, as they enable scholars to critically assess the quality of the information, draw informed conclusions, and ensure findings are robust (Patten, 2013). Analysis involves systematically examining data to identify patterns, trends, and relationships, thereby providing a structured understanding of the subject matter; think of graphs and charts. Interpretation explores the significance, implications, and connections of the data to existing knowledge or theories. In qualitative research, interpretation is crucial for understanding the deeper meanings and contexts of the data, allowing researchers to construct meaningful narratives and insights. Choosing studies that apply similar analysis or interpretation approaches can allow for comparisons across different study designs or different experimental approaches.

Common experimental approaches: Consistent or standardized experimental methodologies facilitate comparisons of studies that may differ in study design (Hedges & Olkin, 1985). Standard methodologies can include sampling techniques, data collection protocols, and statistical analysis procedures. For example, a study using an RCT design to investigate an obesity intervention with 500 participants and a well-documented food diary could be compared with a comparable RCT study on a high blood pressure intervention.

Reproducibility of results: Reproducibility is a cornerstone of scientific comparability, ensuring that findings can be independently validated under similar conditions. Studies that provide comprehensive documentation of experimental procedures, data collection, and analysis methods enable other researchers to replicate the results to establish reliability. Practices like sharing detailed protocols and raw datasets in fields like molecular biology or materials science allow for cross-laboratory validation and strengthen the credibility of conclusions (Ioannidis, 2005). Without reproducibility, inconsistencies or unverifiable findings risk undermining comparisons across studies.

Research question: Selecting articles that directly address the same research question or hypothesis ensures that comparisons are meaningful and aligned with the studies' objectives. For example, a researcher investigating the effects of mindfulness interventions on anxiety could compare two articles that test similar interventions, such as 8-week and 10-week mindfulness-based stress reduction (MBSR) programs. By focusing on studies that share the same core question – such as, "Does MBSR reduce anxiety levels in programs less than 12 weeks?" – the researcher ensures that the findings and conclusions are directly applicable to the overarching inquiry.

Similar arguments from evidence: Comparable articles can emerge when authors construct their arguments using similar types of evidence or make similar arguments based on the different types of evidence. For instance, two studies on renewable energy policy might claim that financial incentives drive adoption, using as evidence case studies in one study and statistical data on subsidies and adoption rates in the other. Alternatively, two articles could evaluate a particular energy policy, the first by demonstrating a policy's short-term effectiveness and the second by examining its long-term impacts.

Study design: Studies that employ similar design frameworks – such as randomized controlled trials (RCTs) in medicine or matched observational studies in social sciences – facilitate direct comparisons by aligning the structure of investigation. A well-defined study design includes clear objectives, consistent inclusion criteria, and systematic protocols, which reduce bias and improve the validity of comparisons (Shadish, Cook, & Campbell, 2002).

Study topic: The primary element of comparison is whether two (or more) articles describe the same phenomena. Even when restricted to a common field (e.g., astronomy), the range of field subtopics can be vast and not directly relatable. For example, research on the visible surface properties of asteroids is very difficult to compare to theoretical particle models for dark matter. On the other hand, disparate fields can sometimes lead to novel insights, such as the application of string theory methods developed for cosmology to studies of condensed matter (Maldacena, 1999).

Beyond these study-level factors, comparable articles can also be identified by considering **publication-level factors**, particularly citations and research impact:

1. **Citation metrics:** Individual papers on a given topic with similar citation counts generally have a similar impact in a field, making them potentially useful comparators. Broader journal measures such as Journal Impact Factor (JIF), Eigen factor, and Article Influence give a quick check of journal reach, although they do not necessarily correlate with the impact of an individual paper (Garfield, 2006).
2. **Field normalizations:** Adjusting metrics based on field standards prevents a pharmacology blockbuster study from being unfairly compared with a pure-math gem (Bornmann & Leydesdorff, 2018).
3. **Subject-category match**: Comparing journals inside the same Journal Citation Report (JCR) category allows for a fair comparison (Clarivate, 2025a).
4. **Citation-culture awareness:** Disciplines publish and cite at different tempos; knowing those patterns keeps cross-field trends in perspective (Waltman et al., 2012).

Both study- and publication-level checks can transform a stack of seemingly disparate articles into a data set you can actually compare.

Box 9.2 Knowledge Check: Question 9.2 Which study-level factor is highlighted as a cornerstone of scientific comparability because it allows other researchers to verify findings under similar conditions?

A. Controlling confounding variables through matched study designs.
B. Ensuring reproducibility of results by providing full protocols and data.
C. Standardizing sampling techniques across geographic regions.
D. Applying robust statistical analyses such as meta-analysis.
E. Normalizing citation behavior with principal component analysis.

(Check your understanding using the Knowledge Check Key at the end of the book.)

9.4 Visual Comparisons

Reading two articles side-by-side becomes far easier when their key elements are displayed rather than described. Visual tools – especially tables, Venn diagrams, and concept maps – convert textual arguments into spatial layouts, letting the eye register patterns the ear might miss. Cognitive-science research on external representations shows that such layouts lighten working-memory load and promote deeper interpretation (Gilbert, Reiner, & Nakhleh, 2008).

CERIC Comparison Template: The CERIC elements are useful for comparing articles at the study level. The CERIC Comparison Template provided in the Supplementary Materials has all the components of the CERIC Basic Template while adding the synthesis task of comparing two related articles. The Comparison Template encourages readers to conduct two (or more) CERIC reviews, draw connections between the CERIC elements using a table or diagram, and construct a comparison summary narrative. Readers may also wish to add additional visual or narrative components – such as creating a Venn diagram or discussing the strengths and weaknesses of the articles – to foster critical thinking and integrate multiple perspectives. We will explore the Comparison Template through a worked example in Section 9.5 of this chapter.

Diagrams and maps: Venn diagrams and concept maps move comparison from simple enumeration to relationship-building. A two-circle Venn diagram (Figure 9.3) forces readers to declare exactly which claims, data types, or theoretical assumptions the articles share and which remain unique. Concept maps extend that idea, linking labelled nodes with arrows that make causal or logical relations explicit. Research on diagrammatic reasoning – whether with Venn or Euler diagrams (Bauer & Johnson-Laird, 1993) or with concept maps specifically (Demirci & Oktay, 2021; Evagorou, Erduran, & Mäntylä, 2015) – shows that visual organization helps learners expose

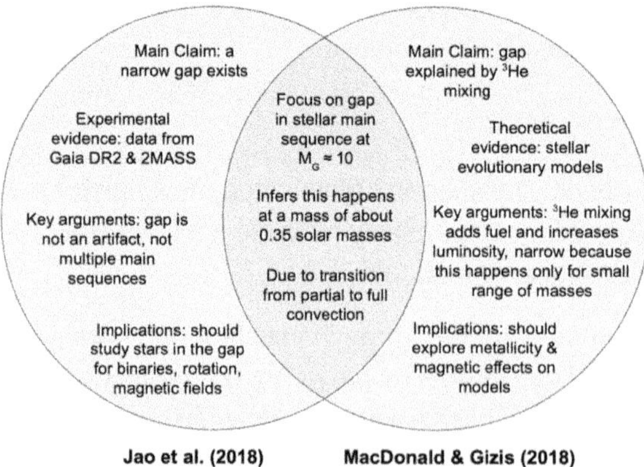

Figure 9.3 Venn diagram comparing Jao et al. (2018) and MacDonald & Gizis (2018). A Venn diagram displays shared and unique themes of two brown-dwarf studies.

hidden assumptions, correct misconceptions, and weave new ideas into existing knowledge structures. A recent meta-analysis of 55 studies even reports a strong, positive effect of concept-mapping practice on science achievement (Anastasiou et al., 2024).

Tables: Tables are the work-horses of comparative reading. A simple grid can line up sample sizes, methods, or effect sizes so that similarities and discrepancies jump out row by row. Systematic reviews and meta-analyses often devote several pages to these matrices as they condense hundreds of data points into a form reviewers can scan quickly (Younas & Ali, 2021). Computational studies that automatically parse tables in life-science articles show that well-structured table grids facilitate machine-based information extraction, a prerequisite for large-scale cross-study comparison (Milosevic et al., 2016; Milosevic et al., 2019). Building tables to categorize evidence before attempting to interpret it is an intellectual move that dovetails with the CERIC workflow.

9.5 Worked Example

Let's build on what we have learned in this chapter and in Chapter 8 by combining the CERIC review of the astronomy article by Jao et al. (2018) with a CERIC review and Comparison to a related article by MacDonald & Gizis (2018), using our CERIC Comparison Template.

We identified our comparison article in Google Scholar using the "cited by" function, applying the **forward citation tracking** strategy described in Section 9.3. We scanned articles published up to two years after Jao et al. (2018), looking for articles that addressed the same topic of investigating anomalies in the Hertzsprung–Russell diagram (HRD) for M dwarf stars. You may recall the worked example in Chapter 8 of Jao et al. (2018), which identified a distinct gap in the Gaia HRD for M dwarfs using observational evidence. As described below, MacDonald & Gizis (2018) present work that aims to explain the gap in the Gaia HRD using theoretical models. This is a great example of a Development relationship between articles as illustrated in Figure 9.1.

For this comparison, we will first conduct a limited CERIC review of the MacDonald & Gizis (2018) article, focusing on the article information, CERIC elements, and CER analysis. Then we will proceed step-by-step through a comparative CERIC review using the CERIC Comparison Template. Take a few moments to open the relevant materials:

- The CERIC Comparison Template in Chapter 9 Supplement,
- The MacDonald & Gizis (2018) article,
- The Jao et al. (2018) article, and
- The completed CERIC review for the Jao et al. (2018) article (from your own analysis or in Chapter 8).

9.5.1 Research Article Information

Table 9.1 Key bibliographic and content information for MacDonald and Gizis (2018), which interprets the Gaia main sequence gap in M dwarfs as a luminosity function dip caused by ^{3}He mixing during the merger of stellar convection zones. (^{3}He is read aloud as *helium 3*).

Key Elements	Content Information
Title	An explanation for the gap in the Gaia HRD for M dwarfs
Lead Author (Year)	MacDonald & Gizis (2018)
Journal, Volume, Page, and URL or DOI	*Monthly Notices of the Royal Astronomical Society, 480*(2), 1711–1714. https://doi.org/10.1093/mnras/sty1888
Abstract	*We show that the recently discovered narrow gap in the Gaia Hertzsprung–Russell Diagram near M_G = 10 can be explained by standard stellar evolution models and results from a dip in the luminosity function associated with mixing of $_3$He during merger of envelope and core convection zones that occurs for a narrow range of masses.*
Keywords	stars: evolution, Hertzsprung–Russell and colour–magnitude diagrams, stars: low-mass, stars: luminosity function, mass function *cool stars, stellar main sequence, Gaia, stellar evolution, stellar mass function, stellar luminosity function*

Note: The first set of keywords are drawn directly from the article; the second set of keywords include some of those we generated and used in the Jao review in order to provide consistency across articles.

Table 9.2 Claims in MacDonald and Gizis (2018): the main claim attributes the Gaia main sequence gap to a luminosity function dip from ^{3}He mixing during convection zone mergers near 0.35 solar masses, while the subclaim emphasizes that this feature arises naturally in standard stellar evolutionary models. (^{3}He is read aloud as *helium 3*).

Notes	Page/Section
Main Claim: The HRD gap in Jao et al. (2018) is caused by *"a dip in the luminosity function associated with mixing of $_3$He during merger of envelope and core convection zones,"* which occurs over a narrow range of masses at around 0.35 solar masses.	Abstract, Section 1, Section 3, Section 4
Subclaim: The feature arises naturally from existing stellar evolutionary models.	Abstract, Section 1, Section 3, Section 4

Note: In this case, the Claim is hinted at in the title but is much more clearly stated in the abstract and main text. We separate the main Claim – the physical explanation for the gap – from a subclaim that highlights that this feature can be explained by current models.

9.5.2 CERIC Elements

1. Claim

A plain language, declarative answer to a scientific question that asserts new knowledge. What is the central claim or hypothesis this article states? Are there additional secondary claims being made? Usually found in the abstract and title (Table 9.2).

Table 9.3 Evidence in MacDonald and Gizis (2018), consisting of stellar evolution modelling across 0.30–0.38 M⊙ with detailed input physics (boundary conditions, opacities, convection treatment, nuclear reaction rates, and equations of state) and use of an initial mass function to generate the predicted luminosity function for comparison with the observed Gaia gap.

Notes	Page/Section
Used their own *"fully implicit stellar evolution code"* to evolve stellar models for masses of "*0.30–0.38* M_{sun} *in increments of 0.005* M_{sun}*,*" from pre-Main Sequence to the age of the Universe (13.7 Gyr), assuming interior abundance Z = 0.0153 (same as BT-Settl models)	Section 2
BT-Settl models for *"outer boundary conditions"* (= bottom of atmosphere; Allard, Homeier & Freytag, 2012a; Allard et al., 2012b; Rajpurohit et al., 2013).	Section 2
OPAL opacities for T > 20,000 K (Iglesias & Rogers, 1996) and opacities from Ferguson et al. (2005), Freedman, Marley & Lodders (2008), Freedman et al. (2014) for lower temperatures	Section 2
Assumed mixing length theory for convection with alpha = 1 with convection *"is treated as a diffusion process"* (Böhm-Vitense, 1958)	Section 2
Assumed nuclear reaction rates from Angulo et al. (1999), with *"electron screening based on Potekhin & Chabrier (2012),"* with careful attention to weak and strong screening effects (electron screening modifies nuclear reaction rates)	Section 2
Equation of state from Saumon, Chabrier & van Horn (1995) with *"additive volume rule,"* includes ionization for C, N, O, Ne, Mg, Si, and Fe; applies "*Coulomb and quantum corrections*" (Iben, Fujimoto & MacDonald, 1992), pressure ionization, and *"Stark effects"*	Section 2
For luminosity function, assumes ages of 0–8.8 Gyr (for the Milky Way's thin disk), uses the initial mass function (IMF) from Parravano et al. (2011), and applies random sampling	Section 3

Note: We have included a lot of detail in the Evidence here, with citations, in order to capture all of the inputs to the modeling.

2. Evidence
What are the data, measurements, observations, models, computations, and other forms of analysis that the authors use to support their claim? Usually found in the abstract, the Results section, and in tables and figures (Table 9.3).

3. Reasoning
What is the logical basis for this argument? What connects the evidence and the claim? Do the authors conduct analysis? This could be a particular analysis methodology or how authors were able to rule out other interpretations. Do the authors use theory or prior research to link the evidence and claim? Usually found in the Analysis section (Table 9.4).

4. Implications
How is this result significant beyond the immediate findings? What has been learned? What further work is needed? Usually found in the Discussion section and possibly in the abstract (Table 9.5).

Table 9.4 Reasoning in MacDonald and Gizis (2018), linking stellar evolution models to the claim: simulations reproduce the observed luminosity dip, show an isochrone pinch in the 0.33–0.35 M⊙ range, and attribute the feature to ^{3}He mixing during convection zone mergers, which occurs only in a narrow mass interval. (M⊙ is read aloud as *solar mass* and ^{3}He is read aloud as *helium 3*).

Notes	Page/Section
Simulated luminosity function shows a clear dip at *"log* L/L_{sun} *= −1.84" or "*M_K *= 6.71,"* which agrees well with the gap reported in Jao et al.	Section 3, Figure 1
Luminosity vs mass diagram shows an *"isochrone pinch"* in the 0.33–0.35 M_{sun} range, caused by a *"faster change of luminosity ... compared to lower and higher masses"*	Section 3, Figure 2
Faster evolution ⇒ less likely to find source at this luminosity ⇒ rare in sample ⇒ dip in luminosity function	Section 3
Cause of pinch is the mixing in of ^{3}He from envelope into core when envelope convection zones merge – increases luminosity (adds fuel)	Abstract, Section 3, Figure 3
If $M < 0.31\ M_{sun}$, fully convective throughout so no merger; if $M > 0.35\ M_{sun}$, merger does not happen; hence, this mixing only happens in a narrow range of masses	Section 3, Figure 3

Table 9.5 Implications in MacDonald and Gizis (2018), highlighting directions for further research, including testing additional stellar compositions and examining the role of magnetic fields in suppressing convection.

Notes	Page/Section
Further research: only considered one composition, could see if this happens for other compositions	Section 4
Further research: have not considered magnetic fields which could suppress convection (important for rapidly rotating stars)	Section 4

Notes	Page/Section
Detection of gap in Gaia HRD by Jao et al. (2018) at MG ≈ 10, spectral type M3, *"where M-dwarf stars are thought to transition from partially to fully convective."*	Abstract, Section 1

Figure 9.4 Context for MacDonald and Gizis (2018), framing the study as a response to Jao et al. (2018)'s detection of a main sequence gap near MG ≈ 10 at spectral type M3, where M dwarfs transition from partial to full convection.
Note: The Context is very spare here, simply stating the existing observational problem before jumping into showing how it can be resolved through stellar models.

5. Context

What is the scientific context of this work? What broader questions or problems are relevant to this work? What is the rationale for conducting this study? Usually found in the Introduction or Context sections (Figure 9.4).

9.5.3 CER Analysis

1. Does the Evidence and Reasoning Support the Claim, and If So, How Exactly?
Yes, the evidence and reasoning support the claim. The authors use stellar evolution models (theoretical evidence) and demonstrate that the convergence of core and envelope convection zones in M dwarfs at around 0.35 solar masses increase their luminosity, "pinching" the luminosity-mass relationship, and leading to a lower probability of observing stars in the luminosity range that aligns with the gap identified by Jao et al. (2018). Moreover, they show that the gap is narrow because it is only a narrow range of masses undergo this mixing.

2. Evaluate the Article's Quality and Significance, Including Argument and Impact
Overall, the article is of fairly high quality with some gaps in key details:

- **Strength of Evidence:** The study deploys an advanced theoretical model that utilizes advanced elements for boundary conditions, opacities, nuclear reaction rates, and equation of state, and deploys well-motivated assumptions in its simulated luminosity function.
- **Strength of Reasoning:** The article provides a clear physical explanation for the existence of a gap as arising from the higher luminosity and more rapid evolution of stars when ^{3}He is mixed into the core at the point of envelope/core merger. It also justifies the narrowness of the gap as the mixing only occurs over a narrow range of masses.
- **Systematic error:** The article notes that the range of stellar parameters tested was limited (e.g., only solar abundances and no magnetic fields), which may be inconsistent with the more general Jao et al. Gaia sample.
- **Consistency:** While producing a gap *"in good agreement with the location of the gap shown by Jao et al. (2018)"* (Section 3), the article does not directly compare the theoretical predictions to the Jao et al. data, making it difficult to assess if the gap is exactly aligned.
- **Transparency:** While the article describes in detail many of the inputs to their model, the "fully-implicit" code is unnamed, not open-source, and some key model choices are omitted, making it difficult to replicate the results.
- **Credibility:** The article is published in a reputable, peer-reviewed journal, ensuring that experts critically evaluated the methods, evidence, and conclusions. The authors have written numerous articles on the subject of the study.
- **Impact:** The article has been cited at least 25 times since publication, giving it the modest impact.

3. Are There Weaknesses in the Article's CERIC Elements, and How Can They Be Addressed?
There are a few weaknesses in the article that can be addressed:

Claim: The Claim is clear but has a relatively narrow scope; this study could be advanced by exploring the influence of metallicity, magnetic fields, and other parameters on the gap.

Evidence: The models used are neither named nor open-source, making it difficult to reproduce the results.

Reasoning: No direct comparison between model predictions and the original Gaia data are provided; assessment of parameter sensitivities (e.g., to mixing length alpha parameter or choice of boundary conditions) is limited and could be quantitatively evaluated.

Implications: While exploration of the impacts of various stellar parameters is proposed, there is no discussion of broader consequences.

Context: Introductory material is very limited and focused entirely on explaining Jao et al. (2018) result. No mention of competing hypotheses or other contemporary studies.

Note. The point of the CER Analysis is not to be "judge and jury" of whether or not the paper is good enough, but instead to evaluate the thread of logic used in the article.

9.5.4 Additional Notes

Next, let's jot a few notes about unclear jargon, key terms, and relevant references that this article cites.

1. *Unclear Jargon and Terms*

Use this space to list any unclear or unknown jargon or terminology and their definitions. Note that each definition includes the citation to its source, in case further clarification is necessary later (Table 9.6).

Table 9.6 Glossary of unclear jargon and terms from MacDonald and Gizis (2018), including definitions and sources for technical concepts related to stellar interiors, energy transport, nuclear processes, and modelling methods.

Jargon/Term	Section/Page	Definition and Source
Convection zone	Section 1	A region inside a star in which energy is conveyed by the mixing of hot material upwards; this contrasts with regions where radiation (photons) carry the energy. Stellar interiors are often divided into convective and radiative layers. Source: Wikipedia, Convection zone.
Luminosity function	Abstract, Section 3, Figure 1	Distribution of stars as a function of luminosity (radiant power), typically expressed as the number or density within a given volume. Source: Wikipedia, Luminosity function (astronomy)
Stellar core and envelope	Abstract, Section 1, Section 3, Figure 3	Interior regions of a star. The core is the hot, dense center of the star where nuclear fusion occurs, the envelope is the shell between the core and the visible atmosphere. These are often distinguished by the mechanism of energy transport (e.g., convection versus radiation). Source: Wikipedia, Stellar
Radiative and conductive opacities	Section 2	A measure of the absorption of energy as conducted by light (radiative) or vibration (conductive). Opacities determine the efficiency of energy transport inside a star. Source: Wikipedia, Opacity

Table 9.6. (cont.)

Jargon/Term	Section/Page	Definition and Source
Mixing length theory	Section 2	In fluid dynamics, a model to describe the motion of convective fluids, in particular the length scale over which a hot rising fluid element can retain its heat before it dissipates in the surrounding (cooler) medium. Source: Wikipedia, Mixing length model
Convective-kissing instability	Section 2	A cyclic interaction in certain low-mass stars where thin radiative layers form between the core and envelope convective zones, "kiss," then separate again, causing luminosity oscillations. Source: van Saders & Pinsonneault (2012), *ApJ*, 751, 98.
Electron screening	Section 2	In the context of nuclear reactions, electron screening describes the reduced repulsion of positively charged nuclei due to the surrounding cloud of negatively charged free electrons, resulting in higher nuclear reaction rates. Distinguished into weak and strong screening depending on whether the screening energy is small or large compared to thermal energy. Source: Salpeter (1954), *Australian Journal of Physics*, 7, 373.
Equation of state (EOS)	Section 2	The thermodynamic relationship between bulk properties of matter, notably pressure, density, and temperature. Source: Wikipedia, Equation of State
Pressure ionization	Section 2	The ionization of atoms caused by collisions in a dense, high-pressure fluid; can change the stellar EOS. Source: Lawrence Livermore National Laboratory, "Experiments shed light on pressure-driven ionization in giant planets and stars"
Bolometric correction	Section 3	Correction made to convert between the absolute magnitude in a specific filter to the total (bolometric) luminosity. Source: Wikipedia, Bolometric correction
Stellar (or initial) mass function	Section 3	Distribution of stars as a function of mass, typically expressed as the number or density within a given volume. Source: Wikipedia, Initial mass function
Isochrone	Section 3, Figure 2	A curve on the HRD (or in Figure 2, luminosity versus mass) that shows where stars of the same age and composition are located. Source: Wikipedia, Stellar isochrones
Magnetic inhibition of convection	Section 4	A process by which strong magnetic fields can prevent convection from arising in the interiors of stars. Source: Gough & Taylor 1966, *Monthly Notices of the Royal Astronomical Society*, 133, 85

2. *Relevant References*

Are there other articles cited in or by this article that may be useful for further investigation? (Table 9.7)

9.5.5 Compare and Contrast

Now, let's draw on the elements of the CERIC Comparison Template to assess the commonalities and differences among the CERIC elements of these two articles.

Table 9.7 Relevant references cited in MacDonald and Gizis (2018), including the original discovery of the Gaia HRD gap (Jao et al., 2018), sources for model inputs such as atmosphere and equation-of-state data, and related studies on magnetic effects in low-mass stars.

Abbreviated Citation (Author, Year, Journal)	Notes on relevance to this article
Jao et al. (2018, ApJL)	Paper that identified the HRD gap, main motivating source for this article
Allard, Homeier, & Freytag (2012a Phil. Trans, R. Soc. A)	BT-Settl atmosphere models, used to set the upper (atmosphere) boundary condition for the stellar interior model presented here
Saumon, Chabrier, & van Horn (1995, ApJS)	Source for H-He EOS (other contributions from heavy elements are added to this)
Feiden & Chaboyer (2012, ApJ)	Demonstrates how strong fields inflate radii and dim luminosities of low-mass stars, offering an alternative or complementary mechanism to the ^{3}He-mixing explanation for luminosity dips.

Table 9.8 Comparative CERIC summary of Jao et al. (2018) and MacDonald and Gizis (2018), contrasting observational discovery and validation of the Gaia HRD gap with theoretical modelling that explains its physical origin, while highlighting shared emphasis on the transition to full convection in M dwarfs.

Jao et al. (2018)	Compare/Contrast	MacDonald & Gizis (2018)
Discovery of a narrow gap at $M_G \approx 10$ in the Gaia HRD, near the transition between full and partial interior convection.	Both validate the reality of a gap in the HRD and associate it with the transition to full convection over a narrow range of masses among M dwarfs.	A physical explanation for the gap as resulting from ^{3}He enrichment during core–envelope convection zone convergence over a narrow range of masses.
Observational – uses Gaia DR2 photometry and astrometry and 2MASS photometry. Theoretical – uses existing evolutionary models.	The studies apply different forms of evidence – observational vs theoretical – to validate and explore the nature of the gap.	Theoretical – Develops stellar evolution models with detailed inputs.
Justifies the reality of the gap, ruling out biases and alternative explanations such as multiple main sequences; argues that convection is important given the mass at which the gap is associated.	Jao et al. focuses on justifying the observational reality of the gap, while MacDonald & Gizis focus on explaining the physical origin of the gap; both articles justify changes in the convection structure of low-mass stars as responsible.	Uses simulated luminosity functions to reproduce the gap; infers the origin of the gap as due to increased luminosity from ^{3}He mixing during convection zone merger, resulting in fewer sources at that luminosity and hence a gap.
Suggests follow-up observational studies to investigate the properties of "gap stars," including metallicity, rotation, and magnetic activity.	Both recommend further research to refine understanding of the gap's underlying physical processes.	Suggests exploring the effects of composition and magnetic fields on the amplitude and location of the gap.
Motivated by advancements in observational precision provided by Gaia DR2, and the discovery of other features in the HRD for other types of stars.	Both articles benefit from the accuracy of Gaia data in the M dwarf regime. Jao et al. specifically builds on this, while Macdonald & Gizis build from the Jao et al. study, a good example of a Development comparison.	Motivated by the need to explain the observed gap and establish a physical origin.

9.5.5.1 CERIC Summary Table

Briefly summarize the main CERIC elements of both articles and compare/contrast the overlapping concepts in the center column, as shown in Table 9.8.

9.5.6 Comparative Narrative Synthesis

Use the compare and contrast results in the previous section to write a short, narrative paragraph comparing and contrasting the key similarities and differences between the articles. Aim for a minimum of 3–5 complete sentences.

Jao et al. (2018) and MacDonald & Gizis (2018) both investigate a narrow gap centered at an absolute magnitude of $M_G \approx 10$ in the Gaia Hertzsprung–Russell Diagram (HRD). This magnitude corresponds to a mass of about 0.35 times the Sun's mass, linking it to the transition from partial to full convection in the interiors of low-mass stars. These studies are motivated by the exceptional accuracy of Gaia data in the M dwarf regime, enabling detection of structures in the HRD that may not have been visible in earlier data. Jao et al. (2018) focus primarily on observational data, using Gaia and 2MASS to identify and confirm the gap's existence. They deploy extensive bias checks to rule out selection effects, biases, and alternative explanations such as multiple main sequences. This study uses previously computed evolutionary models to argue that the gap occurs at mass where stars are expected to transition from partial to full convection. MacDonald & Gizis (2018) build on these results by developing new theoretical models to investigate the physical origin of the gap. Their models demonstrate that mixing of ^{3}He from envelope to core drives a temporary rise in luminosity for a narrow range of masses, resulting in fewer sources near this luminosity in stellar samples. By generating simulated luminosity functions, they can quantitatively reproduce the depth and width of the observed gap in the HRD. Together, the papers form a clear observation-to-theory "development chain," with Jao et al. providing the empirical cue and MacDonald & Gizis supplying the physical mechanism. Both studies advance our understanding of the interior properties of low-mass stars, and advocate for additional studies of the physical properties of stars around the gap, and further theoretical investigation of the impact of metallicity and magnetism on the shape and depth of the gap. Both studies also note open questions, such as Gaia's completeness for low-mass stars, uncertainties in metallicity assumptions, and the timescale of ^{3}He mixing. Beyond stellar interiors, understanding the gap has implications for population-synthesis models and for estimating the occurrence rates of exoplanets around mid-M dwarfs.

This worked example demonstrates how individual CERIC reviews can be merged to create a synthesis of two articles, a first step toward more advanced literature reviews. This case illustrates a specific example of a development chain type of relationship (observation influences theory), where one article's Claim directly motivates the second article's research question. Other article relationship pairs can be approached in a similar manner. We can also add additional information to these articles, such as the quality of the arguments or their differing degree of impact in the broader field. We can further extend our synthesis to three or more related articles. By framing such comparisons in a structured approach that focuses on Claims, Evidence, Reasoning, Implications, and Context, we can quickly identify how articles align or disagree along their core elements, providing a pathway to deeper analyses such as critique, the subject of Chapter 9.

9.6 Chapter Key Takeaways

1. **Comparing research articles is foundational to scientific thinking:** Using CERIC to compare articles equips readers with the critical tools needed to identify patterns, gaps, and contributions across studies. This comparative skill deepens our understanding of how individual findings advance broader scientific conversations and critique.
2. **CERIC provides a structured lens for comparison:** By examining Claims, Evidence, Reasoning, Implications, and Context across two studies, we can systematically evaluate how each article builds knowledge and addresses scientific questions from different angles.
3. **Article comparability depends on both study-level and publication-level criteria:** Selecting comparable articles requires attention to factors such as journal impact, field-specific citation patterns, study design, reproducibility, and statistical rigor. These criteria ensure that comparisons are meaningful and methodologically sound.
4. **Search tools offer complementary approaches:** Citation tracking (forward and backward), "Similar Articles" tools, and keyword-based strategies help us locate related studies efficiently. Understanding the strengths and limitations of each platform supports more precise literature searches.
5. **Visual tools enhance comparative thinking:** Tables, Venn and Euler diagrams, and concept maps allow us to visualize similarities and differences in CERIC elements. These tools improve synthesis, reveal conceptual relationships, and support clearer reasoning.
6. **The CERIC Comparison Template fosters synthesis and higher-order thinking:** By scaffolding comparisons along the CERIC elements, we can conduct efficient side-by-side article analysis, articulate cross-study insights, and contextualize findings within a broader research landscape. This comparison framework builds toward more integrative critique and literature review skills.

10 Using CERIC to Critique and Peer Review Articles

[Science] is intensely skeptical about the possibility of error, but totally trusting about the possibility of fraud.

—Schechter et al. (1989)

10.1 Overview

Chapter 10 synthesizes the insights gained from Chapters 2 through 8 and focuses on the practical use of the CERIC method for critiquing research articles. This chapter starts by discussing the importance and process of peer review, emphasizing that not all published articles are inherently good; and revisits some of the major types of missing information or errors in papers related to individual CERIC elements. Building on the foundation of article review in Chapter 8 and comparison in Chapter 9, this chapter examines how a CERIC review can be used to assess the quality and effectiveness of a research paper's findings, central argument, and impact.

10.2 The Importance of Critical Peer Review

Critical peer review is fundamental to scholarly communication, ensuring the integrity, quality, and credibility of published research (Taylor & Francis, 2025). By subjecting manuscripts to rigorous evaluation, peer review helps maintain high academic standards and supports researchers in refining their work. This process filters out flawed or unsubstantiated studies; and enhances the clarity and impact of valid research findings (Taylor & Francis, 2025).

10.2.1 Define Peer Review and the Role of Critique in the Scientific Process

Peer review is a cornerstone of the scientific process, defined as subjecting research to the scrutiny of other experts before publication (Hoppin, 2002; Kelly et al., 2014). This evaluation helps ensure that published studies meet the field's standards and do not disseminate unwarranted claims or errors (Jefferson et al., 2002; Kelly et al., 2014). Crucially, peer review injects **constructive skepticism** into science: Reviewers question methods and interpretations in a positive spirit, aiming to improve the work rather than simply find faults (Hoppin, 2002). Multiple reviewers bring distinct perspectives and mental maps of the literature, often catching issues that others miss (Akers, 2017; Lucey, 2013). This diversity of expert opinion enhances an article's rigor by providing a well-rounded critique (Hoppin, 2002; Kelly et al., 2014). In essence, peer review acts as a quality filter – one that maintains trust in scientific literature by vetting accuracy and validity (Kelly et al., 2014; Mulligan, 2005).

It is important to address a common misconception: *If it is published, it must be good.* In reality, publication is not a guarantee of high-quality research (Bohannon, 2013). Studies have shown that peer review, while valuable, can fail to detect all errors (Jefferson et al., 2002). For instance, an experiment inserted eight deliberate mistakes into a paper and typical reviewers spotted only two of them on average (Baxt et al., 1998). Many published findings are later refuted or revised, symptomatic of a broader **replication crisis** where many results cannot be reproduced by others (Ioannidis, 2005; Ioannidis; 2019; Mulligan, 2005). Therefore, even published research must be read with a critical mind (Bohannon, 2013). Peer review improves the odds that a paper is sound, but post-publication critique and replication are vital parts of the scientific quality control cycle (Mulligan, 2005). The takeaway is that critical peer review – both pre- and post-publication – is essential for advancing reliable knowledge, by filtering out obvious flaws, fostering constructive dialogue, and instilling in researchers an attitude of healthy skepticism toward Claims (Akers, 2017; Hoppin, 2002).

10.2.2 The Process of Peer Review in Research Journals

Understanding the peer review process is essential for both authors and reviewers. Typically, the process involves several stages (IEEE Author Center Journals, 2025):

1. **Submission evaluation**: Editors assess whether the manuscript fits the journal's scope and meets basic quality standards.
2. **Reviewer selection**: Qualified experts in the relevant field are identified to evaluate the manuscript's methodology, originality, and significance.
3. **Review and feedback**: Reviewers provide constructive critiques, highlighting strengths and areas for improvement.
4. **Revision and resubmission**: Authors address the feedback and may resubmit revised versions for further evaluation.
5. **Final decision**: Based on reviewers' recommendations, the editor decides to accept, reject, or request additional revisions.

Note that steps 3 and 4 may be repeated as a manuscript is improved and reviewers suggest further recommendations. This structured approach ensures that published research is both reliable and valuable to the scientific community.

Adopting a mindset of constructive skepticism is vital for reviewers. This involves critically assessing research while providing feedback that guides authors toward improvement (el-Guebaly et al., 2022). Constructive skeptics question assumptions, evaluate the robustness of methodologies, and consider alternative interpretations of data. Such an approach not only strengthens individual studies but also fosters a culture of continuous improvement in research practices.

Box 10.1 Common Confusion: If It's Published, It's Good, Right?

Many people operate under the assumption that if a study has made its way into a journal, it must be good or least correct. Maybe you have thought this, too. Peer review is often hailed as the gold standard for vetting scientific research. However, peer review has flaws. Let's consider a few examples:

The Hydroxychloroquine Fiasco: Consider the saga of hydroxychloroquine during the COVID-19 pandemic. A study published in the *International Journal of Antimicrobial Agents* by Gautret et al. (2020) suggested that the drug could be an effective treatment. This study led to widespread media coverage and even endorsements by high-profile figures. However, subsequent analysis revealed methodological flaws, and the study was later retracted due to ethical concerns and unreliable data (O'Grady, 2024). This incident underscores that publication does not always equate to quality or reliability.

Can Corn Kill You?: Then there is the infamous Séralini affair. In 2012, a study claimed that genetically modified (GM) maize caused tumors in rats (Séralini et al., 2012). Published in a peer-reviewed journal, the study initially gained widespread attention. However, scientists soon criticized its flawed methodology, including small sample sizes and lack of reproducibility. The journal ultimately retracted the paper (Séralini et al., 2014), illustrating that even peer-reviewed studies can be misleading if not critically evaluated.

The Proliferation of Predatory Journals: In today's academic landscape, not all journals uphold rigorous standards. Predatory journals – publications that prioritize profit over quality – have become increasingly prevalent. These outlets often lack proper peer review, allowing subpar or even fraudulent studies to see the light of day. According to Moher et al. (2017), predatory publishing has "created a credibility crisis in scholarly communication" (p. 211), making it crucial for researchers to critically evaluate the sources they rely upon.

The sharpest critiques of the process argue that "peer review is a flawed process, full of easily identified defects with little evidence that it works" (Smith 2006, p. 178). Others point out that, while peer review plays a vital role in maintaining academic standards, it is not infallible. So, what's a diligent researcher to do?

Deepen critical thinking. Do not accept findings at face value simply because they are published in a reputable journal. Evaluate a study's methodology, sample size, potential biases, and the CER core. As Ioannidis (2005) famously argued, "most published research findings are false" (p. e124), making it imperative to scrutinize even high-profile studies. As budding scholars, it is imperative to approach research with a skeptical mind, recognizing that the path to new knowledge is often paved with both insights and oversights.

10.3 Finding Common Flaws in Research Articles Using the CERIC Structure

To systematically critique research, we can start from the five CERIC components: Claim, Evidence, Reasoning, Implications, and Context. By evaluating each element, one can systematically uncover common flaws that undermine a study's credibility. This categorical approach ensures that no aspect of the paper's argument is overlooked during critique, and transforms a vague and potentially overwhelming task (e.g., "Is this paper any good?") into five clear questions (Bjorn, 2024):

- What is the Claim, and is it supported?
- What is the Evidence, and is it valid?
- What is the Reasoning, and is it logical and sound?
- What are the the Implications, and are they based on the CER core?
- What is the gap and rationale, and is the work logically situated within the Context?

Assessing the answers to these questions based on what is presented in a research article enables a comprehensive critique of that article. In this section, we explore errors typical in each CERIC category.

10.3.1 Common Errors in Claims

The Claim is the central thesis or new knowledge of the article – what the authors assert to be true. A common flaw is when Claims are unsupported or overstated relative to the data. For example, authors might Claim to have "proven" something beyond a doubt – a language that science generally avoids (Hoppin, 2002; Jefferson et al., 2002). Strong conclusions should be drawn only if Evidence is robust; otherwise, speculative Claims must be clearly qualified. An unsupported Claim might also manifest as a hypothesis presented without any rationale or connection to prior research. Studies lacking a clear hypothesis or logical prediction – the "what and why" of the research – raise red flags about the validity of their Claim (Liumbruno et al., 2012). Readers should always ask: Is the Claim explicitly stated and does the paper provide sufficient and appropriate Evidence and Reasoning to back it up? If the answer is no, the Claim may be too weak or too bold given the actual findings.

10.3.2 Common Errors in Evidence

Evidence refers to the data and observations collected in experiments, surveys, and similar methods that are meant to support the Claim. Flaws in Evidence are among the most frequent issues in research articles. One major concern is statistical errors, notably Type I and Type II errors. A **Type I error** is a false positive – claiming an effect or difference exists when it actually does not. A **Type II error** is a false negative – failing to detect a real effect that does exist (Liumbruno et al., 2012; Shreffler & Huecker, 2023). Both errors can lead to faulty conclusions, especially if studies are underpowered (samples that are too small) or use improper statistical tests. Another

Evidence-related flaw is reliance on hidden assumptions or confounding factors (Hoppin, 2002). These occur when an experiment's design does not fully control for variables that could influence the results. For instance, a researcher might conclude a certain treatment affected plant growth, but on closer inspection, factors like soil moisture and container type were not standardized. Such uncontrolled variables undermine the Evidence by introducing alternative explanations.

In reviewing Evidence quality, check for adequate sample size (to avoid statistical errors), proper control groups, and whether the data presented actually align with the Claim. Beware of cherry-picking as well. Authors sometimes highlight thin data that supports their Claim while ignoring data that contradicts it (Morse, 2009). Robust Evidence should include all results, with appropriate analysis of why some data might deviate.

10.3.3 Common Errors in Reasoning

Reasoning is the logical process that connects the Evidence to the Claim. Even with good data, an article can falter in how it interprets or links the data to the conclusions. **Logical fallacies** are common Reasoning flaws, the most prevalent being confusion of correlation with causation (Bohannon, 2013; Colquhoun, 2011). Researchers might observe that two variables changed together and erroneously argue that one caused the other, when no causal link is demonstrated. Researchers must consider other variables that might be causing that correlation. A classic cautionary example is the spurious correlation between global warming and the decline of pirates, humorously pointed out in an article in Forbes magazine by contributor Erika Anderson (2012). It's amusing but not evidence of causation.

Another reasoning error is **fallacy** or mistaken belief, such as anthropomorphism (e.g. treating plants as if they have human-like preferences; Dowden, 2023) or some other misplaced analogy. Fallacies can cause authors to impose inappropriate Reasoning frameworks on their results. Missing Reasoning can also be an issue: Sometimes papers present data but fail to explain how those data address the research question or support the hypothesis (Hoppin, 2002). Every step of the argument should be explicit. If a reader must make a big logical jump, that gap probably hides an unsupported assumption.

Additionally, inconsistencies between different parts of an article can indicate Reasoning flaws. For example, if the Claims do not actually follow from the Evidence presented, the authors may be overinterpreting. Reviewers are advised to watch for overinterpretation or "spin," where authors stretch their conclusions to be more exciting or broad than the Evidence allows (Brewin, 2023; Meadows, 2013; Mohty & Melo, 2025). Good Reasoning requires that conclusions follow inevitably from the data under the stated assumptions; anything else should be acknowledged as speculation or left out.

10.3.4 Common Errors in Implications

Implications are the broader conclusions, applications, or consequences the authors draw from their findings. A frequent flaw in this category is overgeneralization.

Authors might claim their result has sweeping relevance beyond the scope of the study. In reality, a single study often has limitations (sample characteristics, laboratory conditions, etc.) that constrain how far we can generalize the findings. For instance, a psychology study on a small group of undergraduates should not claim to reveal a universal truth about "all humans" without similar findings across diverse populations. Overgeneralized Implications ignore the essential Context and can mislead future research or policy (Brewin, 2023; McNutt et al., 1990). Recent analyses in psychology have highlighted this generalizability crisis, where researchers sometimes draw broad, poplulation-level theoretical inferences from very narrow experimental setups, like 50 undergraduate participants in a behavioral survey (Brewin, 2023). A sound paper will temper its Implications, acknowledging if results are preliminary or only applicable under certain conditions (Fitzpatrick, 2009).

Another aspect of Implications is the ethical and practical consequences. Flawed articles might propose using a finding in real-world practice without addressing safety or ethics. For example, a biomedical study might suggest a new treatment approach but fail to discuss possible side effects or ethical approval – a critical omission in Implications. Moreover, authors need to discuss the limitations of their work from an implementation perspective, and any conditions needed for their findings to be applied (Kelly et al., 2014). *Do they mention if further testing or replication is required? Do they avoid implying causation or effectiveness beyond what was demonstrated?* Articles that acknowledge what their findings do *not* imply are often more trustworthy. Ignoring or painting rosy Implications is a red flag that the authors may be overinvested in a narrative. As a rule, credible research will link Implications back to solid Evidence and Reasoning and it will distinguish between what the study shows directly and what is more speculative.

10.3.5 Common Errors in Context

Context encompasses the background and setting of the research, including the literature review of past work, the theoretical framework, and the motivating research question. Common pitfalls in Context are poor literature reviews or missing foundational research. If a paper does not cite key prior studies in its field, it may be either unaware of them or deliberately ignoring conflicting Evidence (Bohannon, 2013). Both scenarios weaken the article's rationale for why the research needed to be done. A good introduction should present what is already known and highlight the gap the current study aims to fill. When authors omit important references, they risk making Claims that are not novel or that have already been refuted by others. Inaccurate or biased citation is another issue. Sometimes authors cite sources that do not actually support their statements or mischaracterize others' results (Brewin, 2023). For example, citing retracted or discredited studies as if they are valid evidence is a contextual flaw that peer reviewers often catch (Brewin, 2023). Such citation errors can propagate misinformation and give a false sense of the study's grounding.

Context also includes the alignment of the research design with the motivating question. If the study's methodology is not appropriate to address the stated problem, that mismatch is a Contextual and Reasoning flaw.

Box 10.2 Knowledge Check: Question 10.1 Which of these scenarios indicates a Context flaw?

A. Authors cite a retracted study as if it were credible support for their argument.
B. Psychologists generalize results from a handful of undergraduates to "all humans."
C. Researchers state they have "proven" their hypothesis even though the data are modest.
D. A paper claims that because two variables rise together, one causes the other.
E. An experiment's small sample size fails to detect a real effect that actually exists.

(Check your understanding using the Knowledge Check Key at the end of the book.)

Finally, consider the source and venue of the publication as part of Context. Research published in a predatory or non-peer-reviewed outlet might not have undergone rigorous review at all (Falagas, 2007; Retraction Watch, 2015). Pseudoscience is often found in venues outside the mainstream scientific literature, so knowing where a study appears is part of Contextual critique (Beall, 2018).

In summary, a solid Context means the authors situate their work within the existing body of knowledge, credit relevant studies (both supporting and conflicting) and use an appropriate framework. A critique should flag cases where the paper seems to exist in a vacuum, ignoring Context that would challenge or refine its Claim.

10.4 Evaluating the Quality of a Research Article

Critiquing research is not only about finding flaws; it's also about **assessing how strong a research article is overall**. A high-quality scientific study excels in several dimensions that build on the CERIC elements, including a *compelling central argument*, appropriate *scope and context*, and high standards of *rigor and transparency*. A genuine breakthrough paper will be strong on all these fronts and will clearly stand out as a valuable contribution to knowledge. By systematically examining Claim, Evidence, Reasoning, Context, and Implications – and by noting qualities like data transparency, reproducibility, and impact – we can more easily assess what a paper's strengths are and whether it rises to the level of a great study or falls short.

10.4.1 A Clear and Coherent Central Argument: Claim, Evidence, and Reasoning

A good research article presents a well-defined central Claim and supports it with solid Evidence and Reasoning. This means the hypothesis or thesis is stated up front (drawn from the research question identified by the gap), the methodology is capable of testing that hypothesis, and the results are interpreted logically. One should be able to follow the "through-line" of argument from introduction to conclusion without getting lost or encountering contradictions. As part of quality evaluation, check that the conclusions follow directly from the data. *Are the authors' interpretations justified by their results?* A strong article will have conclusions that are well-supported by the findings, without leaps of faith. Reviewers often explicitly validate data-versus-conclusions alignment; if they detect **overinterpretation** or conclusions that stretch beyond the Evidence, they will flag it (Mohty & Melo, 2025). Note that novice readers can have genuine insight when asking loads of basic questions because the field's assumptions are not fully internalized yet.

10.4.2 Evaluating the Scope and Relevance: Context and Implications

Top-tier research clearly situates itself in the broader context and addresses a meaningful question. The study will build upon a thorough literature review, demonstrating the authors' command of background knowledge and ensuring the work is not redundant with past studies. In the introduction and discussion sections, high-quality articles explicitly connect to existing theories and findings, showing how they fill a gap or resolve an open problem (Kelly et al., 2014; Körner, 2008).

Additionally, the paper's *scope* should be appropriate – neither too general to be tractable nor too narrow to matter. **Relevance** is often judged by whether answering the research question would advance the field or have practical importance. The scope of relevance includes a frank discussion of the study's limitations and Implications. Strong research articles acknowledge what remains unknown or uncertain, and they refrain from making grand claims that are not warranted. They also often suggest logical next steps or applications, demonstrating that the authors have thought about where the research leads. In sum, a great research article strikes a balance: It is *important* enough to contribute new knowledge yet *focused* enough to provide credible evidence on its specific question. The article should articulate why the work matters by relating its findings to prior research (context), noting any limitations, and outlining theoretical or practical Implications responsibly (Kelly et al., 2014).

10.4.3 Transparency and Reproducibility

Modern standards for high-quality articles include transparency in methods and data, so that other researchers could, in principle, replicate the work. **Reproducibility** – the ability for an independent team to obtain similar results using the reported procedure – is one of the hallmarks of good science (Resnik & Shamoo, 2017). Authors of strong papers provide sufficient methodological detail, and often supplementary data/code, that readers can verify key analyses or attempt to repeat the experiment. Transparency extends to disclosing any potential conflicts of interest or biases as well. A paper that is vague about how data were collected or that omits crucial steps in analysis invites skepticism. By contrast, clear and complete methodological reporting is a positive indicator of quality (Kelly et al., 2014; Körner, 2008).

The rise of **open science** practices (preregistrations, data sharing, etc.) reflects this criterion. A truly robust study will withstand others looking closely or trying to reproduce it. When evaluating a research article, consider: *Could another competent*

Box 10.3 Knowledge Check: Question 10.2 Which of the following features most strongly signals that a research article meets transparency and reproducibility standards?

A. The article introduces a novel hypothesis but withholds raw data to protect intellectual property.
B. The abstract claims "ground-breaking" results and notes that a prestigious agency funded the work, but the materials section states that proprietary reagents are unavailable for sharing.
C. The discussion section cites many high-impact papers but gives no information about statistical power.
D. The paper discloses all conflicts of interest yet reports its methodology only in outline form because of word limits.
E. The authors preregistered their study, provided complete datasets and analysis code as supplements, and described their protocols in detail.

(Check your understanding using the Knowledge Check Key at the end of the book.)

researcher reproduce these results with the information provided? If yes, that speaks well of the paper's integrity. If not, the paper might lack reliability. Indeed, the scientific community has become increasingly aware that many flashy results crumble under replication attempts, which is why reproducibility is emphasized as a pillar of quality (Resnik & Shamoo, 2017). In practical terms, check for things like clearly described protocols, availability of data or materials, proper statistical power, and adherence to reporting guidelines. These are hallmarks of an article that values rigor over mystique.

In summary, genuinely outstanding research articles do far more than tick off basic reporting requirements. They introduce something new – a question unasked, a dataset unseen, or a method deployed in an unexpected way – while maintaining methodological rigor so thorough that peer reviewers uncover few, if any, weaknesses. The prose exhibits clarity and organization, guiding readers through complex ideas without confusion, and demonstrates significance so that reviewers can assess that it advances the field in novel and important ways (Kelly et al., 2014). Finally, the very best studies display a touch of elegance or creativity – an insightful synthesis or innovative design that elevates them beyond solid science. When these qualities converge, the article achieves broad or deep impact, shaping future agendas well beyond its own pages.

10.5 Applying CERIC in the Peer Review Process

The CERIC framework can serve as a practical checklist for peer reviewers as they critique manuscripts, improving thoroughness and consistency by examining each CERIC element in turn. This section explores guidelines for peer-reviewing using CERIC and considers some practical steps for applying critique.

When you peer-review a draft article, let the CERIC elements guide the flow of your reading:

Is the main research question stated clearly and framed as something novel, testable, and reported in declarative language as a Claim? Proceed to Evidence and evaluate whether the methods, sample size, controls, and statistics are appropriate – and whether the data shown truly match the Claim. Next, trace each logical step of Reasoning. Do the conclusions follow from the results, and have alternative explanations been addressed? Then, review the Implications and decide whether the authors explain the significance of their findings without exaggeration or ignoring ethical concerns. Finally, check the Context for an Introduction Funnel that cites key studies and a genuine research gap.

Keep your feedback concise and actionable. A single numbered list works well, for example:

1. **What is strong:** Note any clear questions, rigorous methods, or thorough analyses.
2. **What needs work:** Point to specific pages or figures and say, for example, "Figure 2 lacks a control, so Claim A is only partly supported."
3. **How to improve it:** Suggest realistic next steps, such as adding another analysis (name it), clarifying a theoretical support, adding comparison to a model, or rerunning a key set of statistics.

By organizing your comments around CERIC and limiting yourself to clear, numbered points, you will give your peer useful feedback or revision while sharpening your own critical-reading skills. You can ask your peers to review your drafts this way, too. Novice readers may feel unprepared to do this level of critique, and that discomfort is okay. This skillset is essential to begin practicing – the sooner, the better.

10.5.1 Practical Notes for Applying CERIC to Critique

In practice, peer reviewers who adopt a framework tend to produce more balanced and thorough evaluations (Akers, 2017; Lucey, 2013; Mohty & Melo, 2025). They catch flaws and note what works well in each category, providing a fair and useful review. CERIC is a powerful tool in the peer review process, guiding reviewers to be comprehensive and systematic in their critiques, ultimately helping to improve the quality of papers that get published. For example:

Claim: well-defined and supported by the Evidence and Reasoning.
Evidence: Methods are strong, but the sample size is small, risk of overgeneralizing.
Reasoning: Be aware of making a logical leap when inferring causation.
Implications: Appropriate to the Evidence but discuss with caveats.
Context: Literature review is clear but missing a relevant rationale for doing the study.

Such a structured critique is extremely valuable to authors (and journal editors) because it pinpointedly identifies where revisions are needed. Furthermore, applying CERIC ensures constructive feedback: by examining each area, the reviewer can offer targeted suggestions (perhaps to add an analysis for evidence, or to rewrite a section for clarity in reasoning, etc.) rather than vague praise or criticism.

10.6 Worked Example: Fatal Flaws Plus Elaborate Fraud

In this worked example, we will apply CERIC to critique a now-infamous retracted study. In 1998, Andrew Wakefield and colleagues published a study in *The Lancet* suggesting a potential link between the measles, mumps, and rubella (MMR) vaccine, gastrointestinal inflammation, and autism spectrum disorders (Wakefield et al., 1998). This publication has been extensively critiqued for its methodological flaws and ethical breaches, ultimately leading to its retraction in 2010 (Editors of The Lancet, 2010) and Wakefield losing his medical license in the UK (Kmietowicz, 2010). Still, its impact on vaccine fear mongering is a profound and lasting Implication. Let's dive into the critique.

- Title: [RETRACTED] "Ileal-lymphoid-nodular hyperplasia, non-specific colitis, and pervasive developmental disorder in children"
- Citation: [RETRACTED] Wakefield, A. J., Murch, S. H., Anthony, A., … Walker-Smith, J. A. (1998). *The Lancet, 351*(9103), 637–641. https://doi.org/10.1016/S0140-6736(97)11096-0
- Abstract: [RETRACTED]

  "*Background.* We investigated a consecutive series of children with chronic enterocolitis and regressive developmental disorder.

  Methods. 12 children (mean age 6 years [range 3–10], 11 boys) were referred to a paediatric gastroenterology unit with a history of normal development followed by loss of acquired skills, including language, together with diarrhoea and abdominal pain. Children underwent gastroenterological, neurological, and developmental assessment and review of developmental records. Ileocolonoscopy and biopsy sampling, magnetic-resonance imaging (MRI), electroencephalography (EEG), and lumbar puncture were done under sedation. Barium follow-through radiography was done where possible. Biochemical, haematological, and immunological profiles were examined.

  Findings. Onset of behavioural symptoms was associated, by the parents, with measles, mumps, and rubella vaccination in eight of the 12 children, with measles infection in one child, and otitis media in another. All 12 children had intestinal abnormalities, ranging from lymphoid nodular hyperplasia to aphthoid ulceration. Histology showed patchy chronic inflammation in the colon in 11 children and reactive ileal lymphoid hyperplasia in seven, but no granulomas. Behavioural disorders included autism (nine), disintegrative psychosis (one), and possible postviral or vaccinal encephalitis (two). There were no focal neurological abnormalities, and MRI and EEG tests were normal. Abnormal laboratory results were significantly raised urinary methylmalonic acid compared with age-matched controls ($p=0{\cdot}003$), low haemoglobin in four children, and a low serum IgA in four children.

  Interpretation. We identified associated gastrointestinal disease and developmental regression in a group of previously normal children, which was generally associated in time with possible environmental triggers."

10.6.1 CERIC Review

Claim: The authors claimed:

> We identified associated gastrointestinal disease and developmental regression in a group of previously normal children, which was generally associated in time with possible environmental triggers.

Wakefield et al. later dubbed this gastrointestinal condition "autistic enterocolitis." Also, in the article, they named MMR as the possible environmental trigger.

Evidence: The authors based their Claim on what they described in the methods as a case series of 12 children (11 boys and 1 girl) who exhibited developmental regression and gastrointestinal symptoms. They conducted a series of tests, including colonoscopy with histology of collected tissue samples, lumbar puncture, blood tests, MRI, and EEG. In the findings section, they report these lab and clinical results. Also, in the article's main text, they described collecting data from parents, and the parents of eight of the children linked the onset of behavioral symptoms to the administration of the MMR vaccine.

Reasoning: They claimed age-matched controls for some of the findings as follows:

> Abnormal laboratory results were significantly raised urinary methylmalonic acid compared with age-matched controls (p=0·003), low haemoglobin in four children, and a low serum IgA in four children.

The authors used connective Reasoning, where they inferred that temporal co-occurrence of MMR vaccination indicated possible causation of the observed conditions, suggesting that the vaccine might trigger both gastrointestinal and developmental disorders.

Implications: The main Implication was layered into the Claim:

> () … which was generally associated in time with possible environmental triggers.

This clause signals a broader concern: something in the environment (which they identified as MMR vaccination in 8/12 cases) might precipitate the newly described syndrome. This language implied the need for further investigation and raising potential public-health questions.

Context: Their stated purpose was:

> We investigated a consecutive series of children with chronic enterocolitis and regressive developmental disorder.

They conflated this purpose and an aim to characterize any underlying pathological syndrome with the rising autism rates in the 1990s.

Next, let's pivot to a full CERIC critique, applying what we have learned in Chapters 2–9.

10.6.2 Problems with Study Design and Methods

Even before its fraud was uncovered, Wakefield et al.'s (1998) investigation could not establish causation: the authors used a 12-patient case series – an inherently descriptive design with no random sampling or control group – so prevalence and causal inference were impossible (Taylor et al., 1999). Later scrutiny revealed that Wakefield had contacted 5 of the 12 families in advance, introducing overt selection bias; and he preselected children who already had vaccination histories, diarrhea, and autism, making the sample unrepresentative and inflating any apparent vaccine link (General Medical Council, 2010). Using this hand-picked cohort, the team then performed colonoscopies, lumbar punctures, and blood tests – procedures that are invasive yet wholly incapable of proving causality.

10.6.3 Problems with Evidence

The study's Evidence has so many flaws that it is a challenge to see if we can find them all. So, we divide this analysis into two groups: problems with the Evidence visible in the article, and problems that emerged later. We found the later flaws by using the "Cited By" function of Google Scholar and then forward tracking.

Here are the major flaws visible in the article:

1. **Single-center, single-clinician case** series of 12 children (11 boys, 1 girl) – far too small and unrepresentative for generalizable conclusions about vaccine safety (Taylor et al., 1999).
2. **No control group** for vaccinated vs. unvaccinated comparison (Godlee et al., 2011).
3. **Parental recall** used to time symptom onset, introducing recall bias (Taylor et al., 1999).
4. **Unblinded assessments** of histology and behavior.
5. **No virological or immunological evidence** (e.g., measles PCR, antibody titres) to link MMR with gut lesions.
6. **Virtually no statistics:** Only a single p-value for urinary methylmalonic acid; even basic descriptive metrics (mean time from vaccination to symptoms) are missing.

Post-publication Evidence revelations:

7. **Undisclosed legal funding** from an anti-MMR lawsuit and other conflicts of interest.
8. **Timeline falsification:** Hospital records showed some children's symptoms arose before or long after vaccination (General Medical Council, 2010).
9. **Questionable pathology:** Independent review found mostly normal tissue despite the paper's claim of "nonspecific colitis" in 11/12 children (General Medical Council, 2010).
10. **No external replication:** Large epidemiological studies failed to confirm any MMR–autism link (Madsen et al., 2002).

This catalog of Evidence shortcomings – flawed sampling, absent controls, unblinded measures, missing mechanistic data, undisclosed conflicts, and subsequent non-replication – reveals that the study failed at every rung of evidential credibility, leaving its core Claim without any valid or reliable empirical footing.

10.6.4 Problems with the Reasoning

Like Evidence, the study has so many flaws in Reasoning that it is a challenge to find all of them. Again, we will divide this analysis into two groups: problems with the Reasoning visible in the article, and problems that emerged later. Again, we found the latter flaws using Scholar's "Cited By" function and forward tracking. Here are some of the glaring ones in the article:

1. **Overgeneralization from an anecdotal sample**: Extrapolates findings from 12 hand-picked children to the entire vaccinated population.
2. **Correlation treated as causation:** No experimental controls, yet MMR implied causal using temporal logic alone (Godlee et al., 2011).
3. **Missing causal tests:** No dose-response or consistency analysis, and major confounders (genetics, perinatal factors) ignored.
4. **Disregard for prior probability:** Autism rates were already rising before the 1988 UK MMR rollout, undermining Bayesian plausibility.
5. **Cherry-picked analogies and no mechanism:** Measles-encephalitis cited without virological data; no plausible pathway offered (Taylor et al., 1999).
6. **Circular reasoning on "autistic enterocolitis":** Nonspecific gut lesions are used both to define and to support the purported syndrome.

Here are more Reasoning problems that emerged after publication:

7. **Post-publication confirmation bias:** Hospital records showed some symptoms pre- or long post-MMR, contradicting the paper's timelines (General Medical Council, 2010).
8. **Unfalsifiable Claim:** Subsequent large epidemiological studies found no autism risk (Madsen et al., 2002; Taylor et al., 1999), yet the authors dismissed those results instead of revising their hypothesis.

These Reasoning flaws – ranging from unwarranted causal leaps in the original article to *post-hoc* confirmation bias after publication – indicate how the authors built their Claim on a chain of logic so weak that each link collapses under even the most basic standards of scientific argumentation.

10.6.5 Problems with Limitations and Implications

Wakefield et al. (1998) never alerted readers that their work had profound limitation: that a descriptive, uncontrolled case series cannot show causation. Nor did they discuss its tiny sample size, selection bias, or methodological weaknesses. Because these limitations were absent, journalists and the public assumed the findings were solid – after all, they appeared in *The Lancet*. Wakefield then magnified that misperception at his press conference, by fear mongering and urging clinicians to "suspend" the combined MMR and give three single-antigen shots a year apart, claiming "sufficient anxiety … about the long-term safety of the polyvalent … MMR" (Wiley, n.d.), even though his article provided no data on the safety or effectiveness of that alternative schedule.

Media coverage of the recommendation was intense, and fear spread like a contagion. UK MMR uptake fell from 91 percent in 1996 to 80 percent in 2003–2004, and measles outbreaks followed between 2006 and 2009 (Nuffield Trust, 2025; Rao & Andrade, 2011). Similar hesitancy spread to Ireland, parts of western Europe, and pockets of the United States, eroding confidence in childhood vaccination programs.

Immediate flaws in the article's Implications:

1. **No risk–benefit calculus**: Implied minimal cost to skipping MMR while ignoring measles-, mumps-, and rubella-related morbidity and mortality.
2. **Unwarranted invasive testing**: By naming "autistic enterocolitis," the authors tacitly encouraged colonoscopies and lumbar punctures in autistic children without validated pathology.
3. **Litigation-driven framing:** Undeclared legal funding steered the study's public messaging toward blame, not science.

Public-health fallout still ongoing, in addition to sustained vaccine hesitancy (found with forward tracking and online news searches):

4. **Research diversion:** autism funds were siphoned into repeated, fruitless MMR replications – still ongoing in the U.S.
5. **Global spill-over:** similar coverage declines and outbreaks in Ireland (2000–2001) and the US (2008, 2025).
6. **Erosion of confidence in the full schedule:** anxiety spread from MMR to other childhood vaccines (e.g., DTaP, Hib), undermining immunization programs.

Each layer of covering up limitations and overinterpretation – from unsupported policy advice to worldwide spill-overs – amplified the public-health damage that originated in a fraudulent evidential foundation. By omitting frank discussion of its limitations and broadcasting a policy change without solid Evidence or Reasoning, the study transformed a weak exploratory Claim into a public-health crisis whose effects – including persistent anti-vaccine activism – are still being felt (Deer, 2011).

10.6.6 But Wait, There's More: Problems with Ethics

The UK's General Medical Council found undisclosed funding and unethical procedures (General Medical Council, 2010), for instance, Wakefield received £435,643 in fees, plus £3,910 expenses (about $725,000) from an attorney, Robert Barr, who was suing the MMR vaccine manufacturer (Deer, 2011). Also, all 12 children were investigated without valid ethical approval (General Medical Council, 2010).

Following multiple independent investigations, including The *BMJ* investigation labelling the work "fraudulent" (Godlee et al., 2011), *The Lancet* fully retracted the article in 2010. The UK General Medical Council revoked Wakefield's medical license the same year (Kmietowicz, 2010), citing professional misconduct and ethical breaches (Editors of The Lancet, 2010).

In summary, this worked example illustrates how initially failing to critique an argument from evidence can cause harm. Despite extensive refutation – including falsified Evidence, faulty Reasoning, distorted Context, overstated Implications, ethics violations, and outright fraud – Wakefield et al.'s original Claim remains a touchstone for anti-vaccine activists today. This example demonstrates how public trust erodes when the essential aspects of a research article are fraudulent and not vigorously critiqued, and transparency is absent.

10.7 Chapter Key Takeaways

1. **Peer review is crucial but not infallible:** Peer review serves as a critical checkpoint in academic publishing, ensuring research quality and credibility. However, it is not a perfect system – errors can slip through, and even published research should be critically evaluated. Studies show that reviewers often miss errors, and the replication crisis has revealed that many published findings cannot be reproduced. Peer review is particularly vulnerable to fraud.
2. **The peer review process involves multiple steps:** The structured peer review process includes submission evaluation, reviewer selection, review and feedback, revision and resubmission, and a final decision. Each step plays a role in strengthening a paper's quality, but the process relies on both the expertise and integrity of reviewers.
3. **Constructive skepticism enhances research quality:** Being a "constructive skeptic" is vital for both researchers and reviewers. Critical evaluation should aim to improve research rather than simply find faults. Effective skepticism questions assumptions, evaluates methodologies, and considers alternative interpretations, leading to stronger, more reliable scientific conclusions.
4. **Diversity in perspectives strengthens research:** Diverse research teams – including those with varied ethnic, disciplinary, and methodological backgrounds – produce more impactful and innovative studies. Peer review benefits from multiple expert viewpoints, ensuring well-rounded critiques that catch potential flaws.
5. **Publication does not equal quality or validity:** The assumption that "if it's published, it must be good" is flawed. Peer review improves the likelihood of a study being sound but does not guarantee accuracy. High-profile cases, such as the retracted hydroxychloroquine study and the Wakefield et al. (1998) MMR-autism paper, highlight the urgency of post-publication critique and replication efforts.
6. **The CERIC framework helps identify research flaws:** Breaking down research into CERIC provides a systematic approach to critique. This method helps to quickly uncover common errors such as overstated Claims, flawed statistical analysis, logical fallacies, exaggerated implications, weak Evidence, and missing Reasoning.

11 Using CERIC for Presentations

> The greatest value of a picture is when it forces us to notice what we never expected to see.
> —John Tukey (p. vi, 1977) in *Exploratory Data Analysis*

11.1 Overview

Chapter 11 demonstrates the use of the CERIC framework to design and deliver compelling presentations of scientific research. It introduces the modified CCERI structure – Context, Claim, Evidence, Reasoning, Implications – as a flexible guide for organizing talks in classrooms, journal clubs, lab meetings, and seminars. In addition to framing a scaffold for scientific presentations, this chapter guides readers on how to choose and discuss figures, explain Reasoning clearly, and incorporate comparison and critique into the flow of a presentation. This chapter also provides strategies for maintaining clarity and engaging audiences across disciplines. It also explores collaborative formats, including live CERIC reviews, seminar preparation sessions, and group discussions, showing how these activities encourage active learning and build scientific literacy. Finally, this chapter offers practical ways to gamify research discussions, making critical reading fun, rigorous, and rewarding.

11.2 How Do We Talk about Primary Literature?

Science thrives on conversation. It is a dynamic, centuries-long discussion tradition that stretches back long before Galileo's telescope. Unlike casual chats over coffee, the scientific conversation follows a formal process of recording, challenging, and revising ideas through primary scientific literature.

This chapter examines the ways scientists – both novice and expert – typically interact with primary literature. We first explore common formats in which scientific research is presented, including class presentations, journal clubs, and lab meetings. We then introduce an approach that clarifies complex research using an adapted CERIC framework: Context, Claim, Evidence, Reasoning, Implications (CCERI). Along the way, we will highlight how structured discussions, collaborative critique, and public outreach enhance scientific engagement. Scientific conversations unfold in many settings. Let's explore where and how they happen.

11.2.1 Classroom Oral Reports

We've all been there: Staring at a dense research paper, wondering how on Earth to understand it, let alone talk about it. It is one thing to read a research article; it is another to explain it to someone else, especially when a grade is on the line. **Classroom oral reports** offer the opportunity to delve deeply into the research literature, either individually or in a team. These assignments are often organized into core research tasks, including choosing one or more relevant research articles, breaking the articles down into their main components, and presenting a summary and analysis of the articles to peers. It is part scientific storytelling, part public speaking boot camp, and part exercise in translation. The classroom oral report helps us grow the early stages of scientific discourse. Research shows that engaging with primary literature increases scientific literacy, critical thinking, and confidence in handling complex material (Kararo & McCartney, 2019; Sloane, 2021).

Class reports often follow a structured, one-way presentation format. The presenting student is in charge, and the audience listens, asks questions, and engages. The structure mirrors real-world academic conferences, preparing students for the kind of presentations they may give at research seminars or public science events. Most instructors guide students through this process using a framework of guiding questions, for example:

- What is the context of this article, and what is the article trying to address?
- What are the article's main findings?
- What data do the authors collect, and how do they analyze it?
- Do the data and analysis support the findings?
- Why does this study matter?

It should be clear from the previous chapters that the CERIC elements provide a natural scaffold for addressing these questions. As we will see, it also helps students avoid getting lost in the weeds and to focus instead on the bigger picture questions.

11.2.2 Journal Clubs

Imagine a book club where the discussion centers on a research paper titled *"The Role of Gut Microbiota in Regulating Neural Function in Mice."* Our scenario mirrors a traditional book club: everyone reads the same work, then gathers to discuss its meaning and significance. In scientific settings, this is known as a **journal club** – a long-standing tradition in graduate programs, research labs, and even undergraduate courses (Sloane, 2021). The process typically unfolds like this:

1. One participant (or a small team) presents the background, methods, findings, and implications of a recent study.
2. The group collectively analyzes the findings, raises questions, and evaluates the article's strengths and limitations.

3. Through discussion, participants develop a deeper understanding of the article's results and quality, while more novice researchers expand their knowledge of the broader scientific field.

The journal club model helps to develop students' ability to read and critique like expert scientists – questioning design, interpretation, and significance. Research shows that journal clubs are active-learning environments that strengthen students' confidence in analyzing and presenting scientific research (Sandefur & Gordy, 2016). Each session brings opportunities to think critically and engage deeply with new ideas. These discussions also reinforce a fundamental principle of science: group scrutiny.

11.2.2.1 Every Claim Must Withstand Scrutiny

If journal clubs are structured academic exercises, **reading groups** offer a more flexible and exploratory approach. With less emphasis on formal presentation, reading groups encourage participants to engage with research in a collaborative and low-pressure setting. Participants read a scientific paper or a selection of papers) in advance, then come together to share insights, clarify confusing sections, and discuss broader themes. Reading clubs often extend beyond a single discipline, allowing students to explore connections between fields. A reading club might focus on astronomy 1 week and genetics the next, fostering cross-disciplinary curiosity. The goal of these discussions is to promote peer learning and scientific confidence, helping students broaden their scientific understanding, refine their ability to analyze research, ask thoughtful questions, and articulate their understanding.

11.2.3 Research Group Meetings

Now, let's take journal clubs into the heart of research settings: the lab. In research groups, literature discussions are often integrated into weekly lab meetings, where students, faculty, and researchers share insights from recent studies. A graduate student might present a newly published quantum materials study to their physics research team, or a marine biologist might introduce a paper on ocean acidification's impact on coral reefs to their research team.

Unlike class presentations, **research group meetings** encourage dynamic interaction. Rather than passively receiving information, participants engage in discussions that connect the paper's findings to their ongoing research questions. In addition, research group meetings are typically more focused on a specific topic of interest to the group than the broader journal club and research group formats. These meetings enhance scientific engagement in three key ways:

- They keep the research group informed about current advances.
- They provide a forum for refining ideas, methodologies, and interpretations.
- They foster collaborative problem-solving, often leading to new research directions.

By participating in lab meetings, students hone their ability to present research, evaluate methodologies, and integrate new findings into their work.

Box 11.1 Knowledge Check: Question 11.1 Which format most closely resembles an academic conference presentation, relying on a structured, largely one-way delivery followed by audience questions?

A. Journal club discussions
B. Reading club sessions
C. Class reports on research papers
D. Weekly lab meetings
E. Informal coffee-break conversations about science articles

(Check your understanding using the Knowledge Check Key at the end of the book.)

Across these formats – class presentations, journal clubs, reading groups, and lab meetings – one truth remains constant: scientific discussion shapes how we understand and advance knowledge, cultivating essential scientific skills from data analysis to critical evaluation. More importantly, these discussions provide an opportunity to go beyond simply absorbing information. Through debate, analysis, and collaboration, students develop the mindset of a scientist – questioning claims, exploring alternative explanations, and seeking deeper insights. A research paper offers much more than just results. It opens the door to inquiry, innovation, and intellectual growth.

11.2.4 Structuring Effective Research Discussions

Every scientific article is a small piece of a much larger conversation, and research presentations are intimately tied to research discussions. A good scientific discussion feels like a lively debate loaded with AH-HA moments – a space where ideas mix, evolve, and refine our understanding of the world. But without direction, these discussions can unravel into meandering conversations, missed insights, or a few voices carrying the weight of the room. The ability to critically engage with primary literature is essential for scientific literacy, but how that engagement unfolds depends as much on how we talk about research as what we talk about. Without structure, research discussions can become academic free-for-alls, where key details get lost, some participants are hesitant to speak, and critique overshadows curiosity.

Unstructured discussions falter for at least three main reasons:

- **Dominant voices and unequal participation:** In every discussion, some voices rise above others. In a classroom or research group, the loudest or most experienced participants can inadvertently steer the conversation, shaping the group's interpretation of an article even when their take is not the most insightful. In unstructured discussions, those who feel less confident in their expertise often remain silent, even when they have valuable perspectives to contribute. Without intentional moderation, these imbalances persist. When one or two voices dominate, the discussion narrows, reinforcing the idea that only certain types of knowledge "count." In reality, the best scientific conversations are collaborative investigations, not solo performances. Structuring discussions to ensure equitable participation enables ideas to challenge and refine one another, leading to deeper insights.

- **Drifting without direction:** Imagine a group of students gathering for a journal club. The article has been sent out, but half the group hasn't read it thoroughly. There's no guiding framework, just a general prompt to "discuss." Someone mentions a figure on page five, someone else jumps to the methods, and soon the conversation is ricocheting between details without ever settling on a central theme. By the end, no one is quite sure what the main takeaway was, except that they feel vaguely exhausted. Unstructured discussions often drift like this. A lack of clear questions or an assigned leader can leave participants navigating a research article without a clear direction, skimming the surface of its findings but never delving deeply enough to truly grasp its core argument. Without a structured plan to guide inquiry, key ideas slip through the cracks.
- **Fault-finding spiral:** Critical thinking is a cornerstone of scientific inquiry, but when discussions focus solely on dissecting a paper's flaws, they can lose sight of what makes the research meaningful in the first place. In one study, students in an informal journal club actively engaged but spent most of their time critiquing the methodology rather than discussing the broader contributions of the study (Atzema, 2004). While skepticism is healthy, a discussion that fixates only on a paper's weaknesses can miss the *why* behind the research – the new avenues it opens up, the questions it raises, and the knowledge it builds upon. Effective critique should be both deconstructive and constructive, questioning the argument and assumptions while also appreciating how a study fits into the larger scientific landscape. Without this balance, journal club discussions can become more about finding what is wrong than exploring what is possible.

Structure and scaffolding are crucial for a robust and inclusive discussion of a research article. A structured preparation might include assigning sections to individual participants or requiring students to submit a ummary of the article before discussion, strategies that are known to improve participation and engagement (Atzema, 2004). When participants come into a discussion with a shared baseline of understanding, discussions gain depth and coherence, enabling more advanced analysis rather than just basic comprehension.

Structured discussions can also help to level the expertise of the group. Scientific papers are dense, and students often struggle with complex methodologies, unfamiliar statistical analyses, or densely packed figures (Abdullah et al., 2015; Hubbard & Dunbar, 2017; Lie et al., 2016). In an unstructured discussion, these points of confusion may never be addressed, leaving gaps in understanding that undermine the entire conversation. The most effective structured discussions slow down where it matters, taking the time to walk through the key findings, examine figures, explain methods, and establish the rationale behind the study's design. Without this scaffolding, a discussion can jump straight to results and conclusions, skipping over the foundational work that makes those results meaningful to the more novice readers.

Finally, without guidance, discussions may focus too much on a paper's minutiae and forget to step back to ask: *Why does this research matter?* Structuring discussions to include questions about Context, significance, and Implications helps ensure that the conversation remains rooted in the larger scientific narrative, rather than just isolated technical details.

In summary, a great discussion, like a well-designed experiment, needs structure. The most effective journal clubs and group discussions employ intentional frameworks to promote engagement, critical analysis, and collaboration. Here are some guidelines to consider when planning for research article discussion, whether it is a class exercise, journal club, or research group meeting:

- **Clear goals:** Participants should understand whether the discussion aims to critique methodology, analyze findings, or connect the study to broader scientific questions.
- **Defined roles:** Assigning responsibilities – such as leading the discussion, summarizing figures, or preparing background context – helps distribute intellectual labor more equitably.
- **Moderation and balance:** Facilitators must ensure that multiple voices contribute, preventing the discussion from being dominated by a select few.
- **Preparation:** Setting clear expectations on prereading, providing guiding questions, or requiring short summaries from participants ensures that everyone arrives ready to engage with the discussion.

A well-structured research article discussion transforms primary reading from an intimidating approach to the vast archive of scientific knowledge into a living conversation, one in which every participant maintains focus and contributes to the collective understanding.

Box 11.2 Knowledge Check: Question 11.2 In journal club discussions of research papers, which strategy most directly prevents the "drifting without direction" problem, where conversation ricochets between details without settling on a central theme?

A. Appoint a moderator who enforces speaking time limits to balance participation.
B. Require each participant to submit a short written summary of the paper before the meeting.
C. Supply a structured set of guiding questions (or a framework like CERIC) that steers analysis step-by-step.
D. Invite an external expert to deliver a detailed critique of the study's methodology.
E. Begin the meeting by cataloguing every statistical test the authors used.

(Check your understanding using the Knowledge Check Key at the end of the book.)

11.3 Presentation Types: Summaries and Critiques

A presentation on a research article typically takes one of two forms: summary and critique. A **summary presentation** focuses on the article's main contribution to the field, breaking down the article's objectives, methodology, findings, and relevance. A critique presentation engages with the study's Claims, Evidence, and Reasoning to assess its accuracy, significance, and limitations. A summary presentation aligns with a typical CERIC review (Chapter 8), while a critique presentation engages the analytical framework of CERIC Critique (Chapter 10). Both can involve comparison, depending on the assignment (Chapter 9).

The approach to a research article presentation depends on whether the goal is to summarize or critique the article:

- **A summary** requires a guiding mindset, which helps the audience understand the key components of the article and its place within the broader literature. Preparing for a summary presentation often involves finding the clearest way to explain the main findings and technical details, while exploring related work to show how the article fits into a broader scientific narrative.
- **A critique** requires an analyst mindset, questioning how well the Evidence supports the Claims, identifying potential weaknesses in Reasoning, and exploring alternative interpretations. Preparation for a critique presentation often involves checking background sources, comparing the study to related work, and considering broader implications.

In journal clubs and group meetings, researchers often engage in both summarizing and critiquing. Knowing when to review and when to analyze can transform a routine summary into an engaging conversation about scientific advances. Let's break down the features of these types of presentations in detail.

Summarizing a research article is an exercise in selectivity and storytelling: The goal of a summary presentation is to guide an audience through the essential aspects of the paper – its motivation, methods, central argument, and implications – without getting lost in unnecessary detail. Many readers initially approach presentations by moving through a paper section by section, summarizing the introduction, methods, results, and conclusion in order. This might seem like a logical way to proceed, but it can lead to a boring information-dense talk that does little more than repeat the text. As one guide bluntly puts it, "This approach is not useful because all that is happening is that the student is reading the paper aloud, forgetting that the audience … has already done so" (Ram, 2025, para 1).

A more effective approach is to structure the presentation around the article's central Claim–Evidence–Reasoning argument (Bjorn, 2024). Instead of treating every section as equally important, focus on the main Claim, the key pieces of Evidence that support it, and the Reasoning that links them. For instance, if the article reports the discovery of an exoplanet in a previously unexpected star system, the presentation should highlight how the planet was detected and how the authors were able to rule out alternate hypotheses to establish their Claim. In addition, the presentation could also provide sufficient Context to justify the unique nature of the detection and Implications that illustrate why this discovery matters. While some technical details may be crucial for understanding the results, others will add complexity without enhancing comprehension. The task of the presenter is to "figure out how to get your audience as quickly as possible to the point where they can understand this idea" (Ram, 2025, para 2). The most effective presentations prioritize what matters most, ensuring the audience leaves with a clear grasp of the research's significance.

A critique moves beyond what a study finds to examine how well it supports its Claims: The purpose of a critique presentation is to evaluate a research study's strengths, weaknesses, and contributions. Approaches include identifying the study's most significant methodological limitations, evaluating its arguments, or highlighting unresolved or newly raised questions. Just as a summary presentation highlights an article's central Claim, a critique presentation highlights its most pressing challenge. For instance, if a study draws a strong conclusion from a small sample size, a reviewer might point out that larger studies are needed to confirm the result. Similarly, if an article proposes a bold interpretation of its findings but lacks direct supporting evidence, a critique might explore alternative explanations.

Experienced scholars often frame critique as identifying the central tension or flaws in a research article, guiding the audience through the significance of these flaws, and highlighting alternative experimental designs or statistical approaches. On the other hand, critique is most effective when it is balanced and constructive. It is easy to highlight weaknesses, but an insightful review also recognizes what the paper does well. Even studies with significant limitations can make meaningful contributions – such as introducing new techniques, framing novel questions, or opening the door for future work. A well-rounded critique acknowledges these contributions while identifying areas for improvement, ensuring that evaluation remains analytical rather than merely dismissive.

In academic settings, instructors often require students to separate summary from critique, for example, dedicating the first ten minutes of a presentation to summarizing the study and the next five to offering a critique. Starting with an objective summary and then transitioning into an evaluation allows the audience to absorb the key facts before engaging in a critical discussion (Atzema, 2004). This separation also ensures that early researchers develop both skills. Over time, as fluency in scientific discussion grows, summary and critique can be blended more naturally. For example, a blended approach presentation could be structured as follows:

1. **Summary of the study's goal and approach:** A brief statement of the study's motivation, questions, and how it aims to address them, in the context of prior or contemporary work.
2. **Overview of the main argument:** A summary of the study's main Claim, Evidence, and Reasoning arc, focused on the key methodological approaches and types of Reasoning and how these interrelate.

3. **Evaluation of the main argument:** A discussion of strengths (such as a rigorous methodology or innovative approach) and limitations (such as a small sample size or unaddressed confounding factors) of the study's CER core argument.
4. **Implications:** A reflection on whether the study provides adequate or inflated Implications of the findings.
5. **Moving forward:** Ideas for next steps, such as new experiments, alternative analyses, or further comparisons with related studies, are needed.

By practicing both summary and critique presentations, we practice and refine our abilities to engage deeply with scientific literature – understanding it, questioning it, and ultimately, contributing to the broader conversation in science.

Box 11.3 Better Practice: Designing Summary and Critique Presentations

As we have seen in this section, summary and critique presentations have different objectives, mindsets, and frames for discussion. Framing these different kinds of presentations requires design considerations around their purpose, focus, and strategies:

Summary Presentation: Highlighting an article's main contribution

- Purpose: Teach the work – make the study's core argument, rationale for Context, and main Implications are clear to listeners, who probably did not read the full text.
- Mindset: Storyteller and guide.
- Focus: The paper's central Claim, the key Evidence, and the Reasoning that links them.
- Strategy: Lead with the breakthrough, then supply only the methods and Context necessary for the audience to grasp why it matters (Bjorn, 2024).
- Avoid: Marching section by section through every detail – always avoid "reading the paper aloud" (Ram, 2025).

Critique Presentation: Highlighting an article's main point of contention

- Purpose: Evaluate the work – probe the CER core argument's validity, rigor, and relevance.
- Mindset: Analytical evaluator.
- Focus: Test each CERIC element:
  - Is the Claim warranted?
  - Is the Evidence sound and sufficient?
  - Does the Reasoning logically connect them?
  - Are the Implications appropriate in scope?
  - Is the Context complete and clear about why the study needed to be done?
- Strategy: Cite strengths, identify limitations, suggest improvements, and compare with related studies to gauge novelty and reliability.
- Avoid: Getting lost in descriptive summary – analysis must move beyond "what" to "how well" and "so what."

11.4 The CCERI Framework for Presentations

If you have ever heard social science researcher Brenee Brown give a TED talk, she does so with a passion, clarity, and logic that is inspiring. Similarly, presenting a research article is more than listing results and summarizing figures. It is **storytelling** – guiding an audience through a logical and engaging journey that reveals what was found and why it matters. Whether in a classroom, journal club, or research group meeting, a compelling presentation transforms raw data into a coherent narrative, drawing listeners into the scientific process. This section explores strategies for structuring a research presentation, integrating essential figures and data, and using the Context, Claim, Evidence, Reasoning, and Implications (CCERI) framework to communicate key findings.

11.4.1 The Narrative Thread: Context, Claim, Evidence, Reasoning, and Implications

A well-structured research presentation follows a logical, temporal arc, much like a compelling documentary or a well-written novel. The CCERI framework, adapted from argumentation models in science (Bjorn, 2024; Toulmin et al., 1984), provides a roadmap for crafting this narrative:

Context: Establishes the background and rationale of the research. This includes the broader problem, existing knowledge, and gaps the study aims to fill.

Claim: Highlights the study's main conclusion or discovery. This is the answer the research provides to the question posed in the Context.

Evidence: Presents the data supporting the Claim. In presentations, this often means selected figures, tables, or key observations.

Reasoning: Explains how the Evidence supports the Claim. This encompasses the underlying theory, methodology, and logic that connect data to conclusions.

Implications: Discusses the significance of the findings. This includes potential applications, unanswered questions, and the study's broader impact.

Using CCERI ensures a presentation forms a **cohesive scientific argument**, where each element flows according to traditional expectations.

11.4.2 Claim-First versus Context-First

The order of these elements can be adapted based on audience needs and presentation style. Both Claim-first and Context-first methods are effective, and presenters often hybridize these approaches. For example, a talk might hint at the Claim early – "This study found something unexpected" – before developing the background; or it may state the Claim in its entirety before examining the Context for justification. The choice of Claim-first versus Context-first depends on:

- **Audience expertise**: Specialists may prefer a Context-first approach with a detailed buildup, while mixed-discipline audiences may benefit from a Claim-first summary before diving deeper.

- **Engagement strategy**: A Claim-first talk offers immediate clarity, while a Context-first talk builds suspense and deeper understanding.
- **Time constraints:** A short talk often benefits from leading with the Claim, while a longer seminar may allow a gradual buildup through Context.

Let's examine each approach in more detail.

Claim-First: Presenters who open with the study's main Claim immediately draw the audience into the new knowledge. This method mirrors the approach used in news headlines or scientific press releases, where the key finding is stated upfront to capture interest. For example, a presenter discussing Kim and Chan's (2004) physics article "Observation of Superflow in Solid Helium" might begin with this paraphrased claim:

This study demonstrates the existence of a new state of matter: a superfluid solid. This kind of direct statement of finding immediately signals what the research is about and why it is significant, prompting the audience to wonder how this Claim was established. The presenter can then step back to provide Context, introduce the supporting Evidence, explain the Reasoning, and discuss Implications. A Claim-first approach helps anchor the audience to the main finding, ensuring that they follow the story without getting lost in background details.

Context-First: Alternatively, presenters may choose to begin with Context, mirroring the structure of traditional publication order. This method builds curiosity and understanding before diving into a study's main findings. For example, the astronomy article "Water vapour in the atmosphere of the habitable-zone eight-Earth-mass planet K2-18 b" by Tsiaras et al. (2019) announced the first robust detection of water vapor in a sub-Neptune/super-Earth planet that orbits within its star's habitable zone. Water vapor is essential because it indicates the presence of water-based clouds, weather cycles, and – under the right surface conditions – liquid water. Yet such signals are faint and had previously been confirmed only in the scorching atmospheres of hot Jupiter-like planets (e.g., Wakeford et al., 2013). Here, a Context-first talk might start with:

> Astronomers had previously detected the presence of water vapor in several hot-Jupiter atmospheres, but by 2018, not a single temperate, Earth-size or super-Earth world in the habitable zone of a star – except for the Earth itself – had yielded a clear detection. What would it mean to discover water vapor in the atmosphere of another Earth-like world?

By establishing the Context and the knowledge gap first, the audience is primed to appreciate the significance of the Claim when it is revealed. Again, the main consideration with this approach is how much background knowledge the audience has.

11.4.3 Organizing the CCERI Elements

CCERI provides a structured yet flexible framework for presenting research, ensuring that an audience understands:

1. Why the research was needed (Context).
2. What the study found (Claim).
3. How the data support the claim (Evidence).

4. Why and how the logic connects the Evidence to the Claim (Reasoning).
5. What the findings point to as the future next steps (Implications).

By explicitly planning for each of these elements, presenters can create a balanced and engaging talk. The CCERI framework also serves as a helpful checklist during preparation or after drafting a slide deck or outline, allowing presenters to review whether each component is clearly addressed. By integrating the CCERI elements directly, presentations move beyond merely listing results to evaluating the main scientific arguments, understanding the motivation of the research, and assessing its broader implications.

As an illustration, let's imagine we are presenting a talk about the article "Molecular mechanisms for tumour resistance to chemotherapy" by Pan et al. (2014), a study on chemotherapy resistance in cancer cells. A CCERI-structured talk might unfold (paraphrased) like this:

Context: Chemotherapy resistance limits the success of treatment in acute myeloid leukemia (AML). Many resistant AML samples overexpress the anti-apoptotic protein BCL-2.

Claim: This study shows that selective inhibition of BCL-2 with ABT-199 (i.e., venetoclax) reverses drug resistance and triggers on-target cell death in AML.

Evidence: The study reports three major lines of evidence

- Figure 2B shows an ≈80% drop in viable primary AML blasts after 24 h of 1 μM ABT-199, whereas treated controls remain unaffected.
- Assays reveal a >10fold reduction in colony formation when BCL-2 is inhibited.
- Finally, genetic knockdown (inactivating or deleting a specific gene) of BCL-2 mimics the drug effect, while inhibiting the similar proteins BCL-XL or MCL-1 does not alter resistance.

Reasoning: Because loss of viability occurs whenever BCL-2 is specifically neutralized – and not when related proteins are targeted – the authors argue that BCL-2 is a key driver of chemoresistance. Control experiments and profiling data strengthen the causal link, as do numerous previously published studies on BCL-2 inhibition in other types of cancer, including primary CNS lymphoma.

Implications: The study positions venetoclax-based regimens as a strategy to enhance standard chemotherapy in resistant AML, suggesting the next steps focus on clinical trials that combine BCL-2 inhibition with cytarabine or hypomethylating agents (drugs that block the BCL-2 protein, which helps cancer cells avoid death).

The structured outline above ensures a logical flow across the main elements, guiding the audience from problem to discovery to real-world impact. More importantly, the CCERI framework helps presenters avoid common pitfalls, such as:

- Spending too much time on background (Context) and rushing through results.
- Focusing only on data (Evidence) without clearly stating the main Claim.
- Presenting results without explaining why they support the Claim (Reasoning).
- Ending without discussing why the study matters (Implications).

11.5 More Guidance on Presenting the Essential Elements of an Article

As we saw in the previous section, the CCERI framework provides a roadmap for crafting a clear and compelling presentation, especially when using structured approaches to convey each element. Let's explore how to present the CCERI chain in a way that captures attention, enhances understanding, and deepens appreciation for the research.

11.5.1 Presenting Context

Context sets the stage by orienting the audience to the problem and describing why it matters. The challenge is to provide just enough background to engage listeners and ensure they understand the key terms and concepts, without overwhelming them. One practical approach is the **Introduction Funnel**: start broadly with a big-picture statement, then progressively focus on the specific research question (see Chapter 7 for more about the Introduction Funnel).

For example, a presenter introducing the Tsiaras et al. (2019) study reporting the detection of water vapor in the atmosphere of K2-18 b might begin with a wide-angle topical statement:

Liquid water is the signature of planetary habitability; it is essential for all life on Earth and drives weather and atmospheric and surface chemistry processes necessary for element recycling. But can we detect water on worlds orbiting distant stars?

Next, the presenter narrows the focus by posing a rhetorical question:

Astronomers have found water vapor in the atmospheres of hot Jupiters, but no one has confirmed it in the atmospheres of smaller, Earth-like planets in the "Goldilocks" habitable zone. What would it mean to see clear water signatures on such worlds?

Next, the knowledge gap can be stated explicitly to bridge into the study's Claim:

Despite rapid advances in space-based spectroscopy, we still do not know whether habitable-zone Earths or super-Earths retain detectable water vapor in their atmospheres, or what such a detection would tell us about their potential to support liquid water and life on the surface of these worlds.

This sequence kindles curiosity, clarifies the open question, and prepares the audience for the paper's central finding.

Visual guides can help clarify key ideas in the Context through a simple schematic illustration or diagram. In the Tsiaras et al. (2019) example, this could be a visual comparison of the relative sizes of Earth, super-Earths, and hot Jupiters, or it can be an illustration of the various states of water on an Earth-like planet, emphasizing that these measurements only probe the upper atmosphere. These guiding illustrations can help the audience grasp the research motivation more easily without excessive detail.

11.5.2 Presenting the Claim

A clear Claim is like a good GPS signal in an unfamiliar location: It tells the audience exactly where the talk is heading. The Claim should be stated clearly and adhere to the article's stated findings. A good practice is to use a clear, declarative sentence of the Claim as a slide title or the entirety of the slide itself. For example, when presenting the Claim for the Tsiaras et al. (2019) article, a simple statement such as:

New telescope data confirm water vapor in an exoplanet's atmosphere. This approach ensures that the audience immediately registers the main takeaway (Garner & Alley, 2013). Framing the Claim concerning expectations makes it even more impactful. For example, the speaker can elaborate by stating:

> Researchers have long hypothesized that water vapor should be present in the atmospheres of super-Earth planets, but observations haven't been sufficient to detect these signatures. Now, new observations obtained with Hubble Space Telescope have confirmed this prediction.

Moreover, if the findings were unexpected, the presentation can add a "superlative" to the stated Claim:

> Since previous attempts at this measurement have failed, the detection was an unanticipated breakthrough.

Visually stating the Claim clearly and succinctly, while orally providing more details, helps audiences understand both the key finding of the article and appreciate the scientific context and significance.

When presenting multiple Claims or findings, it is helpful to enumerate them; for example, the presenter may state:

> This study makes two key claims: (1) detection of water vapor, and (2) characterization of a hydrogen-rich, potentially temperate atmosphere

The slide could list these in bullet form:

Main findings:

- First detection of water vapor on a temperature super-Earth
- First characterization of a hydrogen-rich atmosphere on an Earth-like world

This structured approach ensures that the audience follows along as each Claim is explored with supporting Evidence and Reasoning.

11.5.3 Presenting Evidence

Evidence is often at the heart of a research presentation, helping to translate raw data into scientific insight. The key to presenting Evidence effectively is centering, sequencing, clarity, and relevance. One approach is to introduce each piece of Evidence by explicitly stating what it tests. Instead of jumping into a figure, a presenter discussing the article "Requirement of mammalian DNA polymerase-beta in base-excision repair" by Sobol et al. (1996) might first provide a paraphrased summary of the methodological approach:

To determine whether DNA polymerase-β (Pol β) enhances DNA repair, researchers measured the repair rates in treated and deficient cells.

A more in-depth paraphrased version, delivered orally, might look like this:

Sobol et al. (1996) asked whether mammalian cells lacking DNA polymerase-β can still repair alkylation damage. They compared the rate at which wild-type fibroblasts and Pol β-knockout fibroblasts removed single-strand breaks after methyl-methanesulfonate exposure. Let's look at those kinetics.

The presenter might frame this description with a figure from the article that illustrates the experimental procedures, a task we look at in more detail below.

If multiple lines of Evidence are presented, it helps to emphasize how they reinforce one another. For example, for Sobol et al. (1996):

Re-introducing human Pol β into the knockout cells restores the fast-repair curve, while in vitro incision assays show the same dependence. Together, biochemical and genetic results reinforce the conclusion that Pol β is a pivotal catalyst in base-excision repair.

Engaging the audience with the details of the Evidence can further improve comprehension. Asking a guiding question, such as: *What would it mean if the treatment and control groups had the same outcome?* This invites listeners to think critically before revealing the observed data trend.

11.5.4 Presenting Reasoning

Reasoning connects Evidence to Claims, transforming data into an argument by explaining why the authors interpret the data as they do to make the Claim. This step is often overlooked in a research presentation, yet it is essential to both summarizing and critiquing an article's primary findings. A good approach to presenting Reasoning is to walk through the logical chain that connects data to findings. As an example, let's consider the Reasoning in the article "The Effects of Light Intensity on the Growth Rates of Green Algae" by Sorokin & Krauss (1958), which demonstrated that photosynthetic light energy is the primary driver of algal growth among several species. The inference chain for this Claim can be summarized as follows:

Cultures exposed to 1,700 foot-candles double their cell density in 24 hours, **whereas** cultures kept in complete darkness show virtually no increase. **Therefore**, the energy that drives algal growth **must come from** the light-dependent reactions of photosynthesis.

Here, we've highlighted the logical connecting words. Symbolically, this argument can be summarized as:

- light ⇒ growth (implication)
- ~light ⇒ ~ growth (converse)
- both being true, light ⇔ growth (light is necessary for growth)

(See Chapter 5 for a review of logic symbols). While this seems like a sound enough argument, we also need to explain how and why the experimental design isolates the mechanism:

All flasks contained identical mineral media and were bubbled continuously with CO_2-enriched air, **so nutrient availability and carbon source were held constant**. Light intensity was the only variable that **allowed the authors to discriminate between phototrophic and heterotrophic growth**.

Finally, a good Reasoning discussion must also examine and rule out alternatives:

Could residual organic carbon in the medium support limited heterotrophic growth? Sorokin and Krauss (1958) addressed this by adding acetate to a second dark control; **even with the extra carbon, cell counts remained flat** (i.e., as shown in their Table 1), refuting the heterotrophic explanation.

By highlighting each of these key Reasoning steps – logical chain, isolated variables, and ruling out alternative explanations – the presentation demonstrates that the authors' Claim is scientifically warranted rather than merely plausible.

11.5.5 Presenting Implications

Implications provide the *why it matters* moment, leaving the audience with a clear understanding of the study's significance. A clear set of Implications provided at the end of a presentation broadens the conversation, connecting the research to the larger scientific, ethical, or practical applications.

Let's consider the article "The R136 star cluster hosts several stars whose individual masses greatly exceed the accepted 150 $M_\odot$ stellar mass limit" (a good Claim title) by Crowther et al. (2010). This article reports the discovery of stars in the R136 cluster (part of the Tarantula nebula in the Large Magellanic Cloud, a satellite galaxy of the Milky Way) with initial masses up to about 265 times the mass of the Sun, far beyond the long-assumed 150 solar mass ceiling. Here's how a presenter might frame the Implications of this work:

While it has long been assumed that there is a hard upper mass limit for stars, Crowther et al. (2010) identify several stars in R136 that reach nearly twice the canonical limit. As such, current theories of star formation and evolution, including radiation-pressure feedback during prestellar collapse, are incomplete, while stellar-population models – especially for the early Universe – need revision.

The presentation could frame the key Implications of this study in a series of bullet points. For example, the theoretical Implications may be listed as:

- Star-formation feedback models must accommodate accretion histories that build super-massive protostars without blowing away infalling gas.

- Chemical evolution models must accommodate the death of massive stars, which may include pair instability or hypernova explosions, thereby changing galactic enrichment calculations.
- Cosmological reionization models will need to accommodate the higher flux of ionizing photons from these massive stars.

Or the future directions may be detailed as:

- High-resolution infrared spectroscopy of young super-clusters beyond the Local Group is needed to determine whether 250 solar mass stars are rare or commonplace in metal-poor galaxies.
- Radiation-hydrodynamic simulations that couple disk accretion, stellar winds, and magnetic fields are needed to test how such massive stars assemble.
- JWST time-series photometry could detect early super-luminous supernovae, which would validate the pair-instability end states predicted for progenitors with masses greater than 200 solar masses.

Finally, the presentation could close with a callback:

> So, are we forced to raise the stellar mass cap? Crowther et al. (2010) make a compelling case that nature already did – and now our models must catch up.

By explicitly linking an article's findings to broader theoretical concepts and concrete next steps, the presenter leaves the audience with a vivid sense of why the article is important and what new research questions it raises.

11.5.6 Transitioning between Elements

A smooth flow between CCERI components prevents presentations from feeling disjointed and helps listeners follow the logic. Using transition phrases ensures logical continuity. Here are some examples:

- **From Context to Claim**: "Given this background, the researchers set out to test…"
- **From Claim to Evidence**: "In support of this Claim, the authors conducted three experiments."
- **From Evidence to Reasoning**: "These results indicate a strong effect, but why does this happen? The authors propose …"
- **From Reasoning to Implications**: "With this understanding, we can now ask: what does this mean for the next experiment and the field?"

11.5.7 Including Figures, Equations, and Data Visualizations

Research articles typically include information-rich elements, such as figures, graphs, tables, equations, and other datasets. These are important elements to include in a research presentation. However, a presenter needs to *curate*, *simplify*, and *guide* the audience through the most meaningful figures and data. Data should be presented in a way that the audience can understand the insight; otherwise, their eyes will glaze over (Lamman, 2024). Using visuals such as charts and plots rather than text or tables can make the core message more accessible.

11.5.7.1 Selecting Key Figures

Choosing which figures to present is the first step. Instead of displaying every graph and table in an article, focus on those that directly support the study's main Claim. Typically, these include the "key graph" that illustrates the main Claim or explains the primary argument.

For example, in a study on antibiotic resistance, the most important figure might be a growth curve showing bacterial survival rates before and after drug treatment. A clear graph depicting this trend conveys the key finding far more effectively than a table of numerical values. Similarly, in a materials science study, an atomic-scale microscope image of a newly synthesized nanomaterial provides visual proof of its structure, making the discovery immediately tangible for the audience.

When selecting figures, consider whether the audience needs to see an experimental setup to understand the Evidence and methodology. In a presentation on an astronomical discovery, a series of images illustrating the analysis steps can help articulate how the discovery was detected. In a biochemistry presentation, a schematic of a protein binding interaction could clarify a key step in synthesis.

11.5.7.2 Presenting Figures Clearly

Once the most relevant figures have been selected, the next challenge is presenting them in a way that maximizes clarity, taking into consideration:

- **Legibility:** Ensure labels, axes, and color contrasts are visible, even from a distance.
- **Step-by-step explanations:** Define what is being plotted before interpreting the result.
- **Simplification when necessary:** Remove or fade out extra lines or plot elements that do not contribute to the primary argument.

Copying a figure directly from a paper is not always ideal, especially if the labels are too small or the image contains extraneous details. If needed, reformat the graph for readability by enlarging text, highlighting key trends, or cropping or annotating to focus the audience's attention on the most relevant portions of a figure.

When displaying a graph in a presentation, assume that the audience is seeing it for the first time. Instead of immediately discussing a graph's implications, guide the audience through its construction, step by step.

1. **Orient the axes:** Show and explain the axes, stating both the quantities being plotted and their units.
2. **Highlight control data first**: In a comparison study, start with the control or "expected" trends as the baseline condition.
3. **Then focus on the experimental data:** Shift the focus to the experimental data and highlight key differences from the baseline trend.

4. **Bring in models:** If the study includes comparisons to theoretical or empirical models, these should be introduced last so that the audience can compare both baseline and experimental results to predicted trends.
5. **Link back to the Claim:** Once a graph has been thoroughly described, the presenter can link back to the main Claim of the study, using the comparison of control to experimental data or experimental data to model predictions as the anchor.

This sequence is an example of **progressive disclosure**, a process that reveals data in stages to prevent information overload. As an example, consider the article "HST hot Jupiter transmission spectral survey: detection of water in HAT-P-1b from WFC3 near-IR spatial scan observations" by Wakeford et al. (2013). When describing the key graph in the article (i.e., Figure 14 in the paper), which compares a transmission spectrum to a series of models, a presenter can walk through the plot in the following manner. We have highlighted the key phrases and sentence builders that do the work of progressive disclosure.

This figure shows the Hubble/WFC3 G141 transmission spectra of HAT-P-1 b from Wakeford et al. (2013). **The x-axis is** wavelength and spans 1.1–1.7 μm; the y-axis is the planet-to-star radius ratio. Absorption features in this atmosphere of the planet appear to represent increases in the planet radius, and hence, as positive bumps in the spectrum. **The data points show** a broad bump centered at 1.45 μm, **which aligns with** the H_2O absorption band and is detected **at a level of** ≈250 ppm with > 5 σ significance. **This feature aligns with** water vapor features **present in several different models, indicated by the colored lines**, confirming detection of this feature.

Clarifying the salient features of a figure can be enhanced by layering annotations or call-outs to focus the audience's attention. For example, a color overlay can be used to help distinguish between control and experimental conditions. At the same time, an arrow can be brought in to direct attention to a significant data point or feature in the data.

11.5.7.3 Presenting Data without Overwhelming the Audience

Scientific data can be powerful, but when presented in raw form, it can quickly become overwhelming. The goal of an effective presentation is to communicate the insight inferred from the data, rather than every number or calculation. For example, if a paper includes a table of 10 statistical measures, but only two are critical to the central argument, the presenter could show a bar graph comparing those two measures rather than the whole table.

The phrase "less is more" certainly applies to data-heavy slides. A well-designed visualization often communicates complex results more effectively than raw numbers. If discussing statistical results, always clarify whether a difference is statistically significant or within experimental uncertainty. Simply stating that the treatment group showed higher survival rates than the control is incomplete unless followed by a qualifier, such as "this effect was statistically significant ($p < 0.01$)," indicating a meaningful difference rather than random variation.

Presenters also benefit from posing questions to engage the audience. When showing a bar chart comparing two conditions, a presenter might ask:

What would it mean if these bars were the same height? It would suggest no effect. However, here we see a clear separation, indicating a response to the treatment.

Practicing what to qualify, cut, and keep goes a long way toward helping audiences interpret data actively, making them more engaged participants in the discussion.

How data and text appear together on slides can also be used to support comprehension. One proven strategy is the **assertion-evidence approach** (Garner & Alley, 2013), where generic slide titles – such as *"Experimental Results"* – are accompanied by a short sentence that clearly communicates the point of the figure. For the Wakeford et al. (2013) example, potential slide titles for the key graph could include:

- The transmission spectrum reveals a clear signature of H_2O vapor.
- H_2O vapor is detected in HAT-P-1b at >5 σ significance.
- HAT-P-1b's transmission spectrum matches atmosphere models with H_2O vapor.

Slide titles like these enhance the clarity of the presentation. They ensure each slide delivers a single, clear message. They also remind audience members of the main point of the slide, and keep the presenter focused on the narrative thread. Indeed, studies show that audiences retain and understand content better when slides follow an assertion-evidence format (Garner & Alley, 2013).

11.5.7.4 Presenting Equations

Equations play a crucial role in many quantitative scientific fields, from physics to computational biology. However, too many equations in a presentation can create confusion rather than clarity. The best approach is only to include an equation if it directly contributes to understanding the Claim, Evidence, or Reasoning.

For instance, if a paper presents a new empirical model, it may be necessary to display the equation that defines it. Similarly, in a talk about climate modeling, the Stefan–Boltzmann equation might be shown to clarify how a model uses radiative flux balance to determine surface, ocean, and atmospheric temperatures.

To present an equation effectively:

- Define each variable and its meaning,
- Use animations or progressive highlights to break down the equation step by step, and
- Explain the equation conceptually.

For example, one of the key equations used in fluid dynamics research is the continuity equation:

$$\frac{\partial \rho}{\partial t} + \vec{\nabla} \cdot (\rho \vec{v}) = 0$$

If portions of the audience are not familiar with this equation – which may be the case in a class setting, journal club, or research group presentation – the speaker can take a moment to explain the equation and its conceptual meaning. For example (again, we show the sentence builders in bold):

This is the continuity equation, **which describes** how mass density (ρ) changes in a volume. **The left term** describes the time rate of change of density, while **the right term** describes the net flow of mass (density times velocity) into or out of the volume. Since mass is a conserved quantity – it cannot be created or destroyed – the local mass density can only change with time if there is a net flow of mass into or out of a volume. **Similar equations can be written for** energy, charge, and other conserved quantities.

This approach ensures that every member of the audience has a sufficient understanding of the essential equations in a research study and are not simply lost in algebraic detail.

11.5.7.5 Presenting Interactive and More Complex Data

Research that involves videos, animations, or 3D models can be engaging when used strategically. A simulation showing fluid flow around an airfoil in an aerodynamics study, or a real-time image of a cell dividing, can provide compelling Evidence or Reasoning. However appealing, these media must still serve the core CER narrative. A beautifully rendered molecular animation is useful only if it clarifies a concept that words or static images cannot convey equally well. For example, a figure displaying a complex process through a series of multiple panels can be overwhelming for an audience to view at once. Instead, revealing each stage of the process through a single image at a time helps draw attention to specific features within the process while allowing for an understanding of the overall sequence.

In summary, figures, equations, and data visualizations are powerful tools in scientific presentations. However, their effectiveness depends on how they are utilized. A well-prepared presenter selects figures that reinforce the Claim or key elements of the central arguments, presents them clearly and sequentially, and ensures that every visual element is tied to the narrative. A good presenter also anticipates potential confusion and addresses it proactively through annotation, simplification, and scaffolding. The key is to know what is most important and then cut everything else. Multiple drafts can help to winnow and focus the presentation. By curating key figures and explaining data with clarity, presenters ensure that their audience sees the science as a coherent story, rather than just a collection of numbers and graphs.

11.6 Integrating and Balancing Critique

Presenting research articles involves more than summarizing findings. In academic settings, engaging with a study critically – such as comparing it with other work, evaluating its methods, and comparing its Implications – adds depth to scientific discourse. Using the CERIC framework, presenters can seamlessly embed critical analysis, ensuring that their audience both understands a study and can think critically about its place in the broader scientific conversation.

Presenting a paper as a story in five acts – Context, Claim, Evidence, Reasoning, and Implications – embeds a quick, pointed critique at each stage. Let's return to the study by Tsiaras et al. (2019) reporting the first robust detection of water vapor in the atmosphere of the temperate exoplanet K2-18 b. Imagine we are giving a journal club talk that includes a critique of the article. How do we build it in? A good approach is to consider three actions for each of the CERIC elements: *prepresentation check*, *embed critique*, and *ask the audience*. Let's apply these to our Tsiaras et al. (2019) presentation:

1. **Begin with the Context:** Before 2019, water vapor had been detected only in the atmospheres of hot Jupiters. No one had confirmed its presence in the temperature atmospheres of smaller Earth-like planets residing in the habitable zone of its star, which could be a sign of life.
   a. *Prepresentation check*: What was the state of the field before this study? Had there been any detections of water vapor reported prior to this work? What would such a detection mean?
   b. *Embed critique*: While it is true that no temperate super-Earth had yielded a water-vapor signal before 2019, smaller – and hotter – planets such as the warm Neptune HAT-P-11 b had already shown H_2O absorption in Hubble data (Wakeford et al., 2013). The real novelty, therefore, is water vapor in a habitable-zone temperature atmosphere, not merely planet size.
   c. *Ask the audience*: What is the relevance of finding water vapor in a temperate atmosphere?
2. **State Claim:** The authors assert this is the first clear water signal in a habitable-zone super-Earth.
   a. *Prepresentation check*: Is this Claim novel, specific, and robust?
   b. *Embed critique*: The authors identify K2-18 b as the first habitable-zone water detection, but earlier observations and modeling of HAT-P-11 b have hinted at water vapor in planets outside the habitable zone. There is also a contemporary study by Benneke et al. (2019) on K2-18 b that also reports this feature, as well as clouds. Additionally, other planets may have water vapor in their atmospheres but exhibit weak spectral features due to the presence of clouds.
   c. *Ask the audience*: Given prior detections of water vapor on other worlds, how unique and important is this finding?
3. **Move to Evidence:** Their Hubble Space Telescope/WFC3 spectrum shows a broad 1.4 μm bump with a 3.6σ significance.
   a. *Prepresentation check*: Are the data of sufficient quality to reveal this feature? Are there supporting observations of Evidence that are brought in?
   b. *Embed critique*: The 1.4 μm peak is compelling, but the 3.6σ significance is only just above would be considered significant. Residual features near 1.2 μm indicate unmodelled scatter that could bias the retrieval.
   c. *Ask the audience*: Is this detection sufficiently robust to warrant a Discovery Claim?
4. **Add Reasoning and trace the logic:** The paper links the 1.4 μm bump exclusively to H_2O using atmosphere models.
   a. *Prepresentation check*: Are there alternative molecular species, clouds, or other effects that could give rise to this feature?

   b. *Embed critique*: Attributing the bump solely to water vapor overlooks the possibility of clouds or CH_4–H_2 collision-induced absorption that can change the local continuum.
   c. *Ask the audience*: Have the authors sufficiently ruled out other possibilities to claim the feature is definitely water vapor?
5. **Evaluate Implications:** The authors suggest K2-18 b may be "potentially habitable," yet interior models predict a thick H_2 envelope that could preclude a liquid-water surface.
   a. *Prepresentation check*: Are their statements on habitability proportionate to existing interior models?
   b. *Embed critique*: Calling K2-18 b "potentially life-supporting," but subsequent work by Madhusudhan et al. (2020) indicates that the pressure may be too high at the surface to support a liquid water ocean.
   c. *Ask the audience*: Have the authors made a strong enough case to speculate on the habitability of K2-18b based on the detection of water vapor? What more would we need to know?

Not all these critiques are necessary to show in a presentation. Choosing two or three of the highest-impact critiques – such as signal contamination, model assumptions, and overextended habitability claims – focuses the audience on the major elements of the paper and provides an opportunity to turn a critical lens on these key elements.

11.7 Worked Example with the CERIC Presentation Template

As in Chapters 8 and 9, we have included a CERIC Presentation Template in this chapter's supplemental materials to scaffold the creation of an article review. This template is in Claim–Context–Evidence–Reasoning–Implications order, and provides for spaces to add relevant figures, tables, plots, or other media elements that address each of these elements, as well as a critique section.

To illustrate the use of this template, we undertake a worked example of the article "Genome-wide SNP analysis reveals an increase in adaptive genetic variation through selective breeding of coral" by Quigly et al. (2020). This article investigates whether a novel coral breeding technique can enhance heat tolerance in corals, thereby improving their survival in warming oceans. Let's imagine we are preparing a journal club presentation on this article

11.7.1 CCERI Presentation Structure

To match the Presentation Template, let's adopt a Claim-first structure for our CERIC elements:

Claim: The authors propose that selective breeding can increase coral heat resistance, making reefs more resilient to climate change.

Context: Coral reefs are in rapid decline due to rising temperatures (e.g., Heron et al., 2017; Hughes et al., 2018). Interventions are underway, including select breeding, hybridization, and microbial treatments. There is evidence that corals are adapting to increase, with several candidate genes identified as potential adaptations (Louis et al., 2017).

Key questions: What are the specific genetic changes that are taking place to allow corals to evolve their heat tolerance, and could combining selective breeding with genetic screening further improve coral resilience?

Evidence: Coral eggs and sperm were collected from two colonies in the Great Barrier Reef, bred within colonies and cross-bred, and genotyped. Larvae were raised at two different temperatures, and growth, bleaching, and survival were measured after 70 days. Statistical PCoA (Principal Coordinate Analysis) is used to identify correlations.

Reasoning: Key logic: If breeding increases tolerance, then survival should improve under heat stress. Since the control group (no cross-breeding) had lower survival rates, this supports the hypothesis.

Implications: If cross-breeding can be applied on a large scale, it could be used in reef restoration efforts.

With this in place, let's step through the CERIC Presentation Template slides.

SLIDE 1: TITLE SLIDE

- **Title:** "Genome-wide SNP analysis reveals an increase in adaptive genetic variation through selective breeding of coral"
- **Authors/Journal:** Quigly et al. (2020, *Molecular Ecology*, 29, 2176)
- **Main Claim:** Selective cross-breeding (assisted gene flow) between warm-adapted and cooler-reef coral may enhance coral resilience to warming oceans caused by climate change.

SLIDE 2: CONTEXT

Prior Work:

- Coral reefs are in rapid decline due to rising temperatures (e.g., Heron et al., 2017; Hughes et al., 2018).
- Possible interventions include select breeding, hybridization, and microbial treatments (van Oppen et al., 2015; Chakravarti et al., 2017; Morgans et al., 2020).
- Corals are adapting, and several candidate adaptation genes have been identified (Louis et al., 2017).

Key Questions:

- What genetic changes are taking place to allow corals to evolve their heat tolerance?
- Can selective breeding with genetic screening improve coral resilience?

Figure: Image of coral bleaching event

SLIDE 3: EVIDENCE

Experimental Setup:

- Coral eggs and sperm were collected from two colonies in the Great Barrier Reef
- These were bred and cross-bred, and genotyped, and larvae genotyped
- Larvae were raised at 27.5°C and 31°C for 70 days, and growth, bleaching, and survival were measured

Analysis:

- PCoA revealed 5 families corresponding to 3 bred and 2 cross-bred groups
- Compared alleles and heat-tolerance metrics.

Figure: Figure 1 showing PCoA clusters for five families

SLIDE 4: REASONING

Key Logic: Improved survival aligns with the transfer of genetic alleles from warm-reef alleles

- **Key points:**
  - Genome-wide SNP data confirmed transfer of warm-reef alleles and increased heterozygosity
  - These species had better outcomes in the warmer environment
- **Confounders:** Longer captivity time did not influence results.
- **Limitations:**
  - Short-term offspring study – unclear if effects persist into adulthood.
  - Alternative explanations (e.g., maternal environment and symbionts) were not directly tested and remain open questions.

Figure: Combination of Table 1 showing genetic diversity and Figure 2 showing allele variation

SLIDE 5: KEY GRAPH

Figure: Figure 4 showing the survival, bleaching, and growth distributions for cool and warm temperature conditions.

SLIDE 6: IMPLICATIONS

Key Implication: Assisted gene flow could complement reef restoration efforts, but long-term field trials are needed to confirm its persistence and ecosystem-scale benefits.

- **Potential:** Scaling this method could strengthen reef resilience.
- **Scientific contribution:** Suggests inter-generational acclimatization as a new adaptation strategy.
- **Ethical and logistical concerns:** Could this delay urgent climate action?
- **Policy risk:** Over-reliance on biotechnology may reduce the urgency for CO_2 reduction.

SLIDE 7: CRITICAL EVALUATION

Strengths:

- The study utilized more than 20 unique crosses (fresh and cryopreserved), involving thousands of larvae.
- Many tests yielded significant correlations ($p < 0.01$).

Weaknesses:

- Most assays were short term (<4 weeks),
- Field validation is lacking; larvae were raised in aquaria, and ocean conditions are more complex.
- Adult coral survival was not included.

Subsequent studies:

- Caruso et al. (2021) summarize methods for identifying heat-tolerant coral strains
- Humanes et al. (2021) propose a framework for the selective breeding of tropical corals
- Macadam et al. (2025) extended this study to the Great Barrier Reef, including more colonies, higher temperatures, and progression into adult stages, and found mixed results for hybridized species.

SLIDE 8: TAKE HOME POINTS AND Q&A

Summary: Quigly et al. (2020) contribute to coral conservation efforts by demonstrating how hybridization can increase coral resilience, but follow-up work is needed, and broader climate action remains essential.

- **Further work:** additional field studies and studies with adult corals are needed.
- **A call to action:** Integrating biological and policy solutions is essential to saving coral reefs.
- **Possible Figure**: Image of healthy and bleached coral side by side.

11.7.2 Integrating Critique Throughout

While we used the Presentation Template as is, with a dedicated slide to critique, we could also indicate these critiques throughout the presentation, inviting discussion from our audience on how they interpret these concerns.

- **On Evidence:** Larvae were raised in aquaria; would these results hold in wild conditions where additional stressors exist?
- **On Reasoning:** Temperature tolerance increased, but was this strictly due to genetics, or did controlled environmental factors contribute?
- **On Implications:** The study focuses on assisted gene flow adaptation for a narrow set of conditions and a limited timeframe, but does this breeding approach ensure long-term reef resilience?

11.7.3 Bridging to Discussion

At the end of the talk, the presenter returns to the big picture:

> Quigly et al. (2020) contribute to coral conservation efforts by demonstrating how hybridization can increase coral resilience, but follow-up work is needed, and broader climate action remains essential.

This approach ensures that the audience leaves with a clear understanding of both the study's findings and its place within the broader scientific conversation. By structuring critique within CERIC, presentations become more than a summary and help trigger a more dynamic scientific discussion.

At this point, the presenter could transition to audience Q&A. An effective approach might lead with some guiding questions to stimulate audience discussion. For example:

- How could this study be expanded upon to test for adult survival rates?
- How could we assess the outcomes of coral hybridization in the wild?
- Are there other global coral populations that this approach should be examined on?

Remember that research presentations, in whatever environment they occur, are ultimately about situating research within a broader scientific dialogue.

11.8 Chapter Key Takeaways

1. **Scientific research is ultimately about discourse:** Science thrives on conversation, from informal coffee chats to the primary literature. Hence, the goal primary

goal of a research presentation should be to communicate, critique, challenge, discuss, and revise ideas that emerge in the scientific literature.

2. **Different presentation formats support different skills**: Class presentations emphasize clarity and narrative structure, journal clubs foster collaborative critique, and group meetings connect papers to active research. Each format supports the development of essential scientific practices, such as presenting findings, critiquing reasoning, and framing implications.
3. **Robust scientific discussions are structured and collaborative:** A good scientific discussion feels like a lively debate, but without direction, these discussions can unravel into meandering conversations, an excessive focus on criticism, or dominance by a few voices in the conversation. Setting clear goals for discussions, defined participation roles, proper moderation, and integrating prediscussion preparation can foster productive discussion around a research article.
4. **Summaries and critiques provide different formats for presenting research:** A summary presentation examines an article's contribution to the field; a critique presentation analyses an article's Claims, Evidence, and Reasoning to assess its accuracy, significance, and limitations. These approaches can be used together, and knowing what the expectations of a presentation are helpful to guiding the mindset, preparation, and format for the presentation.
5. **Structured frameworks enhance comprehension**: The CCERI variation (Context, Claim, Evidence, Reasoning, Implications) helps readers and presenters transform complex research into coherent, accessible stories. This framework supports both presenting and critically reviewing research by organizing scientific arguments into a logical flow.
6. **Presentation strategies differ across the CERIC elements:** Context is ideally structured as a funnel (general → specific). Reasoning should highlight the main logical steps and address confounding factors or alternative hypotheses. Implications might be organized by scientific, ethical, or social impacts. Figures, tables, equations, and other data formats are useful when not overwhelming and properly guided by the presenters. Critique can be integrated throughout a presentation or left to the end, with guiding questions used to foster critical research discussions that presentations should aim to inspire.

12 Using CERIC for Literature Review

> Conducting effective literature reviews is essential to advance the knowledge and understand the breadth of the research on a topic of interest, synthesize the empirical evidence, develop theories or provide a conceptual background for subsequent research, and identify the topics or research domains that require more investigation.
>
> —Paré et al., 2015, p. 183

12.1 Overview

Chapter 12 synthesizes the lessons of Chapters 8–10 for the production of a complete and efficient literature review. Literature reviews are distinct from summary works, such as annotated bibliographies, as they tie existing literature into a narrative arc with a distinct Claim addressing the current state of a field or its controversies or knowledge gaps. Literature reviews appear in the introductions of research articles, proposals, dissertations, and as stand-alone reviews. The chapter begins by applying the comparison and critique strategies of Chapters 9 and 10 to larger sets of literature by mapping to a coherent narrative. It also explores practical and ethical considerations in the development of literature reviews, considering readability, tone, and paraphrasing. The chapter concludes with a sample text drawn from the introduction of an article to illustrate how a literature review is used to motivate a new study.

12.2 Defining a Literature Review

A **literature review** is an academic discussion that critically synthesizes published work to explore what a field knows, where it disagrees, and what remains unknown (Saunders & Rojon, 2011). As illustrated in Figure 12.1, a literature review is distinct from an annotated bibliography, which summarizes and evaluates individual articles; a single-article critique, which probes one study in depth; or a comparison study, which focuses on the main themes of just a few related articles. A literature review synthesizes dozens – sometimes hundreds – of articles in a field or a broad topic, across shared themes, various methodological approaches, and potentially conflicting findings, to trace intellectual progress and spotlight gaps ripe for future investigation (Paré et al., 2015). Because a review integrates many studies, it demands more than just cataloguing the research literature. It encompasses a broad contextual base,

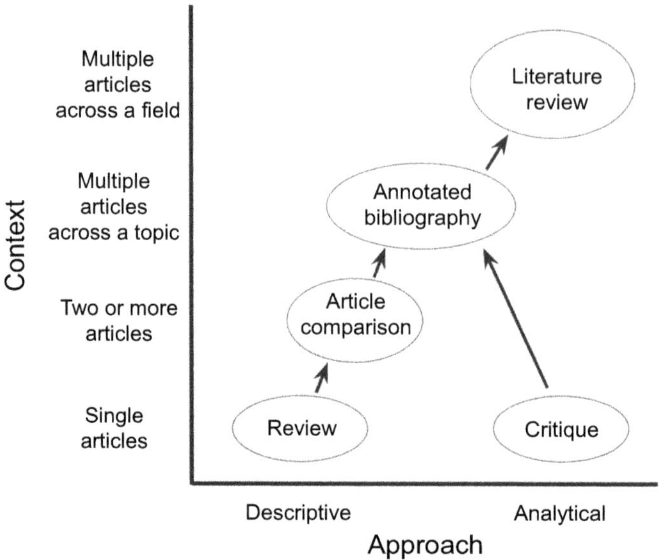

Figure 12.1 Hierarchy of literature overviews organized by approach (descriptive versus analytical) and context (one, few, or many articles).
The diagram ranks types of literature overviews by breadth and analytic depth.

examines interrelated arguments, compares supporting and competing Claims, and delves into a variety of Implications to assess where a field is and where it is headed.

Most importantly, literature reviews integrate disparate studies to generate new, often broad Claims about a field: its intellectual heritage and development, its current consensus and discrepancies, and the knowledge gaps that remain to be explored. Literature reviews establish a direction in which new studies may emerge, justify that path with the Evidence and Reasoning that arises from the existing literature and consider the Implications of these future studies.

As we will see in this chapter, the CERIC framework transforms a literature review from a firehose of information into trackable and focused units of analysis and synthesis.

Box 12.1 Common Confusion: Annotated Bibliographies, Articles Comparisons, and Literature Reviews

We have already seen examples of article summaries (Chapter 8) and article comparisons (Chapter 9). How do these differ from an annotated bibliography or a literature review? Here are a few points of distinction to remember:

- An article summary is a review of a single article to identify its main argument, methods, findings, and conclusions.
- **Article comparisons compare a discrete set of studies**, typically focused on a single issue, common problem, or an immediate development arc. While article

comparisons may reference the broader research through comparisons of the articles' Context and Implications, the scope is strictly on the studies being compared.

- **An annotated bibliography is a set of self-contained summaries**, where each entry provides a short description and possibly an evaluative note. The annotated bibliography is a distillation of individual primary sources and does not go into considerable depth on how the studies relate to one another.
- **A literature review is a full synthesis** that weaves several related articles into a common discourse. Instead of merely listing results in the form of "Smith (2020) found ...; Jones (2021) found ...," a literature review clusters papers around consistent or competing Claims, similar lines of Reasoning, equivalent Evidence, or related Implications, framing all of these elements within a broad Context and creating a unifying narrative for the development of direction of a field.

Here is a short summary to distinguish key features of these common research reviews:

Table 12.1 Comparison of writing features across annotated bibliographies, article comparison papers, and literature reviews, highlighting differences in unit of writing, reader takeaway, and common pitfalls.

Feature	Annotated Bibliography	Article Comparison	Literature Review
Unit of writing	One paragraph per source	One or two paragraphs for all sources	One paragraph per idea or theme
Reader takeaway	What does each article say?	How do these articles compare?	What do these articles lead to?
Common pitfall	Laundry-list of findings	Overly focused, missing broader literature	Can lack a clear direction or narrative arc

12.3 Where Literature Reviews Appear in Science

Literature reviews are prevalent throughout the scientific literature, and in non-peer-reviewed "gray literature," such as proposals and dissertations. They are never one-size-fits-all documents, with scope, depth, and language depending on the audience, the goal of the review, and whether the review is stand-alone or part of a research study. Let's explore the most common places where we can find – and will write – literature reviews.

12.3.1 Literature Reviews in Research Articles

The most common place to find a literature review is in the Introduction (the I in IMRaD) section of a research article. Up to now, we have categorized this information as the Context of a study, but its generation ultimately derives from the practice of literature review. The goal of this type of literature review is to motivate the present study and identify the key knowledge gap that the study seeks to address. A well-developed introductory literature review incorporates all the relevant research related to the study, including competing Claims. By synthesizing this work into a coherent rationale authors can better convey the importance, scope, and need for their study. We will evaluate the introduction of an astronomy article as a literature review in our worked example at the end of the chapter.

More limited literature reviews may appear in other sections of an article. For example, if the study utilizes a new or complex methodological approach, it may include a review of competing approaches and their limitations as justification. If comparative analysis forms a core part of the Reasoning, another limited review may appear in the Discussion section to motivate which prior datasets or models are being used as comparitors and why.

12.3.2 Literature Reviews in Research Proposals

When you ask an organization such as the National Science Foundation (NSF) or the National Institutes of Health (NIH) for research money or access to resources, the request comes in the form of a research proposal. Proposals can take on many forms and lengths depending on the ask, from a single paragraph to requesting a conference talk to a 100-page document to win a major grant. A common requirement is to demonstrate your knowledge of the relevant scholarship. As such, all proposals contain some form of introductory section in which you demonstrate to the reviewers that you understand the state of the field and have identified an essential knowledge or resource gap that you intend to fill.

Successful writers reverse-engineer a proposal solicitation to identify what sphere of research, and possibly what specific knowledge gap, the proposal aims to support. For example, NSF's 2023 solicitation, "Biodiversity on a Changing Planet" (NSF, 23-542, 2023a), requests that applicants "synthesize pattern- and process-based approaches to functional biodiversity and biodiversity dynamics." That single sentence encodes the foundational knowledge on the state of the field ("pattern- and process-based approaches") that should be communicated in the proposal's Introduction section. By concisely and accurately describing the current approaches to studying biodiversity dynamics and tying these to the proposed study design, a proposal has a higher chance of achieving the funder's goals and winning the competitive grant.

12.3.3 Literature Reviews in Dissertations

For graduate researchers, the literature review is both a rite of passage and a mandatory deliverable when completing a research-based degree program. Here, the goal of the review is to prove mastery of the subject, as well as justify the dissertation's purpose, methodological approach, and relevance to the broader field. Doctoral and master's dissertations devote an entire chapter to the literature review because it demonstrates to the dissertation committee that the candidate:

1. Understands the scholarly literature in which the dissertation is framed,
2. Can evaluate that research critically, and
3. Has identified a defensible gap that justifies a new study.

Without a complete and convincing literature review, dissertations that contain elegant research findings might not pass a final defense (Hart, 1998; Randolph, 2009). For example, a study of 30 education dissertations from three US universities by Boote and Beile (2005) found that only two earned a rating of "excellent." The most common weaknesses were superficial treatment of seminal studies, inadequate synthesis, and failure to signal the knowledge gap. Their conclusion is blunt: "A dissertation is only as good as its literature review" (p. 5).

Graduate handbooks echo this verdict. Hart's (1998) classic *Doing a Literature Review* frames the chapter as the "intellectual scaffold" (p. 14) on which every research decision rests. Wisker (2015) urges supervisors to interrogate a draft review with three prompts that echo the CERIC framework:

1. Does it map the field clearly (Context)?
2. Does it reveal a genuine puzzle (Context: Rationale)?
3. Does it justify the proposed method (Reasoning)?

As an illustration, let's wade into the 2013 doctoral dissertation of Dr. Catia Maria Pereira Fraga Caetano of University College London, who examined cell-cycle-regulated transcription and genome stability. Caetano's introductory chapter demonstrates a rigorous literature review that propels an experimental design. In 48 pages, Caetano synthesizes over 100 primary studies on the G1/S transition in *Saccharomyces cerevisiae*, focusing on the DNA-replication checkpoint. Her synthesis converges on a persistent puzzle: Although checkpoint activation stalls the cycle and triggers fresh rounds of transcription, "how this effect is mediated in budding yeast is still not fully understood" (p. 85). Citing work on the yeast species *Schizosaccharomyces pombe*, where checkpoint signaling lifts repression of G1/S genes, Caetano hypothesized that an analogous mechanism operates in budding yeast as well (p. 86). Then, she sketches experiments (pp. 99–102) that will monitor the checkpoint kinase (Rad53) and candidate corepressors *Yox1p* and *Nrm1* during replication stress. Caetano also proposed examining the involvement of specific corepressors and kinases in this process.

This flow – synthesis → gap → hypothesis → method – models what a dissertation literature review must accomplish: map the existing literature, expose what remains unanswered, and build a logical bridge to the research that the dissertation will address. It is an Introduction Funnel (Chapter 7) expanded to encompass the full scope of a doctoral research project.

12.3.4 Stand-Alone Review Articles

In rapidly evolving fields, like astronomy and public health, comprehensive reviews become *de facto* reference works that steer research directions, funding opportunities, and media narratives. Reviews are popular enough to warrant dedicated journals, such as *Annual Reviews of Astronomy & Astrophysics* and *Epidemiologic Reviews*. Review articles succeed when they compress a vast body of literature into a logically argued storyline and conclude with an agenda-setting direction for the field.

Consider the field of gene editing. One of the most influential reviews on CRISPR-Cas9 is Doudna and Charpentier's (2014) *Science* article, cited thousands of times. It opens by framing CRISPR as a paradigm-shifting tool in genome editing, superior in ease of design, scalability, and accessibility compared to older nuclease systems (like ZFNs and TALENs). The authors contextualize CRISPR within a lineage of gene-editing technologies, then summarize studies on specificity and efficiency, including concerns about unintended off-target effects. The review references multiple studies illustrating high efficiency and emerging methods for off-target detection (such as GUIDE-seq and Digenome-seq). Finally, the authors address pressing challenges related to CRISPR's future, including the technical difficulty of therapeutic delivery and the ethical complexities surrounding human genome editing – particularly germline modifications. They emphasize that responsible use of CRISPR requires attention to scientific accuracy and the values and bioethical frameworks guiding its application.

The significance of these ethical concerns is underscored by arguments that the implications of CRISPR extend far beyond gene editing into morally and legally uncharted territory, including experimentation on human embryos (Brokowsky & Adli, 2020). As the technology advances, they call for stronger ethical oversight, writing that "moral decision making should evolve as CRISPR science advances and … () for national and supranational legislatures to consider evidence-based regulation of certain CRISPR applications for the betterment of human health and progress" (p. 88). Indeed, many scholars have argued that the scientific community should not dominate deliberations about moral issues in gene editing because the potential conflicts of interest are too high. Instead, deliberations should include a network of scholars and organizations similar to those established for human rights and climate change (Jasanoff, Hurlbut & Saha, 2019).

Just as Doudna and Charpentier's (2014) review of gene editing pushed the need for bioethical frameworks to the foreground, reviews have long been a mainstay of public policy discussions. Consider the 1,200-page IPBES Global Assessment (2019), which functions as both a literature review and a policy guide on global biodiversity and ecosystems. Its architecture is explicitly CERIC-like. Each chapter states a driving Claim (e.g., land-use change is the largest direct driver of biodiversity loss), collates multiple lines of Evidence (e.g., remote sensing, meta-analyses, indigenous knowledge), and walks the reader through Reasoning chains linking human activities to ecosystem changes. Implications are framed as scenarios for 2030–2050, while Context call-out boxes situate findings in sustainable-development debates. Comprehensive and extensive literature reviews can do more than drive the direction of a scientific field; they can motivate social and governmental priorities relevant to global scientific challenges.

12.4 Preparing for a Literature Review

Preparing a literature review is a **reading-to-write** exercise: you must curate, interrogate, and organize dozens of studies *before* you can weave them into a coherent narrative about the subject. Two complementary techniques make the job more manageable: **structured CERIC-based note-taking** and **visual mapping**.

12.4.1 Structured CERIC-Based Note-Taking

Part I of this book argued extensively for the use of structured reading approaches to read and critique a research article efficiently. With a literature review on the horizon, we apply the same structure to our note-taking. In Chapter 8, we introduced the framework of the CERIC review to organize the summary of one article. At this stage, you may want to simplify that structure to a simple table or spreadsheet to organize your notes for each of the elements. As noted in Chapter 8, when compiling notes, it is good practice to paraphrase the text in your own words, which protects you from accidental plagiarism later. It may also be helpful to write a short narrative about the article at this stage. For example, writing five brisk sentences that capture the essence of each CERIC element will save writing time later.

It is also a better practice to decide early on how to organize your article reviews. Consider storing your notes in one searchable place: a spreadsheet, a reference-manager notes pane, or a shared collaborative database (Bjorn et al., 2022). Using consistent keyword tags as suggested on the CERIC Review Basic Template (Chapter 8) makes it easy to sort or filter entries later on, instantly lining up all studies of, for example, remote-sensing methods or exoplanet detections. When several tagged rows sit side-by-side, converging or conflicting Claims stand out on the screen.

Structured note-taking for each article in your review may feel slow at first, but by the tenth article, the practice becomes second nature. The payoff is significant. Structured note-taking routines and note completeness are positively associated with better quality and more organized writing (Luo & Kiewra, 2019). By the time you are ready to draft your literature review, the relevant studies will already be organized in notes and paragraphs in your own words that contain all of the necessary CERIC elements. It is worth adding here that your notes in your words increase your learning more than any note-taking shortcut, including using generative AI (see more about genAI in Chapter 14).

12.4.2 Visual Mapping Themes and Articles

When your CERIC spreadsheet starts spilling past a few dozen rows, patterns blur, and gaps can hide in plain sight. Translating those rows into pictures is a fast way to recover the "AH-HA!" view of the field. Three complementary mapping styles can be used alone or together – **concept maps**, **mind maps**, and **synthesis matrices**. They are very flexible and let you zoom in and out, from a broad view to details (Eppler, 2006). For instance, you could start with a simple concept map for brainstorming, then build a more detailed mind map that develops themes, and then use those themes to start a synthesis matrix. However, this does not have to be a linear process. If you already know the themes, you can just jump into a synthesis matrix and maybe refine it with additional concept or mind maps as your literature search develops.

- **Concept maps show the why behind links:** Concept maps require every connecting arrow between ideas to be labelled (i.e., causes, tests, and contradicts). For example, if you did a mind map, import the same nodes from your mind map, then write the Reasoning term that connects them. This exercise surfaces shaky logic (e.g., correlation without causation) and pushes you to articulate missing Context or Implications. Concept-mapping grew out of learning-theory research, and learners who build these maps articulate relationships more accurately and score higher on tasks that require the application of new knowledge (Novak & Cañas, 2008).

- **Mind maps spark quick connections**: Mind maps emphasize brainstorming. Start by writing the central research question in the middle of a blank page or white-board application. Use initial branches to map the major themes you spot in your CERIC notes, then divide down to thinner twigs for individual studies or subtopics, as shown in Figure 12.2. Because mind maps are nonlinear, you can drag nodes around until logical clusters emerge, often revealing a hidden bridge paper that links two seemingly distant debates (Wheeldon & Faubert, 2009). Mind maps are powerful as they exploit the brain's associative memory, allowing users to remember relationships 10%–15% better than with lists alone (Buzan & Buzan, 2006). Because of their flexibility, a basic mind map (Figure 12.2) can morph as the review evolves into a more complex thematic map, as shown in Figure 12.3.
- **Synthesis matrices are grids for rigorous comparison:** Finally, when you are ready to draft the literature review, organize the distilled articles for each theme into a synthesis matrix. This matrix can be as simple as a table organized with articles across the top and subthemes down the side. Transfer the relevant Claim, Evidence, and Reasoning notes into the appropriate cells at each intersection, as shown in Figure 12.4. Reading horizontally provides you with thematic paragraphs; scanning vertically allows you to check that you have represented every key paper (Ingram et al., 2006). Empty cells spotlight literature gaps, while crowded rows flag areas ripe for meta-analysis.

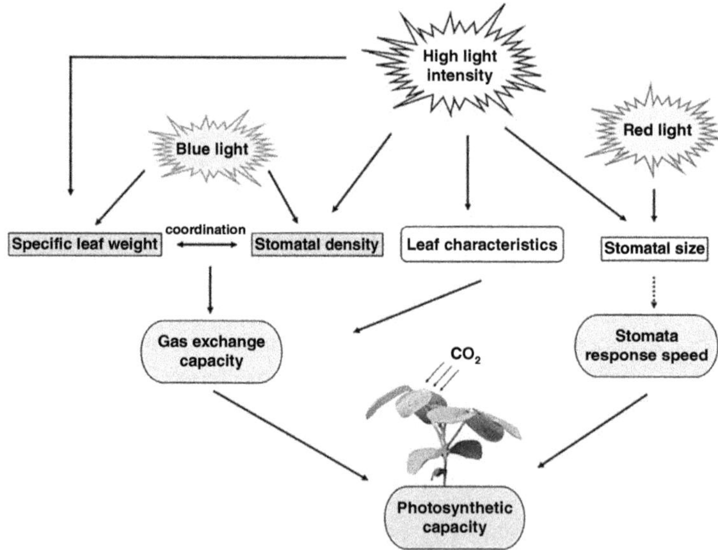

Figure 12.2 Basic concept map.
A simple concept map links core ideas using arrows.

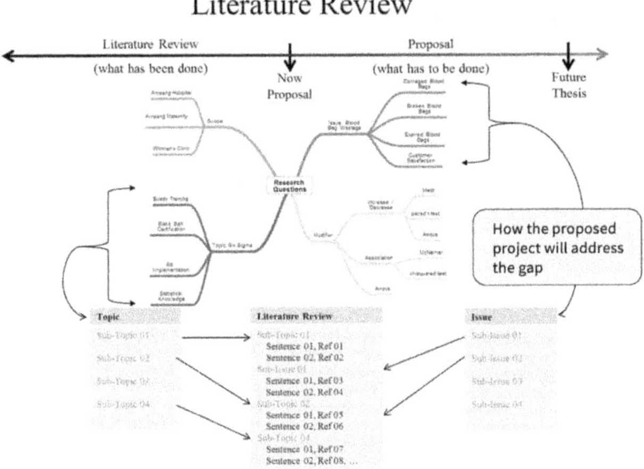

Figure 12.3 Mind map.
Radial mind map in preparation for a literature review of branches and subtopics from a central node, suggesting organization and themes.

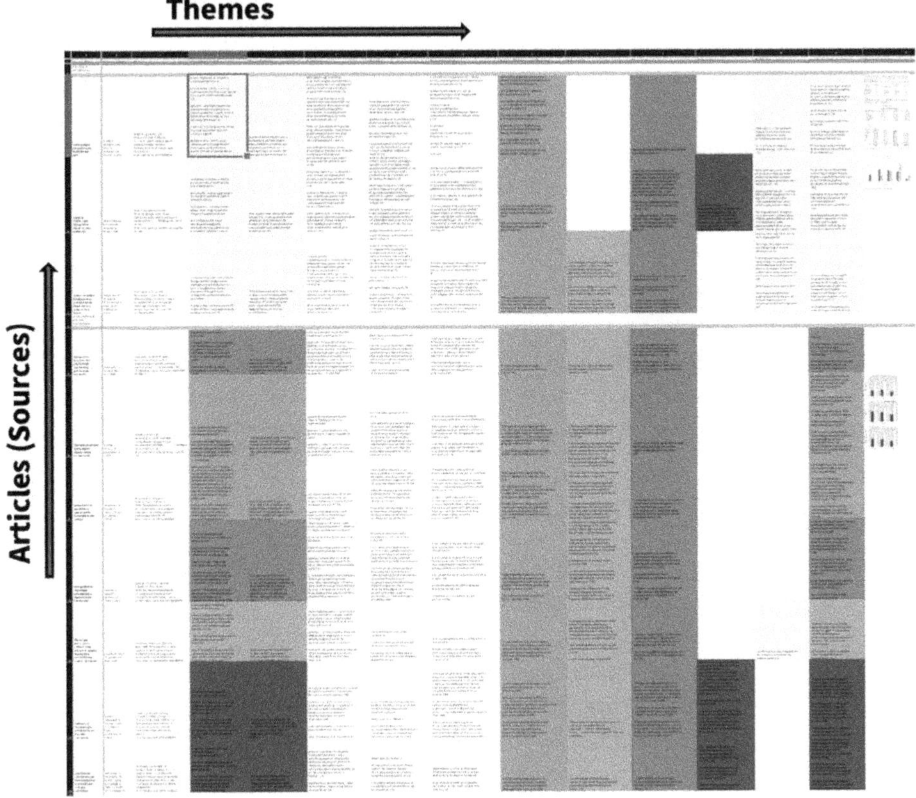

Figure 12.4 Synthesis Matrix.
Table headers themes (column headers) and sources (row headers). The themes include the CERIC elements plus any other key disciplinary and project categories, such as theoretical or conceptual framework, methodology, and related articles. Shading shows articles that overlap.

Box 12.2 Better Practice: What Is the Scope and Purpose of My Review?

Before proceeding with your literature review, ask yourself these questions to keep your reading focused and your writing on track:

1. **Focus:** What thesis, problem, or research question will this review clarify?
2. **Review type:** Are you surveying *theories*, *methods*, *policy debates*, *quantitative* findings, *qualitative* insights – or a blend?
3. **Scope:** Which sources count (journals, books, reports, and media)? Within which discipline(s)?
4. **Coverage:** Is your search broad enough to capture all key studies yet selective enough to skip the noise? Do the number of sources match the length and depth required?
5. **Critical lens:** Are you comparing and evaluating studies by looking at CERIC elements, strengths, weaknesses, and points of agreement or tension?
6. **Balance:** Have you acknowledged Evidence that challenges your perspective?
7. **Relevance:** Will readers see this review as timely, useful, and clearly connected to your overarching research goal?

Keep this checklist visible while you gather literature because it guards against scope-creep, confirmation bias, and last-minute panic.

Used together, these tools transform the vast troves of information in your articles into an articulate and evolving "atlas" of the field – one that keeps each CERIC element visible and retrievable while you synthesize a literature review.

12.5 Designing the Literature Review

Literature reviews must do a lot more than list results; they must explain how a body of work evolves, where it collides, and what future studies are needed. Developing this storyline hinges on two critical elements: the **narrative arc** and **thematic synthesis**.

12.5.1 Selecting the Narrative Arc

Many literature reviews falter at peer review for a straightforward reason: structural incoherence, whereby readers cannot see why topics unfold in the order they do (Booth et al., 2016). The remedy is to establish a single organizing storyline upon which the main themes and studies are anchored. A helpful metaphor is to imagine the review as a journey, and the narrative arc is the road that best reaches our destination (the goal of the review) and supports our fellow travelers (the readers).

Several recurring narrative arcs provide reliable routes for literature reviews. Each arc is driven by a dominant CERIC element, and each evokes a visual metaphor that signals how readers will "travel." Let's look in detail at the three most common of these arcs: the research gap, the historical development, and the methodological evolution.

12.5.1.1 The Research Gap (Spiral) Arc

This arc is by far the most common for literature reviews. Imagine an hourglass that narrows ever tighter toward a gap leading to expanding future work. Reviews built on the research gap arc start from a broad perspective, surveying the established, consensus findings of the field. The narrative then tightens through controversies and missing knowledge to expose that something significant – perhaps even critical – is missing or unresolved (Green, Johnson & Adams, 2006). The CERIC element that drives the research gap arc is Implications. Every turn of the spiral asks, "What remains at stake if we do not solve this problem?" Here is a typical pattern for the research gap arc:

- **Begin wide** by summarizing the consensus of a field (high-level Context + Evidence).
- **Tighten the spiral** by comparing results until unresolved contradictions or missing observations surface (Reasoning).
- **Name the gap** explicitly, showing its practical or theoretical stakes (Context + Implications).
- **Bridge** to the proposed methodology or resource that will allow researchers to span the gap (Claim + new Evidence to be developed).

A clear literature review follows this pattern. Begin wide by reviewing the many successes of, for instance, CRISPR-Cas9 applications, and then tighten the spiral to identify obstacles and gaps. By documenting various failed approaches – perhaps viral vectors, nanoparticle toxicities, and immune response – name the main gap now, for instance, that gene editing is now among the field's main limitations. Finally, the review must bridge to proposed research directions, such as transient ribonucleoprotein complexes that may would allow researchers to address the gap.

12.5.1.2 The Historical Development (Ladder) Arc

Imagine a ladder whose rungs mark the field's major milestones. The historical development arc works when research has unfolded through clear phases or "eras" – early discovery, consolidation, expansion – and the review's goal is to trace this stepwise progress. This is a chronological narrative exploring how knowledge has evolved, highlighting key milestones, gaps, and turning points. Context is the driving CERIC element: each rung comes alive only when situated in its social, technological, or theoretical origin. A danger of the historical development arc is nostalgia bias: romanticizing formative studies while overlooking recent work.

A typical pattern for the historical development arc is to divide the review into sections focused on a particular era. Then, within that era we address:

- What was the dominant idea or paradigm? What toppled it?
- Did a particular technique or methodology flourish at that time? What replaced it and why?

For example, a review on the technological advances of artificial intelligence (AI) by Toosi et al. (2021) explores the development of the field through three distinct eras: early AI, characterized by symbolic logic; the AI Winter, when funding crashed; and the deep-learning Summer, when GPUs and big data algorithms became dominant. The transitions between eras are key, and these are often driven by technological advances and breakthroughs.

12.5.1.3 The Methodological Evolution (Pathway) Arc

Some debates revolve around, not what was found, but how. Visualize parallel lanes, each representing a different research technique. The narrative moves laterally, contrasting the assumptions and motivations that drive, for example, randomized trials versus modelling studies. The storyline typically follows methods, instruments, and techniques. Readers are given a comparative view of the relative merits of different approaches, their strengths, limitations, and blind spots. Evidence and Reasoning drive this narrative arc.

A common hazard of the methodological evolution arc is drowning in technical or statistical detail. It is easy to lose sight of the big ideas and technical breakthroughs that result in new methodological approaches. As a countermeasure, Petticrew and Roberts (2008) advise spotlighting core logic, not every coefficient-level of detail.

A typical pattern for the methodological evolution arc is to assign competing techniques to individual sections, possibly in chronological order, highlighting strengths, weaknesses, and critiques of various applications. A summative assessment at the end can identify the existing shortcomings of present methodologies and propose specific advances needed to move the field forward. For example, Rao et al. (2024) review progress in the development of adaptive optics (AO) in astronomical observations, starting with an overview of the key technologies and then progressing through specific advances such as laser guide stars, deformable secondary mirrors, extreme AO, and wide-field AO, giving examples of instruments in each category. The authors conclude with an assessment of future advances needed for next-generation large ground-based telescopes and machine learning algorithms for rapid AO corrections.

In addition to these common narrative arc types, there are several less common narrative arcs that are used in literature reviews. Let's briefly review these.

12.5.1.4 The Crossroad Weave Arc

Interdisciplinary reviews can resemble two colored fabrics interlaced. They stage a dialogue between, say, psychology and computer science, showing the Contexts or Implications that overlap, making cross-fertilization fruitful. Success depends upon true integration. A compelling illustration of a "tight weave" crossroad arc appears in literature on Computer-Supported Collaborative Learning (CSCL), a field that is interdisciplinary from conception. Crucially, arguments in CSCL studies – such as the dynamics of peer interaction, affordance of digital tools, and learner cognition – are consistently structured across disciplines, revealing a truly interdisciplinary dialogue rather than a mere juxtaposition. For example, Hmelo-Silver and Jeong (2021) analyzed more than 700 publications spanning psychology, computer science, cognitive science, and educational research. They found that, rather than functioning as separate strands, CSCL integrates psychological theory with computational design and group cognition in shared conceptual and methodological frameworks. Indeed, the review's crossweave arc reflects the interdisciplinarity of its subject.

12.5.1.5 The Foresight Horizon Arc

When a funding call or policy roadmap demands forward vision, reviewers often trace a horizon line – a mountain silhouette far into the future. This arc interleaves current findings into alternative Implication scenarios (best-case, worst-case, and most-likely), projecting possible developments without drifting into pure speculation. Because the exercise flirts with prediction, the risk is "crystal-ball" thinking: bold but unanchored. However, when done well, a foresight arc invites decision-makers to act.

The National Academies' (2023) *Decadal Survey for Earth Science and Applications from Space* exemplifies this balance. The report outlines science priorities and mission recommendations for the next 10 years, but grounds every forecast in a transparent chain of reasoning. Rather than speculating broadly, each proposed trajectory – such as continuing measurements of surface deformation or refining atmospheric composition models – is explicitly tied to past mission outcomes, technological feasibility, and unresolved questions articulated by the scientific community. The structure of the survey shows how forward-looking vision can be built step-by-step from what is already known, integrating observational Evidence, technical readiness, and stakeholder input. Its strength lies in its clear vision and also how that vision is justified through many lines of Reasoning.

12.5.1.6 The Novice On-Ramp Arc

For audiences new to a research field, a well-structured on-ramp is essential. This arc scaffolds understanding by foregrounding widely accepted findings and standard methodologies before introducing open questions or emerging debates. It provides just enough conceptual grounding to support critical engagement, without overwhelming the reader with technical density too early. However, the on-ramp arc risks slipping into textbook prose – overexplaining foundational content at the expense of synthesis and critique.

A clear example of this arc appears in Feng's (2010) review of dark matter candidates in particle physics. Aimed at readers with preparation in college physics but who may not specialize in the field, the review begins with a concise but accessible overview of the Standard Model and its limitations, establishing why dark matter remains one of the most compelling puzzles in contemporary physics. Then, the review introduces historically dominant candidates, such as WIMPs and axions, along with the basic principles of direct and indirect detection. These early sections operate as conceptual on-ramps, enabling newcomers to understand key terminology, experimental goals, and baseline assumptions. After that foundation is established, the review transitions toward unresolved tensions in the literature: Why leading candidates have so far eluded detection, what newer models (e.g., hidden-sector particles) propose, and which experimental strategies might close the gaps.

Again, when designing a literature review, choose the arc that matches the goal of the review (such as identifying gaps, tracing development, and looking toward future possibilities) and the intended audience (such as experts in the field, policymakers, or novices). With a clear narrative metaphor in mind, the literature review becomes a guided road trip rather than a maze of citations.

Box 12.3 Knowledge Check: Question 12.1 Which of the following literature review topics would best be served by a Methodological Evolution (Pathway) arc?

A. Showing how CRISPR delivery remains an unsolved bottleneck despite abundant cutting-efficiency studies.
B. Comparing manual, automated and AI-assisted radiology image classifiers to reveal their respective error profiles.
C. Tracing the historical acceptance of plate tectonics from Wegener's hypothesis to GPS confirmation.
D. Weaving ecological resilience theory with political-economy models to propose a new sustainability metric.
E. Projecting likely trajectories for quantum-computing hardware over the next two decades.

(Check your understanding using the Knowledge Check Key at the end of the book.)

12.5.2 Structuring Thematic Synthesis

While narrative arcs provide a compelling overall storyline, literature reviews often use **thematic synthesis** to structure the main sections. In this approach, sources are grouped according to key concepts, themes, or central issues rather than organized chronologically or by individual studies (Webster & Watson, 2002). This structure can generate a more compelling argumentative review (UCLA Undergraduate Writing Center, 2016). Practically, this means dividing the review into subsections, each exploring a distinct theme that emerges from the analysis of multiple sources.

The thematic synthesis approach is especially effective when comparing different methodological approaches, theoretical frameworks, or recurring patterns across a diverse body of research. Scholars widely recommend this concept-focused strategy because it simplifies and enriches synthesis, allowing reviewers to discuss multiple studies together under unified themes (Webster & Watson, 2002). Common thematic structures or "lenses" include grouping studies by methodological approach, theoretical perspective, or scientific controversies.

1. **Methodological lens**: In this lens, studies are grouped into sections based on the tools that they deploy; for instance, "in-situ sensors" versus "remote-sensing satellites" in glaciology, or "finite-element" versus "computational-fluid-dynamics" models in biomechanics. Because different tools draw from different premises, Reasoning becomes the core CERIC lever; we are constantly asking how one method handles uncertainty, scale, or bias differently from its alternates.
2. **Theoretical lens:** In some fields, it is rival ideas rather than tools that delineate divisions in a field. A behavioral-ecology review might allot one section to niche theory, another to neutral theory, and a third to hybrid models. The reviewer's job is to set up each camp's central Claim, anchor to its intellectual Context (who

proposed it, which assumptions it inherits), and then weigh how convincingly subsequent Evidence and Reasoning have sustained or weakened the idea.

3. **Scientific controversy lens:** Some controversial topics demand a debate format centering on a key question. For example: – Does sodium drive hypertension? Does the ALH84001 meteorite indicate there was life on Mars? In this controversy lens, each subsection presents one side of the argument with its full C→E→R→I→C sequence before the next side gets equal time. The symmetry keeps bias at bay and brings the argumentative structure into sharp relief for the reader.

Choosing the right lens is largely a matter of where the intellectual tension resides – in tools, theories, or disputes. Reviewers often find it helpful to combine a lens with a broader narrative arc. For example, we might organize your review theoretically while still arranging the themes in a historical order – chronologically or by major milestones – creating a sense of progression. Whichever lens and arc works best for the review, CERIC tags ensure that every paragraph does analytic work rather than slipping into digressive summary.

Box 12.4 Better Practice: Appropriate Use of the Controversy Framework

The controversy lens is a common approach to literature reviews. In many fields, researchers disagree – on everything from technical details to fundamental interpretations, competing theories, and conflicting results. Resolving these tensions may even be central to the progress of the field. Rather than avoiding debate, a well-crafted literature review uses them to its advantage by organizing around the major positions or strands of Evidence. This approach makes for a compelling narrative and invites readers into the heart of scholarly debate of why researchers disagree and how conclusions are drawn. This kind of narrative opens up opportunities to apply the full power of the CERIC framework.

A controversy review is balanced when the following considerations are made:

- **Provide equal space**: When organizing by controversy, each major viewpoint should be treated in its own subsection or set of paragraphs. By applying the CERIC framework impartially to each side of the debate, the review maintains clarity and neutrality, enabling readers to understand what the disagreement is and how each conclusion is constructed from the literature.
- **Weigh the merits**: The reviewer must weigh conflicting interpretations, analyze methodological differences, and ask why the Evidence and Reasoning support one Claim over another. This deep engagement strengthens the review's analytical rigor.
- **Maintain objectivity**: While the existing research may favor one perspective, the purpose of a literature review is not to take sides but to synthesize – to reveal where the weight of Evidence lies and what Claims are justified. If one position clearly dominates, that conclusion can be presented with appropriate caveats grounded in the literature. CERIC helps keep this balance by anchoring each section in the articles' arguments rather than in personal opinion.

12.5.3 From Trees to Forest: Paragraph-Level Synthesis

With the narrative arc and thematic lens chosen, and an appropriate outline constructed, it is time to start writing the review. Come back to your synthesis matrix (Figure 12.4). Adding subthemes as rows is analogous to adding tress to a growing forest: synthesis gives readers a path through the forest, helping them to interpret meaningfully. The following steps outline this process (Webster & Watson, 2002).

1. **Group the thematic clusters:** Each of the primary themes of your review should include a cluster of 5–15 studies; fewer gives thin evidence, more becomes unwieldy (Tranfield, Denyer & Smart, 2003).
2. **Compare Claims:** Create a new mini-matrix: rows = studies, columns = CERIC. Color-code "agreement" (e.g., green) and "contradiction" (e.g., red) in the Claim column. Patterns leap out, for example, three green boxes followed by two red boxes warrant a subheading about conflicting findings.
3. **Critique Evidence and Reasoning:** Along the Evidence and Reasoning columns, ask three questions for each study (Hall et al., 1994):
   - Does the Evidence and Reasoning logically support the Claim?
   - Are statistics and analysis appropriate?
   - Have authors addressed confounding factors and rival explanations?

   Annotate weak Evidence or Reasoning in yellow; later, they form the basis of *limitations* sentences.
4. **Connect Implications:** Draw arrows on the matrix margins linking studies that push the field forward. Those arrows often form the topic sentences of your synthesis paragraphs. For example, "Preliminary findings from high-resolution LiDAR studies (e.g., Author X, 2019; Author Y, 2020) demonstrate that…."
5. **Contextualize:** Finish each paragraph by zooming out. Which subdiscipline, ecological niche, or time period is not covered? Absence becomes explicit groundwork for your concluding *future research* section.
6. **Prototype:** Draft one full paragraph from a single cluster. Circle each CERIC element. If you cannot find all five, the paragraph is either too thin (add Evidence) or too bloated (split it). Repeat until every theme heading contains at least one polished prototype; then weave prototypes together with forward-linking transitions, such as, "Having established why accuracy remains contested, the next section examines…."

Finally, check your writing rhythm. Many expert reviewers like the 1–3–1 pattern inside a thematic subsection:

1 – Introductory overview paragraph;

3 – Evidence- and Reasoning-rich synthesis paragraphs, each focused on a micro-theme; and

1 – Section conclusion paragraph that foreshadows the next section.

12.6 Practical and Ethical Considerations for Scientific Literature Reviews

Rigorous synthesis is only part of a literature review's job. It is also essential to make the review easy to read, easy to find, and easy to trust. This section addresses two intertwined duties in service to these goals:

1. **Reader-friendliness:** Clarity, inclusivity, and accessibility of prose and layout. How do we make the text welcoming, clear, and easy to navigate?
2. **Research integrity:** Transparent attribution, faithful paraphrasing, and balanced synthesis. How do we ensure every idea is attributed, paraphrased, and represented accurately?

Let's examine strategies to fulfilling each of these duties

12.6.1 Writing Accessible and Reader-Friendly Literature Reviews

A review that dazzles with insight but frustrates the reader with dense jargon or vague language is unlikely to be cited – or even read. Three intertwined craft-of-writing skills – clarity, inclusivity, and document accessibility – ensure that the hard work reaches its intended audience. These writing skills also help to clarify and refine thinking through the process, as reading, writing, and researching are coconstructed skills (Kwan, 2009).

- **Appropriate tone**: Critique the ideas, not the researcher. Instead of saying," Smith's absurd theory ignores basic thermodynamics," write "Smith's model assumes homogeneous mixing, a simplification that conflicts with empirical temperature gradients." The latter is specific, testable, and collegial (Medvecky & Leach, 2019). Adopting this stance reduces confrontational peer-review cycles and signals that your synthesis weighs Evidence over ego.
- **Clear structure:** Most scientific reviews adopt the common logical scaffolding of a three-tier outline: introduction, body, and conclusion. The introduction states the purpose, scope, and organizing principle of the review (e.g., chronology, method, or research gap) and may refer to contemporary, related reviews. The body walks the reader through the bulk of the review, organized by thematic lens into sections that are themselves suborganized into introduction–synthesis–conclusion paragraphs, all building toward the central Claim of the review based on its narrative arc (e.g., identifying research gaps, development of a field). A concise conclusion restates the central Claim and summarizes the Implications as next steps for the field.
- **Conduct a self-audit for readability:** A good writing practice in general is to read your draft out loud or use a "Read Aloud" tool; stumbling often points to awkward phrasing or syntactic overload. Running a Flesch-Kincaid readability check can also help (Kincaid et al., 1975): target grade 13 or lower for cross-disciplinary journals.
- **Inclusive language:** Inclusivity is not political window dressing; it is a professional obligation recognized in major style manuals. The American Psychological

Association (APA, 2020) stresses that bias-free language "acknowledges diversity, conveys respect, and promotes equal opportunity." In practical terms, that means avoiding gendered generics ("mankind" and "chairman"), defaulting to the singular they when a person's pronouns are unknown, and separating individuals from conditions ("people with epilepsy," not "epileptics"). Such adjustments may look minor, but scholarship in nursing shows found that prose using people-first language is perceived as more trustworthy (Jensen et al., 2013; Snow, 2007).

- **Know your audience:** Different review audiences require different considerations for field-specific language. If the review is aimed for new researchers (e.g., a short topical review aimed at graduate students), it will be necessary to define more terms, reduce unnecessary jargon, and stick to general topics. If the review is aimed for experts in the field (e.g., a conference session review), the review can be more concise and delve into specific points of discussion.

12.6.2 Ethical Considerations in Literature Review Writing

If clarity invites a reader into the work, integrity preserves the trustworthiness of it. Three inter-related hazards – **plagiarism**, **patch-writing**, and **misrepresentation** – are major reasons not to trust a work. For example, a large-scale study of 10,700 retracted research articles found that research misconduct of all kinds (e.g., fabrication, falsification, image manipulation, and plagiarism) accounted for 57% of cases, with plagiarism itself responsible for roughly 18% (Brainard & You, 2018). Research journals use similarity-checking tools as a first screen, typically investigating manuscripts whose overall similarity index exceeds 20–25% (IEEE, 2025). A better practice is to develop a trustworthy research process at the student level before needing it at a professional journal level.

Patch-writing is another hazard that occurs when a writer attempts to paraphrase, but ultimately, adheres closely to the source's wording or sentence pattern (Howard, 1995). Like plagiarism, a CERIC-based workflow can be protective as patch-writing tends to crumble in the Reasoning category. Explaining in plain language how authors link their Evidence to their Claim forces us to rebuild the logic, not the sentences. A useful strategy is to speak the explanation aloud first – as if teaching a class – and type from the spoken version. Studies of graduate writers show that this "oral paraphrase" strategy cuts patch-writing incidents significantly (Pecorari & Shaw, 2019).

Misrepresentation is usually an error of omission where we tout a conclusion but neglect the study's caveats. The twin columns of Implications and Context make such shortcuts glaring. For every takeaway recorded – for instance, "CRISPR off-target effects appear minimal in vitro" – our structured notetaking template should prompt, "Under what conditions? Which cell lines? Who funded the work?" When a Claim reaches the drafting stage with an empty Context box, a well-constructed synthesis matrix flags it. Then, we must either revisit the source or qualify the prose. This approach balances enthusiasm with the integrity conditions readers deserve.

Box 12.5 Common Confusion: Copy Now, Rewrite Later

The seemingly innocent temporary copy-paste is where most plagiarism starts. Cognitive-psychology research shows that once text sits intact in your draft, you are 60% less likely to rework its syntax and diction on revision (Lee, Hitchcock & Casal, 2018). Two traps follow:

1. **Ownership blurring**: Days later, we often misremember whose words they are, omitting quotation marks or a citation.
2. **Surface tinkering**: Under deadline pressure we swap a few synonyms, producing patchwriting that still mirrors the original structure and fails integrity checks.

CERIC to the rescue! The CERIC-based workflow forces us to paraphrase the article's main Claim immediately, in our own prose, and to park any genuine quotation in a clearly flagged Evidence cell. Because the wording and the idea live in separate boxes, the copy-and-paste temptation disappears and the draft grows from our own language, not the source's. Borrow ideas? Absolutely – with attribution. Steal wording? Never, and structured notetaking with CERIC makes that much harder.

12.7 Worked Example: A Methodological Evolution Arc

The methodological evolution arc traces how researchers confront an essential question, demonstrate the failure of standard methods, and devise a new toolkit or strategy that addresses the issue. This arc is common for students doing graduate dissertations, where the work often involves developing new methods to address research questions.

Here, we illustrated an example of the methodological evolution arc in the introductory literature review of an astronomy article by Burgasser and Mamajek (2017). This article aimed to determine the age of a low-mass star known TRAPPIST-1, which hosts seven Earth-size planets. The age of the star is a critical factor in determining the habitability of its planets, but as the Introduction section notes, standard age-dating methods for Sun-like stars do not work for low-mass stars, leading to large uncertainties and conflicting estimates in the literature. So, the authors use the methodological evolution arc to motivate a new approach to determining the age of this important exoplanet system. To help identify the main review elements, we reproduce the entirety of the Introduction section of the article here, indicating the citations reviewed in boldface font.

Title: "On the age of the TRAPPIST-1 system"

Citation: Burgasser, A. J., & Mamajek, E. E. (2017). *The Astrophysical Journal, 845*(2), 110. https://doi.org/10.3847/1538-4357/aa7fea

Introduction:

TRAPPIST-1 (2MASS J23062928–0502285; Gizis et al., 2000) is an ultracool M8 dwarf 12 pc from the Sun that was recently identified to host at least seven Earth-sized planets, three of which orbit within the star's habitable zone (Gillon et al., 2016;

> Luger et al., 2017). The planets were identified by both ground-based and space-based transit observations, and span orbit periods of 1.5–19 days and semimajor axes of 0.011–0.062 au (21–114 stellar radii). These observations and others have provided important constraints on the physical parameters of the star, as summarized in Table 1.
>
> One crucial parameter of TRAPPIST-1 that is currently poorly constrained is its age due to the weak empirical age diagnostics available for ultracool M dwarfs. While rotation (gyrochronology), activity diagnostics, and lithium depletion are standard age-dating tools for solar-type stars (Soderblom, 2010), the physical properties of ultracool dwarfs restrict their application at the bottom of the main sequence. The low degree of ionization of ultracool dwarf photospheric gas reduces the star's coupling with magnetic winds, resulting in spin-down timescales that can exceed the age of the Milky Way Galaxy (West et al., 2008). Depletion of lithium, which provides approximate ages for solar-type stars up to 1–2 Gyr (Sestito & Randich, 2005), is complete for fully convective low-mass stars by~200 Myr, and is only useful for young ultracool dwarfs (Stauffer et al., 1998). Spectral age diagnostics, such as surface gravity-sensitive features, are also limited to stars younger than~300 Myr (Allers & Liu, 2013). While the kinematics of TRAPPIST-1 suggest that it is an "old disk" star (Leggett, 1992; Burgasser et al., 2015), such labels are insufficient to firmly constrain an age. Filippazzo et al. (2015) report a wide age range of 0.5–10 Gyr, while Luger et al. (2017) adopt a more constrained, but still broad, range of 3–8 Gyr. In contrast, several studies in the literature have argued that TRAPPIST-1 may be young based on the strength of its nonthermal magnetic emission (Bourrier et al., 2017; O'Malley-James & Kaltenegger, 2017).
>
> Age is necessary for understanding the formation, orbital evolution, stability, and surface evolution (including habitability) of the planets orbiting TRAPPIST-1. Here, we present an analysis of several empirical age diagnostics for this source that allow us to more precisely quantify its age. In Section 2, we analyze in turn the color-absolute magnitude diagram (CMD), average density, lithium abundance, surface gravity features, metallicity, kinematics, rotation, and magnetic activity of the star. In Section 3, we combine these into a concordance age of 7.6 ± 2.2 Gyr, and discuss implications on the stability and habitability of the planetary system.

The Introduction contains 14 separate references, not all of which are relevant to the methodological evolution literature review used to motivate the study. Let's examine the primary components of this arc, extracting and paraphrasing the relevant text, beginning with the Context of why age matters:

> TRAPPIST-1 (...) hosts seven Earth-sized planets ... These observations and others have provided important constraints on the physical parameters of the star ... age is necessary for understanding the formation.

So, we need the age to understand the system's formation. That's the key Context. The authors then build their methodological argument that we can breakout into several major steps.

1. Identify existing methodological approaches for measuring stellar ages:
   - **Quoted text:** "... () rotation (gyrochronology), activity diagnostics, and lithium depletion are standard age-dating tools for solar-type stars (Soderblom, 2010)."
   - **Paraphrased text:** Ages for Sun-like stars can be determined from their rotation periods, degree of magnetic activity, and the presence of lithium in their atmospheres.

2. Identify limitations for these approaches in low-mass stars:
   - **Quoted text:** "weak empirical age diagnostics available for ultracool M dwarfs … The low degree of ionization of ultracool dwarf photospheric gas reduces the star's coupling with magnetic winds, resulting in spin-down timescales that can exceed the age of the Milky Way Galaxy (West et al., 2008) … Depletion of lithium … is complete for fully convective low-mass stars by ~200 Myr, and is only useful for young ultracool dwarfs (Stauffer et al., 1998) … surface gravity-sensitive features, are also limited to stars younger than~300 Myr (Allers & Liu, 2013)."
   - **Paraphrased text**: Normal stellar age metrics fail for low-mass stars because:
     - They don't spin down enough to show a meaningful change over the age of the Milky Way
     - Lithium depletion only applies for ages less than 200 Myr
     - Gravity-sensitive features only apply for ages less than 300 Myr
3. Describe how these limitations result in large uncertainties and conflicting ages for TRAPPIST-1:
   - **Quoted text:** "One crucial parameter of TRAPPIST-1 that is currently poorly constrained is its age … Filippazzo et al. (2015) report a wide age range of 0.5–10 Gyr, while Luger et al. (2017) adopt a more constrained, but still broad, range of 3–8 Gyr … kinematics of TRAPPIST-1 suggest that it is an "old disk" star (Leggett, 1992; Burgasser et al., 2015) … several studies in the literature have argued that TRAPPIST-1 may be young based on the strength of its nonthermal magnetic emission (Bourrier et al., 2017; O'Malley-James & Kaltenegger, 2017)."
   - **Paraphrased text:** Current age estimates are extremely broad for TRAPPIST-1 (0.5–10 Gyr or 3–8 Gyr), and there are conflicting indications that the star could be very old (based on kinematics) or young (based on magnetic emission)
4. Summary of how authors use these limitations to motivate a methodological approach to improve the age of TRAPPIST-1:
   - **Quoted text:** "Here, we present an analysis of several empirical age diagnostics for this source that allow us to more precisely quantify its age … the color-absolute magnitude diagram (CMD), average density, lithium abundance, surface gravity features, metallicity, kinematics, rotation, and magnetic activity … we combine these into a concordance age of 7.6 ± 2.2 Gyr"
   - **Paraphrased text:** The authors evaluate several stellar age indicators, including location on the CMD, measured average density, lack of lithium, surface gravity features, metallicity, kinematics, rotation, and magnetic activity. By combining these, they infer an age of 7.6 ± 2.2 Gyr, a significant improvement over prior age estimates that indicate the system is old.

The short literature review provided in the Introduction of this article establishes a methodological evolution: how star ages are measured now, why they can be used for this source, and hence the motivation to apply a new technique. By replacing ineffective aging methods with a combination of diagnostics, they narrow what was initially a 10-giga year (Gyr) uncertainty to ±2.2 Gyr, a significant improvement that resolves the arc. This example illustrates how a short literature review can focus a paper's Context and motivate the study's adopted approach. More extensive reviews could be used to justify larger research efforts (e.g., a large funding proposal), advocate for a new direction for a field (e.g., review articles), or motivate changes in governmental policy (e.g., policy reviews).

12.8 Chapter Key Takeaways

1. **A literature review synthesizes many sources into a cohesive narrative**. Unlike an annotated bibliography that merely summarizes each study independently, a literature review identifies common themes, articulates areas of consensus or controversy, and finds essential gaps in knowledge.
2. **Literature reviews are found in many forms**. Practical applications include the introductions of research articles, a proposal, the introductory chapter of a dissertation, and a stand-alone review. Each format conveys a specific purpose, scope, and audience.
3. **The CERIC framework provides an analytical structure for developing literature reviews**. Well-structured and organized CERIC reviews of individual articles can be combined with mind mapping, concept mapping, and synthesis matrices to synthesize a large body of research into a focused narrative.
4. **Frame literature reviews around a narrative arc and thematic lens**. To avoid an unfocused maze of research findings, identify which narrative arc your review serves, and organize the structure around themes of methodology, theory, or controversy.
5. **Adopt strategies for reader-friendliness and ethical standards**. Developing a clear structure, using inclusive language, adopting an appropriate tone, knowing your audience, and checking readability help create a reader-friendly literature review that will be read. Using rigorous CERIC practices, including regular paraphrasing and checking element completeness, can reduce plagiarism, patch-work, and misrepresentation.

13 Using CERIC with Social Collaborative Annotation (SCA)

We were all learning together, and I felt supported, [which] helped me calm down and get things done.

—Doctoral participant, Bjorn (2023)

13.1 Overview

Chapter 13 extends the use of CERIC to peer-based discussion and social collaborative annotation (SCA). This chapter begins by discussing the unique benefits of SCA as a pedagogical tool, highlighting how it fosters active learning, structured discourse, and collective meaning-making in academic settings. It then offers guidance on selecting platforms that align with research and learning goals. This chapter also presents best practices for incorporating CERIC into SCA discussion groups, demonstrating how structured annotation frameworks improve analytical skills, reading comprehension, and peer collaboration. Specific strategies – such as role-based annotation, color-coded highlights, and guided discussions – can increase the benefits of SCA on learning in research environments.

13.2 What Social Collaborative Annotation Brings to Research Education

Social Collaborative Annotation (SCA) develops the essential scientific practice of discourse and transforms reading from a solitary activity into a shared, real-time conversation (Brown & Croft, 2020; Ghadirian, Saleh, & Ayub, 2018). Using web-based tools, such as Hypothes.is or Perusall, readers anchor comments, questions, and even emojis to specific passages, then reply to one another in the margins. The result is an "interpretive community" (Fish, 1980) in which meaning is negotiated socially rather than individually (Cui & Wang, 2024) This chapter defines SCA, reviews its academic benefits, and explains why we recommend it for coursework as well as informal study groups that extend beyond the semester calendar.

13.2.1 Defining SCA

One entry point to developing critical reading is using annotations. Annotations are defined as "notes that mediate reading and writing" (Kalir & Garcia, 2021, p. 182).

They can be as small as a highlight or as complex as a paragraph-length critique. When those notes are shared in a common digital space, they become social annotations. Add synchronous chat, searchable threads, and group tagging, and you have SCA: a collective problem space in which readers set joint goals, test tentative interpretations, and co-construct knowledge (Enyedy & Stevens, 2006). Because annotation bridges reading and writing, it functions as "first-draft thinking" (Bjorn, 2023; Lee et al., 2023), especially when paired with other supports, such as role assignments or reflection journals (Ghadirian et al., 2018; Sigmon & Bodek, 2022). SCA supports critical reading and thinking by enabling us to synthesize ideas, critique each other's interpretations, and practice academic discourse skills (Bjorn, 2023).

13.2.2 Why SCA Matters for Critical Reading and Critical Thinking

SCA helps students become better critical readers by turning reading into an interactive, question-driven activity. Instead of reading alone and passively taking in information, readers using SCA can engage actively with the text, ask questions, challenge ideas, and respond to what others are saying. This process strengthens key aspects of critical thinking – like analysis, evaluation, and reflection – by encouraging students to think more deeply and collaboratively (Bjorn, 2024; Yeh et al., 2017).

- **Active reading is an essential part of critical thinking, and SCA makes active reading easier to practice:** Social annotation motivates students to read more intentionally by introducing multiple perspectives and encouraging them to take a position on what they read (Porter, 2022). Active reading – not just reading more – was the strongest predictor of higher course grades, and that social annotation helped students become more active readers (Clinton Listell, 2023). When readers annotate together, it is possible move beyond just understanding the text and begin constructing meaning as a group.
- **Quality of thought and reflection also improves with SCA:** Even if readers write fewer annotations over time, they remain mentally and emotionally engaged when annotation is part of the regular course structure (Chen et al., 2024). The depth and insight in those comments also matter. For example, when instructors follow up annotations with writing prompts that ask us to summarize or synthesize ideas, we gain practice in connecting small details to larger arguments – a key academic skill (List & Lin, 2023). Some groups also use role-based prompts – such as questioner, clarifier, or devil's advocate – to help readers examine texts from different perspectives and maintain a rich and focused discussion (Kaendler et al., 2015; Xu & Dai, 2024). These roles enhance Reasoning and aid in managing the mental effort required for complex texts (Eryilmaz et al., 2014; Ghadirian et al., 2018).
- **Reading comprehension and metacognitive awareness improve with SCA:** Metacognitive awareness is the ability to think about and monitor our own

Box 13.1 Knowledge Check: Question 13.1 What is the most plausible reason early SCA interventions sometimes fail to improve critical-thinking scores – even when comprehension and metacognition increase?

A. Readers had already reached a "ceiling" in critical-thinking ability before the study began.
B. Teams focused on discrete text fragments, so they rarely practiced synthesizing ideas across a whole article.
C. Annotation platforms lacked social-interaction features, such as replies or upvotes.
D. Researchers measured improvement after only one semester, whereas critical thinking develops over several years.
E. Most participants found color-coding distracting, which increased cognitive load.

(Check your understanding using the Knowledge Check Key at the end of the book.)

understanding, and these skills also benefit from well-designed annotation activities. For example, readers using SCA in graduate-level science classes were more adept at recognizing when they were confused and utilized peer input to clarify their thinking (Bjorn, 2024). Other studies have linked SCA with improvements in critical-thinking test scores, especially when paired with reasoning-based tasks (Johnson, Archibald, & Tenenbaum, 2010; Gil-Flores & Torres-Gordillo, 2012).

To sum up, SCA provides us with a structured approach to developing the essential habits of strong, critical readers: paying close attention to ideas, thinking critically, and engaging in meaningful dialogue with others. These skills support academic success and are also essential for professional and lifelong learning.

13.2.3 The Role of SCA in Higher Education

SCA is gaining attention as a powerful tool in higher science education, as it promotes deeper comprehension, a low-stakes learning environment, active learner engagement, and shared knowledge building (Bjorn, 2023; McCormack & Keating, 2024). Rather than learning in isolation or relying entirely on the instructor to lead discussions, readers in SCA-based discussion and study groups participate in a shared inquiry. Research shows that this shift – from traditional lecture formats to more participatory learning environments – can foster collaborative problem-solving and meaningful conversations among peers (Bjorn, 2023; Brown & Croft, 2020; McCormack & Keating, 2024).

When we use social annotation tools to read and discuss an article together, we begin to coconstruct meaning as a group. This kind of participation turns reading into a collective process, where students learn by comparing interpretations and refining their thinking in response to others (Porter, 2022). SCA also engages us in critical reading, writing, and thinking, transforming them from passive readers into more active, community-oriented learners (McCormack & Keating, 2024). Let's review some of the evidence about SCA in higher education and why it matters to natural science researchers.

13.2.3.1 Brief Review of the Literature: SCA in Higher Education

SCA creates low-stakes, peer-based learning environments: An environment that promotes psychological safety boosts motivation and engages all learners in the dialogue (Brown & Croft, 2020; Zhu et al., 2020). When SCA accompanies research discussion groups, it helps lower barriers to participation – normalizing uncertainty, increasing equity and motivation, improving flexibility, and accelerating both social and cognitive growth. There are several key factors:

1. *Psychological safety* is the belief that one can ask questions, express confusion, or offer a tentative interpretation without judgment – emerges as a central mechanism for SCA's benefits. Organizational research consistently demonstrates that environments characterized by high psychological safety foster learning behaviors, collaboration, innovation, and peer engagement (Newman, Donohue & Evan, 2017; Zhu et al., 2020). Learners in SCA settings observe peers admitting misunderstandings, which helps normalize fears and uncertainties and encourages honest inquiry.
2. *Equity* in SCA is supported by inclusive features, including written prompts, roles, optional anonymity (Surmiak, 2018), and asynchronous participation, which help balance contributions across language proficiency and status differences (Brown & Croft, 2020; Sun et al., 2023; Cui & Wang, 2024). These designs support raciolinguistic equity (Rose & Flores, 2017), ensuring that minoritized speakers, as well as those who are quieter or non-native, contribute meaningfully alongside their more vocal peers.

   This design also supports flexible participation, allowing learners with different schedules (e.g., caregivers, international students) to contribute comfortably. Over time, well-structured environments can sustain motivation, as norm-setting around contributions, private subgroup channels, and guidance help reduce overload (Bjorn, 2024). Flexibility also contributes to less pressure.
3. *Metacognitive lift* occurs when we can compare our thinking with our peers' thinking. Reflecting on our own thought processes helps us to eventually transfer those habits to independent study (Sun et al., 2023). Reflective practice helps us feel more confident in questioning our interpretations and less pressured to appear immediately certain (Bjorn, 2024).). More thoughtful annotation is correlated with better exam performance (Hwang, Wu & Chen, 2007), and readers who actively respond to peers over time exhibit deeper cognitive engagement (Xu & Dai, 2024).
4. *Motivation* receives a boost in peer-centered environments where grades matter less than meaningful exchange. Learners often cite peer reactions – not assessments – as the key driver for continued annotation (Nokelainen et al., 2005). Removing the

expectation of a single "correct" response encourages exploration, risk-taking, and a sense of confidence that we can do the work (Bjorn, 2024; Porter, 2022).

SCA helps build supportive learning communities: A supportive learning community increases motivation and academic performance. Readers who regularly engage in social annotation often report a stronger sense of connection with others and remain more engaged cognitively and emotionally (Chen et al., 2024). SCA can also reduce anxiety about reading and improve students' confidence in doing research, which researchers refer to as *self-efficacy* (Bjorn, 2024).

A shared problem space is one of SCA's key strengths, as it creates a common area for learners to explore the meaning of complex texts together. This is a process in which students contribute ideas, interpret each other's thoughts, and refine their understanding through back-and-forth conversation (Enyedy & Stevens, 2006). This kind of dialogue helps us strengthen our analytical reading skills and become better at expressing our ideas (McCormack & Keating, 2024). SCA can improve both individual understanding and group-level learning (Bjorn, 2024).

Engagement needs ongoing support: Engagement with SCA is not always consistent. Chen et al. (2024) used time-series clustering to track how engagement patterns change, and they found two main groups: one that stayed consistently low in participation and another that started strong but dropped off over time. Interestingly, the learners in the declining group showed higher initial levels of cognitive, emotional, and social engagement – writing more comments and replying more often to peers. These findings suggest that while SCA can spark strong initial interest, group facilitators may need to support engagement throughout the project including by changing roles or meeting occasionally in person.

Community building is an effective strategy. Sustained, low-pressure conversation – especially in large classes or online settings – can help us feel less isolated and more connected to peers (Kalir, 2020a; Yang, 2011). This sense of connection matters: research shows that feeling like we belong can boost academic motivation and persistence. In a review of 78 studies, found that a strong sense of belonging is linked to better grades and higher rates of persistence – especially for students from historically marginalized groups (Fong et al., 2024). Learners who feel a greater sense of belonging tend to perform better and are more likely to stay in school (Gopalan & Brady, 2019).

These findings align with *self-determination theory*, which suggests that when we feel a sense of belonging (or "relatedness"), we are more likely to stay motivated and engaged (Ryan & Deci, 2000). SCA supports this kind of motivation by giving us opportunities to exchange ideas and learn actively with others (Bjorn, 2024). In addition, learners who consistently respond to others' annotations and participate in group dialogue tend to show stronger cognitive and social engagement over time (Xu & Dai, 2024).

To keep this kind of engagement going, it helps when groups follow clear expectations. *Structured group norms* – like assigning specific roles (e.g., questioner, clarifier) or using guided prompts – can support deeper conversation and help all members contribute meaningfully (Zhu et al., 2023). Whether groups are chosen by learners or assigned by an instructor, these strategies give everyone a way to stay active and connected during the reading process.

These findings underscore the importance of structured SCA groups in higher science education, especially where class sizes are large and disengagement can be high. Boosting connection helps to sustain high levels of learner engagement across multiple reading activities. If your instructor does not assign an SCA discussion, please use the takeaways from this chapter to form your own.

13.2.3.2 Ethical Guidance for SCA Usage

SCA promotes active engagement with research articles and peer-to-peer meaning-making. However, the tension between collaborative learning and individual assessment becomes crucial when final evaluations and checkpoints – such as comprehensive exams, theses, or qualifying exams – must reflect individual mastery. Cooperative learning theory emphasizes this need: effective group learning should combine positive interdependence with individual and group accountability (Johnson & Johnson, 1987), Ultimately, the goal is for each student to be evaluated on individual performance. Several strategies can enhance the ethical use of SCA, particularly in journal clubs and study groups:

1. **Set dual-phase work as the default study mode:** For example, group members collaborate to annotate and then produce a solo submission (e.g., a written critique or summary). This approach mirrors cooperative learning models, such as Student Teams–Achievement Divisions, where group study is followed by individual quizzes or exams that determine final grades based on each student's individual understanding.
2. **Build in a transition from group to individual learning:** After setting dual-phase work as the default, create a deliberate shift in each session from group SCA to independent written reflections or concept maps, in which each group member interprets and expands upon the group-generated ideas by themselves and in their own words. This ensures personal accountability while supporting metacognitive development.
3. **Encourage group attribution and individual reasoning**: Ask group members to reference each other's annotations in the individual reflections, explaining why they adopted, modified, or rejected those ideas. This enhances critical thinking and personal authorship while preserving the benefits of the group.

The idea is to capture the benefits of SCA while upholding the integrity of individual assessment. This may be particularly helpful in high-stakes academic contexts, such as comprehensive exams and theses. However, we suggest that learners do not wait for instructors to initiate these moves, and instead, proactively facilitate better practices within your study groups.

13.3 Choosing and Setting Up Digital SCA Spaces

Digital annotating is a skill you can develop deliberately – especially when you are learning to search for academic sources and evaluate their credibility (Ghadirian et al., 2018; Gil-Flores et al., 2012). As we have previously reviewed in this chapter, SCA enhances this skill by transforming how you engage with readings. Instead of reading alone, you and your classmates turn the text into a shared discussion space – highlighting, asking questions, and building a running commentary (Bjorn, 2024; Brown & Croft, 2020). The result is stronger comprehension, sharper analysis, and better awareness of your own thinking.

The SCA platform matters only in that it should match the systems already in use at your school. If your SCA tool does not integrate well with your school's learning platform or course workflow, the benefits can get lost in tech problems (Bjorn, 2024; McCormack & Keating, 2024). A good fit supports real-time collaboration and makes it easier to carry your work into lab settings, field sites, or interdisciplinary projects.

If your school does not use an SCA platform or you want to know more about some of the options, we will briefly review a few here. Most SCA platforms fall into three broad categories. Each category supports a slightly different way of working:

- **Dedicated SCA tools** (Hypothes.is and Perusall): These are designed specifically for collaborative reading. These platforms let you anchor comments to exact phrases, respond in real time or later, and save the discussion for future review.
- **LMS-integrated tools** (Canvas LMS, Blackboard, and Moodle): These systems often include built-in annotation tools or offer plug-ins like Hypothesis. They provide a streamlined experience where readings, grades, and annotations stay in one place.
- **General annotation and collaboration tools** (Diigo, Google Suite, HyLighter, Microsoft OneNote, Microsoft Word, and PDF viewers): These let you annotate almost anything – helpful for reading lab protocols, technical memos, or field documents that sit outside your school's LMS. They do not always support threaded discussion, but they are flexible and widely accessible.

Each category has its strengths. Purpose-built tools are ideal for text-heavy discussions, LMS tools are more integrated with for coursework, and general tools are great for informal sharing or working across platforms or institutions.

13.3.1 What the Research Says about Platform and Engagement

Hypothesis and Perusall are two of the most widely studied SCA platforms, used in both U.S. and international classrooms. These tools support synchronous and asynchronous annotation, and allow students to see and respond to each other's ideas in real time (Cui & Wang, 2024; Porter, 2022; Sun et al., 2023). Several studies show that students who annotate before class – especially when prompted to engage with the text – score higher on related quizzes and exams (Sun et al., 2023; Cui &

Wang, 2024). These tools also integrate directly with LMS platforms like Canvas and Blackboard, as long as your institution supports them.

General annotation tools may not offer live discussion, but they shine when used for annotating PDFs, reports, or technical documents that do not need structured discourse. They are especially useful for working in lab groups or research settings, where you may need to mark up a shared protocol or article outside of class.

In all cases, SCA supports more than just highlighting – it enables multilayered interactions, such as:

- Commenting on specific lines or figures in real time.
- Responding to peers' ideas to test or refine your own.
- Revisiting discussions across multiple readings to see how your understanding evolved (Sun et al., 2023).

Hypothes.is and Perusall, in particular, are widely used for their ability to facilitate real-time engagement and discourse (Bjorn, 2024; Cui & Wang, 2024; Kalir, 2020a; Porter, 2022; Sun et al., 2023). They dominate U.S. and global classrooms because they anchor every comment to the exact sentence you mean, support both synchronous and asynchronous discussion, and archive the discourse for future study (Bjorn, 2024; Cui & Wang, 2024; Kalir, 2020a; Porter, 2022; Sun et al., 2023). Canvas and Blackboard offer integration with Hypothes.is, if your institution subscribes to it.

The general tools typically let you layer highlights and comments on almost any document you already use. While they lack some of the real-time chat bells and whistles, they shine when you need to annotate a lab protocol, a policy memo, or any file that sits outside the LMS. Together, these three categories give you options: a purpose-built social reader, a one-stop LMS solution, or a flexible toolkit that travels wherever your coursework takes you. One sign that SCA is a good match is that the tool eventually fades into the background, and conversation takes center stage, evolving into deeper discussion.

13.3.2 The Makings of a Strong SCA Tool

Once your SCA group (or instructor) selects the platform, it is helpful to know which features matter most. A strong SCA platform should:

- Integrate with your LMS, so documents, discussions, and grades (if relevant) are centralized
- Be accessible, with features like screen-reader support, keyboard navigation, and adjustable font sizes
- Support flexible timing, so both night owls and 9–5 students can participate equally
- Work across academic tools like Google Docs, OneNote, or reference managers
- Encourage metacognition, by including comment tags, reflection prompts, or conversation frameworks that help you think about how you're thinking (Ghadirian et al., 2018)

Student experience also matters. As one graduate student put it:

> My advisor and I share key articles by sending PDFs through Slack. We don't annotate them – we just say, "Here's the article and here's what they found." It's not really a conversation. I think SCA could really improve our communication. (Bjorn, 2022, p. 217)

This example captures a common problem: Passive file-sharing without active discussion. SCA, when used effectively, bridges that gap by turning marginal notes into meaningful dialogue (Bjorn et al., 2022; Hollett & Kalir, 2017). When SCA tools are well implemented, learners report:

- Stronger peer engagement, through both commenting and dialogue
- Better retention of concepts, as ideas connect across readings
- Higher-quality feedback, where peer comments and instructor notes become part of your learning, not just one-off corrections

Ultimately, every annotation becomes an entry point for shared inquiry. You get to see other readers' thinking in real time, reflect on your own ideas, and join a shared space where everyone is asking questions, annotating, and learning together – regardless of experience level or confidence.

13.4 Integrating SCA in CERIC Reviews

Blending SCA with the CERIC framework is most effective when clear, purpose-driven prompts guide annotations. Without structure, the comment thread can drift off-topic or stall entirely. But with the right setup, SCA becomes a clear, intellectual workspace where we build toward deeper insights together.

13.4.1 Using CERIC Prompts

In this section, you will find a set of starter prompts designed to jumpstart your annotations – and your conversations about research articles. These draw directly from the CERIC review approach introduced in Chapter 8, but with a twist: each prompt includes a metacognitive check to help you tune in to your own thinking, your energy level, and your focus while you read. Whether you are solo-reading or working in a group, these gentle nudges can help you reflect and refocus in the moment.

PROMPT 1: IDENTIFY THE PAPER'S PRIMARY CLAIM

Look for a clear, plain-language statement of what the authors discovered or proved – usually in the abstract or title. It is their answer to the "what's new here?" question.

Metacognitive check: Are you distracted right now? Do you need to refocus?

PROMPT 2: IDENTIFY THE EVIDENCE THAT SUPPORTS THE CLAIM

Zoom in on the key data – measurements, observations, model outputs – that the authors selected to support their main Claim. This usually shows up in the abstract, results, and methods sections.

Metacognitive check: Are you feeling overwhelmed? Would a short break help?

PROMPT 3: IDENTIFY THE REASONING USED TO CONNECT THE EVIDENCE TO THE CLAIM

This is the logical glue. Look for causal logic, comparative reasoning, or model-based interpretation in the results or analysis section.

Metacognitive check: Are you confused right now? What do you need to understand better?

PROMPT 4: IDENTIFY THE IMPLICATIONS OF THESE FINDINGS

Go beyond the findings and the future trip. Why do these results matter? Check the abstract and discussion for forward-looking statements about significance, applications, or future research.

Metacognitive check: Are you motivated right now? If not, what could help spark interest?

PROMPT 5: IDENTIFY THE CONTEXT OF THIS STUDY, SPECIFICALLY THE RATIONALE

Live in the past (for a moment). The rationale behind the study is usually in the introduction. What prior work left a gap? What need or unanswered question launched this project?

Metacognitive check: Are you able to focus right now? What's competing for your attention?

PROMPT 6: ANALYZE THE MAIN ARGUMENTS OF THIS STUDY

Does the Reasoning hold up? Are there assumptions that go untested? What Evidence would make the Claim stronger? This is where you shift from reader to reviewer.

Metacognitive check: Are you aware of the strategies you are using right now – like comparing, summarizing, or questioning?

PROMPT 7: IN ADDITION, COMPARE THIS PAPER TO AT LEAST ONE SIMILAR

Choose at least one other article for comparison. What are the relative strengths and limitations of each? Does this article have stronger Reasoning or a higher impact (citations per year)? Does the topic matter more in your field?

Metacognitive check: Can you think of multiple ways to approach this comparison?

You do not have to use all seven prompts every time, but as a study group, using even a few can shift the tone of your critical reading from passive to engaged. You will notice your annotations evolving – from simple tags or highlights to deeper questions, critiques, and connections. That is the real goal: To move together from basic reading toward collaborative analysis and critique in a nutshell, this is what expert researchers do.

Figure 13.1 gives examples of how these prompts can play out in an actual annotation thread. Expect variation: some comments will stay surface-level, while others will dig in. That mix is healthy. Over time, and with practice, your group can build toward deeper discussions that feel more like science – and less like homework.

13.4.2 Better Practices for SCA

SCA becomes most effective when grounded in thoughtful design – beyond a good platform. Annotations scattered in the margins will not magically build understanding. What matters is how you engage: whether your comments build on others' ideas,

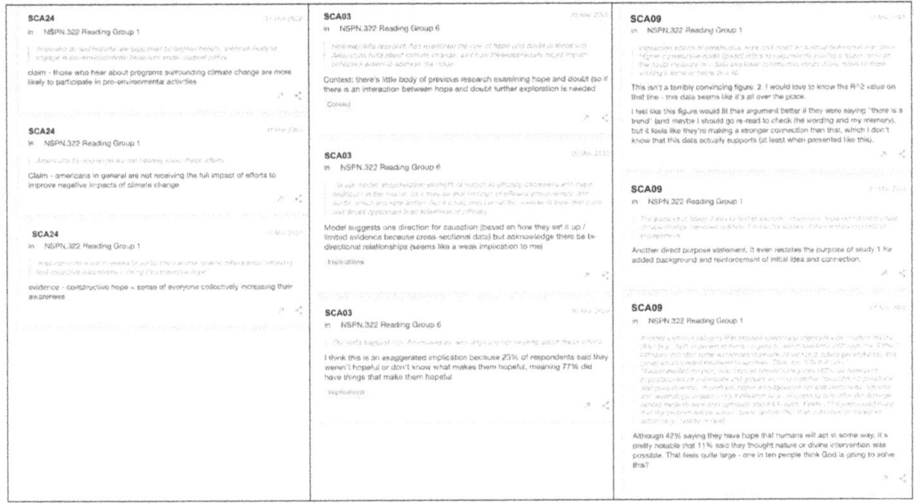

Figure 13.1 Examples of SCA with CERIC prompts.
Article annotations matched to CERIC prompts and tags in SCA practice.

Box 13.2 Better Practice: Advanced CERIC Prompts for SCA

After practicing with the beginning prompt set, level up by challenging the group with advanced prompts. These prompts assume a higher level of familiarity with research articles and SCA. They encourage deeper critical engagement, exploration of broader research contexts, and potential lines of inquiry.

- **Critical evaluation of Claims (C):** Compare the authors' main Claim with related studies you have encountered. Does this Claim align or conflict with previous research, and why?
- **Assessing quality of Evidence (E):** Critique the methods used to gather Evidence. Could the experimental design, sample size, or analysis technique affect how reliable or generalizable the results are?
- **Dissecting complex Reasoning (R):** Identify parts of the text where the authors' argument might rely on assumptions or gaps in logic. Suggest alternative explanations or perspectives.
- **Exploring multilevel Implications (I):** Extend the findings beyond the article's stated Implications. Consider policy impacts, interdisciplinary relevance, or ethical concerns.
- **Contextual shifts (C):** Imagine the study under different conditions – such as a larger sample, a different geographical setting, or a different theoretical framework. How might those contextual changes alter the research question and results?
- **Synthesize across readings:** Draw connections between this article and other research you have annotated. How do the Claims and Evidence in this study reinforce, challenge, or extend previous findings? What Reasoning sets it apart?
- **Propose future research directions:** Based on your CERIC analysis, suggest a follow-up experiment, a new research question, or a more refined approach to address remaining gaps or uncertainties.

whether you are asking questions that matter, and whether the group is structured in a way that keeps the conversation moving. Engagement needs fuel. So, some facilitators and groups add extra layers:

- **Short tutorials** on how to annotate effectively (Porter, 2022; Bjorn, 2023). These can answer questions and model better practices.
- **Participation credit** may be possible (if course related), and research shows it can nudge baseline participation upward – especially for students who tend to stay silent in ungraded conversations.
- **Readers will engage in different** ways. Some will annotate instinctively. Others will not unless there is a clear prompt or incentive. Variation in participation is normal. One reason SCA is effective is that it meets readers where they are, while providing a gentle nudge toward deeper thinking.

Over time, the nature of your own annotations may change. Initially, you may want to highlight or tag simple definitions or results. That is a good start. As you gain confidence, you will likely move toward analytical comments that connect ideas, question assumptions, or even compare across studies. Researchers describe this as "annotation development" – a shift from surface noticing to metacognitive awareness and critical engagement (Sun et al., 2023; Kalir & Garcia, 2019). One graduate student in an SCA study put it simply:

> I moved from "this is interesting" to mapping how each idea connects to the next. (Bjorn, 2023)

Annotation practices improve with practice, as does critical reading. A few simple habits can upgrade the experience from scattered commenting to a powerful shared resource:

- **Set a shared timeline:** Instead of just knowing the due date, agree on which pages to annotate by a specific date. This keeps the group moving together – and reduces last-minute overload.
- **Rotate micro-roles:** Try rotating weekly roles like group facilitator, lead annotator, connector, skeptic, summarizer, or resource finder. These help distribute effort and bring quieter voices into the mix.
- **Use threaded replies**: Instead of starting new comments, reply to existing ones. It keeps ideas together, and you can even tag key threads ("e.g., Evidence Thread 2") to track them.
- **Add helpful links:** Datasets, videos, or articles can add real value – especially if you include a one-sentence note explaining why the link helps.
- **Export and summarize:** After each week, have one member, perhaps the summarizer, download the annotation thread (in PDF or CSV format) and post a bullet-point summary. This becomes a built-in study guide.
- **Be mindful of privacy:** Use anonymous mode or pseudonyms if anyone prefers it, and never include private information in public comments. FERPA and GDPR rules also apply here.
- **Debrief live:** Take 10–15 minutes at the end of each cycle – either on Zoom or in person – to surface key takeaways and raise questions for class discussion.

Practicing these habits will strengthen your group's use of SCA and help convert reading time into productive, collaborative, and low-pressure learning. To sum up, these strategies support better comprehension and retention and will also prepare you for higher-stakes challenges like lab reports, exams, and team presentations.

13.4.3 Common Pitfalls and Fixes

Even the most diligent critical readers sometimes drift off the CERIC track during SCA. Momentum stalls, study sessions drag on, and participation fizzles. These issues are common. However, diagnosing and fixing them early saves mental energy and

Box 13.3 Knowledge Check: Question 13.2 Which statement best captures this chapter's argument that sustained student engagement in SCA depends on both structure and autonomy?

A. Weekly graded prompts plus mandatory role assignments (e.g., summarizer, skeptic) with no free-form responses.
B. Open, ungraded annotation with no instructor guidance, relying solely on peer curiosity.
C. Low-stakes environment plus structured prompts and optional roles, allowing group members to experiment while meeting clear expectations.
D. Instructor-led demonstrations only, followed by individual quizzes; annotations are optional extra-credit.
E. Strict rubric that awards points only for the quantity of highlights, irrespective of comment quality.

(Check your understanding using the Knowledge Check Key at the end of the book.)

turns margin commentary back into thoughtful analysis. Let's consider some of the most common pitfalls and how to fix them.

PITFALL 1: TECHNICAL FRICTION

- *What happens:* Login issues, browser glitches, or access delays disrupt the annotation process.
- *Fix:* Use tools integrated into your campus LMS (e.g., Hypothes.is embedded in Canvas). When technical access is streamlined, we annotate more consistently and the focus stays on learning (Jarrett, 2024).

PITFALL 2: ANNOTATION OVERLOAD

- *What happens*: In large groups, too many overlapping comments overwhelm readers. You may skim or ignore your peers' thoughts entirely.
- *Fix*: Use curated visibility – such as threaded replies or limited comment filters. A controlled visibility interface in a study of large online cohorts reduced overload and improved discourse quality (Almahoud et al., 2025).

PITFALL 3: SURFACE-LEVEL ENGAGEMENT

- *What happens*: Highlighting without interpretation (e.g., "interesting") stays at the surface and stalls critical thought.

- *Fix:* Rotate roles such as summarizer, skeptic, or connector (add more as needed). Rotating these roles deepens reasoning and strengthens collective understanding (Xu & Dai, 2024; Schellens et al., 2005). Scaffolded prompts further help to produce richer commentary (Brown & Croft, 2020).

PITFALL 4: UNCLEAR EXPECTATIONS OR TOOL CONFUSION

- *What happens*: Group members are unsure how to annotate, which sections matter, or how depth is measured.
- *Fix:* Begin with short, whole-group sessions on annotation strategy: What quality annotations look like and why they help learning. Clear group norms tend to lead to better engagement and comprehension.

PITFALL 4: QUIET OR SILENT PARTICIPATION

- *What happens:* Some people dominate the conversation, while others remain silent, especially those who are second-language readers or shy. The loud ones take others' silence as a cue to talk more, while the quiet ones cringe.
- *Fix:* Include equity features such as optional anonymity and rotating roles. While empirical data on SCA-specific anonymity effects are limited, broader studies on anonymity in group settings show it can balance participation and reduce status hierarchies (Brown & Croft, 2020; Surmiak, 2018; Wilkinson & Wilkinson, 2017).

PITFALL 5: ENGAGEMENT DECLINES OVER TIME

- *What happens:* Students start strong, but participation fades after a few sessions.
- *Fix:* Embed prompts that invite comparison across texts, critique assumptions, or apply theories to practice. Longitudinal research finds that prompts focused on reasoning sustain deeper cognitive involvement over time (Zhu et al., 2023).

Early diagnosis of these issues preserves your cognitive bandwidth for genuine analytical thinking. Fixing them early means you spend less time untangling tech glitches or frustration – and more time analyzing Claims, weighing Evidence, and sharpening your Reasoning.

13.4.3.1 Annotation Fatigue

Annotation fatigue sets in when marginalia morphs from surgical notation into dense graffiti. A group member complains about sore wrists, crowded pages, and – most telling – annotations that blur rather than clarify. The root cause is often unfocused, bored, or indiscriminate highlighting masquerading as engagement. Here is a quick diagnostic:

- **Signal:** Fluorescent highlight blankets half the page; notes capture facts and neglect claim or reasoning; annotation responses are one-word yes/no/emojis.
- **Likely source:** Sentence-by-sentence or spotty reading instead of CERIC hunting; anxiety about missing something.
- **Rapid fix:** Apply a two-pass approach. First pass – underline verbs that reveal CER. Second pass – highlight phrases that supply Evidence or Reasoning. If a sentence offers neither, leave it blank. Average page time decreases, and annotations regain diagnostic clarity.

Facilitation tip: This could be a responsibility of the skeptic role. Demonstrate the two-pass rule in a live, whole-group session before members start a new article. Compare a "good enough" page to an over-inked one so everyone sees efficiency rather than perfection as the target. Ask for input about the two-pass approach before it becomes a rule.

13.4.3.2 Off-Topic Tangents

Lively discussion can veer from an article about exoplanet detections to whether octopuses dream. Digressions feel delightful in the moment but corrode the chain from Evidence to Reasoning. Three derails appear repeatedly: anecdotal personal stories, future tech, and déjà-vu debates imported from other courses. Here is a quick diagnostic:

- **Signal:** Connections start with "This reminds me of …" or the margins fill with emojis or meme tangents not related to the article.
- **Likely source:** Unclear discussion goal, participants seeking relevance through autobiography, or the beginning of burnout.
- **Rapid fix:** Park the tangent in a "sandbox" – a reserved corner of the shared SCA space, and label them "curiosities." Acknowledge its value, add a tag, then pivot back with a question that uses article language verbatim ("How does the author's Evidence justify that Claim?"). Return to the sandbox in the final five minutes of a live session if time allows.

Facilitation tip: This fix could be a responsibility of the connector role. Rotate roles frequently as needed to prevent anyone from becoming bored or burnt out.

13.4.3.3 Citation Creep

Critical readers can sometimes treat citations like insurance policies: the more, the better. The manuscript bloats with parentheticals, while Reasoning thins. Annotators lose the through-line, and ironically, group trust erodes. Here is a quick diagnostic:

- **Signal:** Two or more citations appear at the end of every sentence, or paragraph topics shift to a summary of sources rather than analysis.
- **Likely source:** Fear of plagiarism, misunderstanding of synthesis versus compilation.

- **Rapid fix:** Apply the "1–3–1" ratio. Open a paragraph with one framing sentence that establishes the Claim. Spend up to three sentences weaving Evidence from other authors. Conclude with one sentence of Reasoning that names what the evidence does, not what it is. Anything that does not advance the ratio is moved to a footnote or falls away entirely.

Facilitation tip: This could be a responsibility of the lead annotator role. During synch sessions, ask the group members to highlight Reasoning sentences. If highlights are scarce, it becomes obvious what everyone needs to work on.

13.5 Worked Example

Even well-prepared SCA groups can encounter difficulties. An overload of highlights, vague commentary, or drifting discussion can cloud the article's argument and weaken collective analysis. This worked example follows a self-organized graduate study group as they navigate – and recover from – common SCA pitfalls.

Scenario: In a graduate seminar on Earth-system science, five students elect to form as an SCA group and annotate the classic research article by Alvarez et al. (1980), "Extraterrestrial Cause for the Cretaceous–Tertiary Extinction." They choose Hypothes.is as their SCA platform and implement a CERIC tagging system:

- Claim = green
- Evidence = blue
- Reasoning = yellow
- Implications = orange
- Context = pink

Each group member agrees to post five annotations asynchronously before their synchronous weekly check-in.

13.5.1 Early Signs of Overload and Distraction

Within a few hours, the shared document is flooded with 241 highlights and 138 comments. Many are clustered in the Introduction, where entire paragraphs are tagged as Evidence. Reactions like "Iridium here too – wow!" appear alongside Evidence references, but without clarification of their relevance. The group begins to feel overwhelmed. One member mentions sore wrists and blurred focus – a classic sign of annotation fatigue.

Simultaneously, enthusiasm for the topic sparks some off-topic comments. A conversation on iridium anomalies turns into a joking exchange about a road trip to the Chicxulub crater. While lighthearted moments can build rapport, the thread begins to drift from analysis to informal speculation.

In the group's collaborative summary, another pattern emerges: citation creep. Paragraphs end with strings of parenthetical citations (e.g., ... citation issues described by earlier work of Greenburg, 2009; Fister & Perc, 2016; Fowler & Aksnes, 2007; Simkin & Roychowdhury, 2005; Wilhite & Fong, 2012) with little explanation of how those studies connect to the argument. Instead, a better practice is to name the key point and the relevant citation. For example, on the topic of citation creep, we can consider citation distortion (Greenburg, 2009), citation cartels (Fister & Perc, 2016), excessive self-citation (Fowler & Aksnes, 2007), inaccurate citation (Simkin & Roychowdhury, 2005), and coercive citations (Wilhite & Fong, 2012).

These challenges – annotation fatigue, conversational drift, and citation overload – are common in SCA groups and rarely stem from apathy. Instead, they often result from:

- Anxiety about missing important points
- A lack of shared strategy
- Well-meaning enthusiasm that temporarily overshadows focus

If not addressed, these issues can worsen the overload and cause the group to fall apart.

13.5.2 Self-Correction through Member-Led Micro-Interventions

With no instructor in the loop, the group holds a short video meeting. A rotating facilitator leads a screen-share walkthrough to surface pain points and propose targeted fixes. They use familiar SCA roles – skeptic, connector, lead annotator, facilitator, and summarizer – to distribute facilitation. They identify several pitfalls and discuss fixes.

Pitfall 1: Annotation Fatigue → Fix: Two-Pass Rule

- First pass: Circle verbs that articulate the main Claim (green).
- Second pass: Highlight only phrases that provide key Evidence (blue).
- Result: Highlight density drops, and visual patterns emerge. Fewer comments feel more purposeful.

Facilitation Tip: The skeptic role could demonstrate a before-and-after comparison to model clarity over coverage.

Pitfall 2: Conversational Drift → Fix: Sandbox Tag

- A new tag, "curiosities," captures interesting but tangential insights (e.g., speculation about dinosaurs or crater tourism).
- This preserves energy and curiosity without derailing the analytical thread.

Facilitation Tip: The connector role can help bridge back to core CERIC prompts using argument-based language: "How does this Evidence support the Claim?"

Pitfall 3: Citation Creep → Fix: 1–3–1 Ratio

- (1) One sentence of Claim; (3) up to three Evidence sentences; and (1) one Reasoning sentence.

- Reasoning sentences are tagged in yellow to visualize where synthesis is (or is not) occurring.

Facilitator Tip: The lead annotator encourages group members to highlight yellow when they spot Reasoning – not add more sources, unless they specifically connect to CER. Also, the 1–3–1 ratio can flex to any variation that better meets the research, such as 1–5–1 or 1–3–2.

These interventions are modest and experimental – not formal policy. The group agrees to try them for one article and reassess next week.

13.5.3 Reannotation and Recalibrated Focus

Over the next week, the group trims redundant comments and revisits core ideas with a fresh perspective. Here's what improves:

- **Evidence tagging becomes more selective.** Blue tags (E1–E3) now appear under three key data points about global iridium spikes. A yellow tag follows: "Global distribution of iridium challenges volcanic origin – supports impact hypothesis."
- **Reasoning prompts richer discussion.** One comment – "How do trace-metal ratios rule out terrestrial sources?" – generates six threaded replies unpacking isotopic logic.
- **Contextual tagging gains precision.** Pink tags mark stratigraphic layers and geochronological benchmarks, clarifying what sets this paper apart from earlier extinction theories.

When the group meets again, the conversation shifts from note density to argument structure. They weigh whether the iridium anomaly alone warrants the authors' conclusion or whether later-discovered shocked quartz is the stronger evidence. This pivot – toward evaluating Claims rather than cataloging passages – marks a meaningful shift.

13.5.4 Reflection and Transfer

As group members reflect, they ask themselves, *what changed?* A consensus arises: Once we tried the 1–3–1 ratio, we realized how much summary was crowding out actual reasoning. Now it feels okay to annotate – not just collect quotes. A new thread begins, linking to the learning research on SCA:

- **Cognitive load decreases** as annotation clutter falls away (Chen et al., 2024).
- **Participation broadens,** and one quieter group member volunteers for the role of lead annotator on the isotopic analysis thread, facilitated by role rotation (Sun et al., 2023).

- **Strategies transfer,** and another group member applies the two-pass rule unprompted during a later annotation session on an ice-core study.

To sum up, these strategies worked for this group because the members were already motivated and willing to experiment. Not every group will respond the same way. Still, several insights generalize with care:

- **Demonstrate fixes visually:** Cleaning one page is often more persuasive than listing best practices.
- **Link each common pitfall to a micro-skill:** Matching problems to habits makes the group toolkit grow intentionally.
- **Close each session with a "next move" reflection:** Articulating what you'll try next helps consolidate metacognitive awareness.

These peer-developed strategies – like the two-pass rule, sandbox tagging, and the 1–3–1 citation framework – are not silver bullets. However, they can structure clearer thinking and more focused annotating when used flexibly. Their strength lies in adaptability because these micro-interventions are platform-agnostic, discipline-neutral, and compatible with both instructor-led and peer-organized study.

13.6 Chapter Key Takeaways

1. **SCA transforms reading into a collaborative process:** Rather than reading and annotating in isolation, SCA changes reading into a communal endeavor. SCA fosters deeper comprehension, shared inquiry, and more meaningful dialogue than solitary reading by enabling real-time text interaction and peer engagement.
2. **SCA facilitates critical thinking, metacognition, and evolving expertise:** Observing peers' annotations and thinking processes helps us question our assumptions and refine our interpretations. Over time, we can develop heightened metacognitive awareness, more robust critical thinking, and the confidence to explore complex materials more thoroughly.
3. **Low-stakes, peer-based environments enhance participation and motivation:** Readers who operate within a supportive, low-pressure framework – free to ask questions, challenge assumptions, and admit confusion – become more motivated to collaborate and learn. This inclusive environment decreases performance anxiety and promotes healthy risk-taking, strengthening engagement and retention.
4. **Effective SCA implementation relies on clear structure and scaffolding:** Research across multiple studies highlights the importance of structured guidelines, targeted prompts, and thoughtfully designed tools. Facilitators who incorporate explicit norms and expectations, role assignment, color-coding, and tagging of annotations help the SCA group transition from surface reading to deeper analysis.

14 Using CERIC with Generative Artificial Intelligence (GenAI)

> Heavy reliance on AI may hinder students' development of essential academic skills (e.g., reading, writing, analysis, and problem-solving), which ultimately reduces their capacity for independent thinking.
>
> —Taslim et al., 2025, p. 315

14.1 Overview

Chapter 14 explores the integration of CERIC with generative artificial intelligence (genAI) tools, discussing a range of emerging issues, from cognitive off-loading and overreliance to critical thinking development. This chapter begins by defining genAI and its relevance to academic research, then delves into how CERIC and genAI interact, highlighting both benefits and limitations. This chapter discusses the integration of genAI into CERIC review exercises, examining the potential for enhancing learning while avoiding overreliance and cheating.

14.2 What GenAI Brings to Research Education

Since the introduction of generative artificial intelligence (genAI) tools in 2022, such as ChatGPT (Generative Pre-trained Transformer) and Claude, genAI has evolved from a novel concept to a nearly ubiquitous tool in higher education – appearing in classrooms, labs, and libraries, and reshaping how we engage with knowledge (Rudolph et al., 2023). GenAI tools can appear to summarize dense readings in seconds, explain complex ideas in plain language, and even draft practice questions tailored to a student's needs. As a result, genAI is restructuring the way we study and learn in higher education. Its rapid rise invites reflection on learning goals and the creation of knowledge. What is an original work if we draw from the entire Internet to make it? Where does academic support end and cheating begin?

This chapter traces the foundations and expanding role of genAI in higher education and research, focusing on the mix of opportunities and tensions it raises for critical reading and thinking. If used skillfully and with skepticism, genAI can become a support – never a substitute – for deeper inquiry, sharper reasoning, and more expansive intellectual work. However, overreliance and cheating are real and pernicious risks with emerging consequences for how our minds think and learn.

14.2.1 Defining GenAI

At its core, the original genAI is a large language model (LLM) that can be considered as a probabilistic storyteller. Drawing on vast swaths of written text – including millions of books, articles, images, and recorded conversations across the internet – genAI models learn to predict what word, code fragment, or pixel should follow another (Bubeck et al., 2023; Rudolph et al., 2023). The model inside ChatGPT-4o, for instance, can "think productively," seeming to reason and can almost instantly generate seemingly fluent text, functional code, or compelling imagery (OpenAI, 2024). Current (as of 2026) public tools such as ChatGPT, Bard, and Claude demonstrate this capacity on demand, instantly producing outlines, debugging scripts, or generating study questions with striking coherence.

It is also true that these models are evolving rapidly. Claude, for instance, is now capable of acting like a coworker to whom you assign tasks, and with some comfort using a terminal window, a person can assign a Claude agent to the supervise other agents running various sub-tasks.

Across university campuses, genAI already streamlines core information workflows. A controlled evaluation showed that ChatGPT and Bing-AI retrieved 88% of gold-standard citations while executing systematic-review search strings 45% faster than PubMed (Gwon et al., 2024). A global survey of >23,000 university students reported that their most frequent academic uses of ChatGPT are brainstorming ideas (46%), summarizing readings (41%), and locating research articles (38%) (Ravšelj et al., 2024). Parallel polling of North American university students confirms the pattern: 34% use genAI to summarize or paraphrase text and 17% use it for on-the-fly translation (Tyton Partners, 2023). Additionally, 30% of students surveyed believe that they need genAI skills to gain employment (Lee & Low, 2024). Complementing those self-reports, an experiment reported ChatGPT's Spanish and Russian translations of clinical teaching leaflets clinically usable in 92% of cases, underscoring its growing role in language access (Dzuali et al., 2024). AI-based academic search tools are growing, and the list includes Ellicit, Scite Assistant, Scopus AI, Undermined, and Web of Science Research Assistant (Tay, 2025).

In the natural sciences, the influence of genAI is evident in the research literature. Early discipline-specific pilots are appearing. For example:

- **Bioinformatics:** The scGPT model adapts a transformer algorithm to detect cell-type-specific gene-expression patterns (Cui et al., 2024).
- **Materials science:** Deep generative models are being applied to inverse design problems in materials science, including variational auto-encoders, Generative Adversarial Networks, and diffusion models (Lu, S. et al., 2022).

Although today's genAI chatbots may feel brand-new, they are built on decades of work with adaptive-learning systems and intelligent-tutor-research (Holmes & Tuomi, 2022; Miao et al., 2021). What sets the current generation of models apart is their ability to generate new text, code, or images on demand and reflect on task performance while operating, not merely select from pre-written templates. Whether that generative power translates into real learning depends, in part, on thoughtful teaching strategies that keep our minds actively involved. The Probing Chain-of-Thought framework (ProCoT) embeds reflection checkpoints that require us to question, test, and refine a model's output (Adewumi et al., 2023). Similarly, the Critical AI Literacy framework builds human oversight into working with LLMs (Joseph, 2023).

14.2.2 Why GenAI Matters for Critical Reading and Critical Thinking

Critical reading and critical thinking drive scholarly inquiry, but they operate at different levels:

- **Critical reading** stays close to the page – What is the Claim? What counts as Evidence?
- **Critical thinking** compares and evaluates ideas and arguments across texts or fields. Recent studies show qualified support for the notion that genAI can support critical thinking (e.g., Gearing, 2024; Larson et al., 2024; Lee et al., 2025; Schulz et al., 2024). Others argue it could inhibit it (e.g., Lee et al., 2025).

LLMs can accelerate a first assessment of either a single text or a comparison of texts by flagging Claims and summarizing Evidence. However, LLMs work best when paired with conventional search skills (Gwon et al. 2024). Outputs, like auto-summarizations, taken at face value often miss methodological caveats or hallucinate sources (Ji et al., 2023). The risk of outsourcing Reasoning grows because LLM output appears very fluent, making it tempting to adopt as is, especially under deadline pressure (Gerlich, 2025). Indeed, "higher confidence in GenAI is associated with less critical thinking, while higher self-confidence is associated with more critical thinking" (Lee et al., 2025, p. 1).

Our countermeasure is the **ABC habit – Always Be Critical**. ABC is a default mental mode that we have highlighted throughout this book, encompassing the analysis, evaluation, and comparison of ideas within and between texts. ABC plus skillful usage can be a way for genAI to scaffold the development of scientific discourse skills. Used lazily, it erodes them. The balance point is to do the slow and challenging work of developing critical reading and thinking skills amidst LLMs' many shortcuts, made especially tempting when viewed as necessary to academic survival. While the genAI shortcuts may help with short-term academic survival (like getting tomorrow's assignment done tonight), the emerges research points to harm for independent thinking – putting at risk the most crucial skill for becoming a scientist.

Box 14.1 Common Confusion: How Is Critical Reading Related to Critical Thinking?

Imagine looking at a masterwork painting. Critical reading is the moment you lean in to inspect the brushstrokes, pigment choices, and tiny repairs. Critical thinking is when you step back, scan the entire gallery, and ask what the collection as a whole is about. These skills work in tandem – one zooms in, the other zooms out – and both are essential for scientific research.

- **Zoom-in work – critical reading**: Spot the article's Claim, check whether the Evidence and Reasoning actually backs it, and note any hidden assumptions or leaps.
- **Zoom-out work – critical thinking:** Compare that article's Claim with other articles, weigh rival explanations, and identify what gaps exist and what the field could do next.

Typical points of confusion:

- **Evidence vs. significance**: Jumping to "Is this important?" (thinking) before verifying "Is it even supported?" (reading).
- **Single paper vs. conversation**: Treating one study as the final truth instead of one voice in an ongoing debate.
- **Method details vs. big-picture judgment**: Either drowning in statistics tables or skipping them entirely. Both extremes miss how methods shape broader conclusions.
- **Acting too soon**: Rushing to apply findings – cite, share, design experiments – without first vetting the article's internal logic.

Key takeaway: Read critically first, think critically second. Check a painting's brushstrokes before judging the whole gallery.

14.2.3 GenAI in Higher Education

LLM tools can definitely lighten the reading load. Experiments with ChatGPT-4 show that it can compress a multi-page article into bullet-point summaries in seconds (Bubeck et al., 2023) and can be prompted to check those summaries for factual consistency almost as quickly. Yet, taking an auto-summary or its fact-check at face value risks skipping the critical interrogation that scholarship demands. Reviews of AI discovery platforms find that these tools are most useful when experienced searchers layer them onto conventional techniques, not when they are used as standalone replacements (Tay, 2025). Common pitfalls include:

1. **Surface engagement:** A chatbot may omit that a clinical trial used a convenience sample, giving the study a false aura of generality.
2. **Overreliance:** LLMs sometimes blend or invent findings. These "hallucinations" are well-documented (Ji et al., 2023). Accepting flawed or fabricated output and relying on it is where overreliance begins.
3. **Fabricated citations:** Fluent prose can cloak non-existent references; a citation and DOI that looks perfect yet resolves nowhere is a classic tell-tale sign. For instance, ChatGPT produced this result when we prompted it for "GenAI cognitive load articles":
   a. FAKE = Kohnke, L., & Moorhouse, B. L. (2025). How generative-AI-assistance impacts cognitive load during academic writing tasks. In K. Smith & J. Doe (Eds.), Advances in Artificial Intelligence in Education (pp. 345–360). Springer. https://doi.org/10.1007/978-3-031-71385-9_31← **Too perfect!**
   b. REAL = Kohnke, L., & Moorhouse, B. L. (2025). Enhancing the emotional aspects of language education through generative artificial intelligence (GenAI): A qualitative investigation. Computers in Human Behavior, 167, 108600. https://doi.org/10.1016/j.chb.2025.108600← **What those authors really wrote**.
4. **Eroded metacognition:** Outsourcing summaries dulls habits like questioning and cross-checking. The evidence is mixed: Calculators did not erode basic arithmetic (Hembree & Dessart, 1986; Ellington, 2003), but heavy GPS use *does* impair way-finding skills (Speake & Axon, 2012; Ishikawa, 2018).

The challenge with genAI in reading-intensive settings is not whether to use it, but how. Pairing the speed of LLMs with the skepticism of a trained reader remains the central task. Line-by-line engagement – testing each Claim against its supporting Evidence and underlying Reasoning – may feel laborious next to an instant summary, but that rigor is what separates insight from illusion (Gearing, 2024; Larson et al., 2024; Lee et al., 2025). The ABC practice reminds us that the appearance of machine fluency is not human understanding, and persuasive prose is not proof that an argument is sound.

When guided properly, genAI can lower barriers to comprehension and support deep engagement, especially for novice critical readers. Several studies highlight how AI tools can scaffold early-stage reading. Let's review a few of these.

Chen and Leitch (2024) and Kasneci et al. (2023; position paper) report that LLM "reading companions" can clarify jargon, condense dense scientific prose, and pose helpful follow-up questions – providing immediate scaffolding without oversimplification. Stadler et al. (2024) found that AI assistance during complex reading tasks reduced cognitive load and improved persistence. In a classroom pilot, Rudolph et al. (2023) showed that students who used ChatGPT-generated explanations alongside primary research articles reported higher confidence and longer engagement with difficult texts – but *only* when instructors required them to verify at least two claims per summary. Similarly, in a qualitative study, Kohnke and Moorhouse (2025) found that first-year university students described genAI as reducing anxiety and creating a more supportive environment for tackling challenging readings.

However, even as genAI shows promise, mounting evidence indicates it can undermine critical reading when left unchecked. Speed without scrutiny tempts us to trade comprehension for convenience.

Several large-scale surveys show that most undergraduates hold positive attitudes toward GenAI and that roughly one-third already use AI summaries instead of reading the assigned text (Alonso, 2024; Atewell, 2025; Moutlon-Tetlock & Shah, 2025). In addition, there is some evidence that students increasingly substitute AI-generated summaries for assigned readings, often skipping the original text (Alonso, 2024). Schulz et al. (2024) demonstrate experimentally that reliance on instant synopses reduces mental effort and harms subsequent recall, signaling an erosion of reading stamina. Van Dis et al. (2023) warn that habitual reliance on instant synopses (aka shortcut science) leads to shallow processing. For example, when student participants built essays directly from LLM-generated outlines, they frequently omitted key methodological details – for example, control groups or sampling frames – leading to flawed arguments (Lee et al., 2025). The same report showed that student participants who trusted genAI's first draft tended to use fewer independent checks and exhibited lower levels of self-monitoring. Complementing these data, Gerlich (2025, early-access) warns that habitual use of turnkey AI weakens the very reasoning processes genAI purports to augment.

These findings point to a paradox: GenAI can act as both an ally and an adversary in the reading process. Which role dominates depends on the reader's stance. When we continue to ask, "Does the Evidence truly support this Claim, and what Reasoning proves it?," we center human critique and limit genAI to a research partner. If we become over-reliant and stop asking critical questions, the tool becomes a thinking shortcut that weakens the skills we have been working so hard to develop in higher education and as researchers-in-training.

14.2.4 Brief Review of the Literature: GenAI in Higher Education

The evidence around genAI usage in higher education is starting to build. In the previous section, we reviewed some of the research about genAI and reading. Next, let's take a brief tour of current studies, focusing on the usage and outcomes for undergraduate and graduate students in science and technology. We will organize this min-review around three themes: higher-order thinking, unchecked use, and how to balance risks and benefits.

GenAI can sharpen higher-order thinking: When we embed GenAI within structured learning environments, users appear more likely to question, refine, and expand ideas. These benefits are particularly evident when GenAI use is paired with metacognitive prompts, peer discussion, and iterative revision tasks.

In a study of technology-education majors, Hakiki et al. (2023) asked technology-education majors to incorporate ChatGPT into their weekly project work; a brief self-check rubric prompted students to question and verify every AI suggestion. On the final test questions that required explanation and evaluation rather than mere recall, the ChatGPT-plus-rubric group outperformed the control section ($t = 5.42$, $p < .001$). Students also reported feeling more confident about their analytical and problem-solving skills, reinforcing the idea that AI can help if learners stay in control of their thinking.

Joseph (2023) showed that LLM tools can bolster students' ability to synthesize and critique information. Similarly, de la Puente Pacheco et al. (2024) found that debate sessions mediated by ChatGPT – and bracketed by faculty verification prompts – lifted critical-thinking scores in an ANCOVA analysis ($\eta^2 = .15$) compared with traditional debates. In an introductory chemistry course, Guo and Lee (2023) required students to critique and refine AI-generated essay drafts. As a result, participants came away with sharper questions, explored alternative explanations, and reported stronger confidence in their Reasoning.

These studies suggest that GenAI can enhance higher-order thinking *if* students are required to stay cognitively active. Without guided reflection, however, the technology alone does not sustain meaningful intellectual engagement.

These gains disappear when genAI use is left unchecked: While structured GenAI use can promote higher-order thinking, these gains often vanish when the technology is introduced without guidance.

Without scaffolds that encourage reflection or verification, students may begin to treat GenAI output as definitive, bypassing the reasoning process entirely – a major problem called overreliance. Lee et al. (2025) and Gerlich (2025, early access) both warned that unstructured GenAI use can lead to cognitive offloading: Students rely on the model's answers instead of constructing their own arguments. When prompted to explain their reasoning, many deferred to ChatGPT's phrasing without evaluating its logic or relevance. In these cases, the tool becomes a shortcut, not a scaffold. Passi and Vorvoreanu (2022) likewise found that learners often struggle to determine how much trust to place in AI-generated text. Church (2024) extended this finding in a study of undergraduate fact-checking behavior, where students reviewing GenAI summaries overlooked key inaccuracies and rarely traced outputted claims back to primary sources. When we cannot discern what to trust, we cannot build strong Claims using an argument from Evidence, with the potential to seriously hinder a career in science.

However, even well-intentioned integration can fall short. At Georgia Gwinnett College, chemistry faculty introduced ChatGPT into peer discussion sessions to boost engagement. While the tool prompted more students to speak up, Guo and Lee (2023) reported that roughly one-third of instructors saw no improvement in argument quality. Participation increased, but Reasoning did not.

These findings point to a cautionary lesson. GenAI can spark curiosity, but it will not build intellectual muscle on its own. Without deliberate prompts, clear expectations, and space for reflection and critique, the technology may encourage surface engagement without supporting the deeper work of academic thinking.

Aim to keep the benefits while curbing the risks: To preserve the intellectual benefits of GenAI while avoiding its cognitive shortcuts, researchers consistently emphasize the need for structured verification (Wang, Wang & Su, 2024).

When students are asked to interrogate the AI's output – rather than accept it at face value – the technology becomes a tool for reasoning, not a replacement for it. Well-crafted prompts, reflective scaffolds, and structured models, such as ProCoT and the Critical AI Literacy framework, suggest some guardrails. These frameworks require learners to analyze the prompt, evaluate the context, and assess the task alignment of the AI's response (Adewumi et al., 2023; Miao et al., 2021).

Open-ended assignments further strengthen this effect. When students are asked to spot flaws in ChatGPT's summary of a research article or rewrite an inaccurate explanation, the AI shifts from solution generator to intellectual sparring partner (Wang et al., 2024). These kinds of challenges force students to articulate why a response is weak, defend an alternative interpretation, or justify their revisions using disciplinary knowledge.

The key point emerging from the research is that structured critique is not optional. As Gearing (2024) cautions, simply layering GenAI into coursework does not guarantee deeper or critical thinking. Without deliberate design, it is too easy to act from academic survival mode and produce more text with less insight – defaulting to fluent but shallow and unchecked responses.

The takeaway is that genAI does not build critical analysis by default. Deliberate course policies, structured rubrics, and self-assigned reflection tasks are essential to ensure that GenAI serves scholarly thinking rather than short-circuiting it. When well integrated, the tool can support greater human cognitive effort. When left unstructured, it risks displacing it.

14.2.5 Ethical Guidance for GenAI Usage

When applying the CERIC framework, ethical scientific practice centers on transparent Reasoning. For every Claim we articulate, explicitly link the supporting Evidence, lay out the Reasoning that connects them, and label any Implications or Context drawn from secondary sources. If we enlist genAI to speed up literature searches or draft provisional summaries, we want to stay on the right side of ethics. This means:

Box 14.2 Knowledge Check: Question 14.1 Which statement best reflects the caution about letting genAI create quick summaries of scientific articles?

A. Large language models have already achieved artificial general intelligence (or AGI), eliminating the need for human review.
B. GenAI's speed can hide logical weaknesses, so readers must still perform deliberate CER scrutiny.
C. Benchmark studies have proved genAI summaries are largely error-free, making output critique unnecessary.
D. Because genAI writes so fluently, traditional critical-reading instruction should be phased out of higher education.
E. Researchers' only remaining role is to supply data that future genAI systems can learn from.

(Check your understanding using the Knowledge Check Key at the end of the book.)

- Disclosing its use in a methodology section,
- Cross-checking each factual statement against the primary article, and
- Citing and checking the original source – not the chatbot – as exemplified in rigorous analyses of foundational studies.

This practice comes with a key caveat. Before beginning any CERIC-based assignment, whether working alone or in a group, review the most current guidance issued by your program, instructor, and institution on Research Ethics or Academic Misconduct. While many are still emerging, local academic rules supersede any general advice in this chapter. It is always our individual responsibility to verify and seek clarification when in doubt.

Finally, remember that CERIC is a tool for critique, not a shortcut to conclusions. Apply the better practices in Part II of this book – avoid copying any prose verbatim, refrain from inserting unverifiable references, and never allow unvetted summaries to stand in for your own examination of figures, tables, and methods. Also, remember the ABCs: LLM summaries of research articles are likely incorrect, even if they are convincing. By pairing your institution's academic integrity guidelines with the disciplined transparency baked into CERIC, it is possible to uphold scholarly integrity while tapping the efficiency benefits that modern genAI can provide.

14.3 Integrating GenAI into CERIC Reviews

Let's examine a skillful application of genAI for analyzing a research article. We will show how an LLM chatbot – often a quick-fire summarizer prone to hallucinations – can become a disciplined contributor when its answers are subject to scrutiny. By incorporating prompts to question genAI output and remaining skeptical, our workflow can utilize genAI in a manner that fosters stronger scientific literacy (Adewumi et al., 2023).

Let's step through an example of using genAI to expand a forward-looking literature search. We begin by drafting a workflow; note that the full transcript interaction with prompts is provided in the Supplement.

14.3.1 Setup a Workflow

1. **Identify a starting article:** Pull the full text version and annotate the article for specific elements the article for specific elements, such as the methodology, or if broader, run a manual CERIC review.
2. **Pose targeted genAI queries:** Ask the model to fill in specific gaps; for example, ask it to generate explainers for each lab technique used in the study.
3. **Reconcile genAI output with the text**: Return to the article, confirming or rejecting each genAI suggestion; fact-check each and discard anything unsupported.
4. **Expand Context (as needed):** Use a follow-up prompt to locate real follow-on studies, then verify the references and integrate them into your literature map.

14.3.2 Run the Workflow

We applied this workflow to Jinek et al.'s (2012) article "A Programmable Dual-RNA–Guided DNA Endonuclease in Adaptive Bacterial Immunity" about the revolutionary CRISPR–Cas9 gene-editing methodology. A ChatGPT-4o (July 2025 build) query of "Which post-2012 assays now test CRISPR off-target effects?" returned three validation assays, which we then prompted GPT-4o to define in plain language:

- **Validation assay #1 – PCR/Sanger sequencing:** A basic method to amplify and read short DNA sequences to check for specific edits.
- **Validation assay #2 – Deep sequencing with Cas-OFFinder/CRISPOR:** High-throughput tools that detect off-target edits by comparing many DNA reads to predicted cut sites.
- **Validation assay #3 – T7 endonuclease screening:** A fast test that uses an enzyme to detect mismatches in DNA strands caused by genome editing.

Next, we ran a manual CERIC pass of the Jinek article, confirming that none of these assays appeared in the study. Then, we switched over to Google Scholar because GPT hallucinates and searched the literature for these assays. We located many post-2012 articles that implemented each of them. We narrowed that search to four studies of *validation methods* that we could add to our literature pool:

- Bae et al. (2014) – Cas-OFFinder and CRISPOR
- Concordet & Haeussler (2018) – Cas-OFFinder and CRISPOR
- Dehairs et al. (2016) – PCR/Sanger sequencing to identify INDELS (insertion or deletion of bases into DNA)
- Sentmanat et al. (2018). – Cas-OFFinder and CRISPOR

Box 14.3 Knowledge Check: Question 14.2 Which statement reflects the best use of CERIC plus genAI in critical reading practice?

A. To declare that CERIC is obsolete because genAI already captures every CERIC element in its summaries.
B. To urge instructors to ban genAI tools altogether and rely solely on manual reading strategies.
C. To predict that genAI will soon embed CERIC automatically, eliminating the need for human analysis.
D. To show that CERIC and genAI always reach identical conclusions, proving redundancy in running both.
E. To highlight that CERIC supplies the Evidence and Reasoning tracing genAI may overgeneralize or miss, cautioning readers against uncritical genAI use.

(Check your understanding using the Knowledge Check Key at the end of the book.).

14.3.3 Evaluate the Results

If we were writing a literature review, we would delve deeper into each article, conducting a comprehensive CERIC review and comparison and a synthesis matrix (see Chapter 12). The exercise illustrates the value of genAI in broadening the Evidence base, while CERIC safeguards accuracy by forcing us to:

- Verify each suggested method against the focal article;
- Confirm that the follow-up citations are legitimate; and
- Assess whether the newer studies genuinely strengthen – or merely update – the original research question guiding our review.

In summary, a CERIC-first, genAI-second workflow captures machine speed and supports rigor, assuming the ABC reader always verifies and cross-checks output against source.

14.4 Worked Example

Now let's match off an automated genAI summary against a full CERIC analysis to show, in concrete detail, where machine speed helps and where human reasoning remains essential. We fed Humata.ai (July 2025 build) – an LLM that generates article digests – the Jinek et al. (2012) CRISPR-CAS9 study that we paired earlier with ChatGPT. We prompted it to identify the article's CERIC elements. Then, we compared the LLM's output with a manual CERIC review conducted by an experienced critical reader (see Chapter 8). We share short verbatim excerpts from each review to highlight key similarities and differences between the reviews. A full transcript of the LLM interaction appears in the Chapter 14 Supplement.

For context, Humata.ai is a genAI tool that is designed to automatically summarize research articles (Nst & Dewi, 2025). While genAI can efficiently extract surface-level information, it often overlooks important evidentiary details, methodological limitations, and nuanced Reasoning. We demonstrate here that such tools must be paired with independent analysis to obtain a robust article review. Let's dive in.

14.4.1 Draft a Workflow

First, we set up a workflow:

1. **Run a baseline CERIC pass manually:** Annotate the article's Claim, Evidence, Reasoning, Implications, and Context; note any uncertainties.
2. **Pose targeted GenAI queries:** Feed the article in Humata.ai and prompt it with targeted questions about each of the CERIC elements ("*What is the article's main claim?*")
3. **Reconcile GenAI output with the text**: Return to the article, confirming or rejecting each genAI suggestion; discard anything unsupported.
4. **Expand Implications (as needed):** Use a follow-up prompt to locate real subsequent studies, then verify the references and integrate them into our literature map.

Here is the focal article:

> **Title**: "A programmable dual-RNA–guided DNA endonuclease in adaptive bacterial immunity"
>
> **Citation**: Jinek, M., Chylinski, K., Fonfara, I., Hauer, M., Doudna, J. A., & Charpentier, E. (2012). *Science*, 337(6096), 816–821. https://doi.org/10.1126/science.122582
>
> **Abstract**: Clustered regularly interspaced short palindromic repeats (CRISPR)/CRISPR-associated (Cas) systems provide bacteria and archaea with adaptive immunity against viruses and plasmids by using CRISPR RNAs (crRNAs) to guide the silencing of invading nucleic acids. We show here that in a subset of these systems, the mature crRNA that is base-paired to trans-activating crRNA (tracrRNA) forms a two-RNA structure that directs the CRISPR-associated protein Cas9 to introduce double-stranded (ds) breaks in target DNA. At sites complementary to the crRNA-guide sequence, the Cas9 HNH nuclease domain cleaves the complementary strand, whereas the Cas9 RuvC-like domain cleaves the noncomplementary strand. The dual-tracrRNA:crRNA, when engineered as a single RNA chimera, also directs sequence-specific Cas9 dsDNA cleavage. Our study reveals a family of endonucleases that use dual-RNAs for site-specific DNA cleavage and highlights the potential to exploit the system for RNA-programmable genome editing.

14.4.2 Generate a Glossary of Key Terms

This article is very jargon-heavy, and we realized immediately that we need to define some key terms. We initially generated this list using ChatGPT and then refined it by trimming and verifying the content. Here's our glossary:

3′ end = The tail end of a DNA or RNA strand. The seed region that guides Cas9 binding is found here.

Adaptive immunity (in bacteria) = A bacterial defense system that stores viral DNA snippets and uses them to recognize and destroy future invaders.

Blunt double-strand break (DSB) = A clean, even cut through both DNA strands – Cas9's default cleavage pattern.

Cas9 = A bacterial enzyme that cuts DNA at precise locations when guided by RNA.

Chimeric guide RNA = A single engineered RNA that merges crRNA and tracrRNA, simplifying Cas9 programming.

crRNA (CRISPR RNA) = Provides the address for Cas9, guiding it to the correct DNA sequence.

Dual-RNA guide = Two RNA molecules – crRNA and tracrRNA – work together to direct Cas9 activity.

Endonuclease = An enzyme that cuts DNA at internal points rather than at the ends.

Functional genomics library = A collection of guide RNAs is used to test gene function across the genome systematically.

HNH / RuvC-like domains = Cas9's two catalytic regions; each cuts one strand of the DNA.

Knockout model = An organism or cell in which a specific gene has been disabled to study its function.

Loss-of-function mutation = A genetic change that disables a gene or protein – used to test Cas9's mechanism.

Miniaturization = Simplifying biological systems (like CRISPR) to make them easier to use in research or therapy.

NGG = A short DNA motif (any base followed by two guanines) is required for Cas9 to bind and cut.

PAM (Protospacer Adjacent Motif) = A short DNA sequence (e.g., NGG) next to the target site is essential for Cas9 activity.

Plasmid = A circular piece of DNA is often used in lab experiments to test Cas9 cutting.

Programmable = Describes Cas9's ability to be re-targeted to different DNA sequences by swapping its guide RNA.

Protospacer = The DNA sequence that matches the guide RNA and gets cut by Cas9.

Reasoning = The logic connecting observed evidence (e.g., DNA cuts) to the claim that Cas9 is programmable.

Seed region = A critical 13-nucleotide stretch in the crRNA that must match the DNA exactly for Cas9 to cut.

Single-strand nick = A cut in only one strand of the DNA is often used to test which part of Cas9 is active.

Specificity of Cas9 = Cas9's ability to target and cut only the intended DNA sequence.

Streptococcus pyogenes = The bacterium from which the Cas9 used in this study was initially discovered.

TAL effector nucleases (TALENs) = Custom-designed proteins that cut DNA by recognizing specific nucleotide sequences. Before CRISPR, TALENs were widely used for genome editing, though they required more effort to design and construct.

tracrRNA (trans-activating crRNA) = The partner RNA that activates and stabilizes the crRNA–Cas9 complex.

Zinc-finger nucleases (ZFNs) = Engineered proteins that bind and cut DNA using zinc-finger domains linked to a cutting enzyme. They were among the earliest genome-editing tools but required complex protein engineering for each new target.

After generating the glossary, we manually compared the Humata.ai review to the human CERIC review. In each paired excerpt below, we underline the information noted by only one source and leave as unformatted text information identified by both sources.

14.4.3 Run the Workflow

Here's what happened next.

14.4.3.1 Claim

Humata.ai announced:

> The CRISPR-Cas9 system functions as a programmable, RNA-guided DNA endonuclease capable of precise, sequence-specific genome editing.

The human reviewer, using CERIC, framed the Claim more tightly:

> Cas9 is a site-specific DNA-cutting enzyme that needs both a crRNA and a tracrRNA, along with an NGG protospacer-adjacent motif (PAM), to make blunt double-strand breaks three base pairs upstream of the PAM. These two RNAs can be joined into a single chimeric guide, which makes Cas9 easy to program for targeting different DNA sequences.

Both statements capture programmability, dual-RNA guidance, and PAM dependence, yet only the human quantifies **the motif (NGG), the cut geometry (-3 bp), and RNA fusion** – details that later anchor the evidence.

14.4.3.2 Evidence

When asked to list supporting data, Humata.ai offered the following line:

> The study shows that the mature crRNA alone is insufficient for cleavage, but when combined with tracrRNA, Cas9 can be programmed to target and cleave any dsDNA sequence efficiently and specifically.

Efficiency aside, that sentence is largely accurate but omits where the result appears and how specificity was measured. The human CERIC pass instead cited the relevant figure and crRNA alone result.

> Plasmid and oligonucleotide substrates were cut only when both crRNA and tracrRNA were included (**Figure 1A–B, p. 816**), confirming that Cas9 requires both RNAs to function; using **mature crRNA alone resulted in no detectable cleavage**.

Humata wrote:

> Identification of a short PAM sequence adjacent to the target site as essential for recognition and cleavage further supports the claim.

The human review drilled deeper:

> Single-nucleotide scanning mutagenesis identified a **critical 13-nucleotide "seed" region** at the **3′ end of the protospacer**, along with an NGG PAM sequence located on the non-complementary strand (**Figures 3D–E and 4A–C, pp. 818–819**).

By anchoring to figure panels and noting precision details, the human review adds crucial details and nuance.

14.4.3.3 Reasoning

Humata's core inference read:

> The demonstrated ability of Cas9, guided by dual-RNA structures, to introduce precise double-strand breaks at specific sequences shows that the system is inherently programmable.

The human reviewer linked cause to effect more explicitly:

> **Because disabling the HNH or RuvC domains causes Cas9 to cut only one DNA strand (a single-strand nick)**, it suggests that each domain is responsible for cleaving one strand. When the full, functional Cas9 is used, it restores clean double-strand breaks – showing that these cuts are guided by the RNA's match to the DNA sequence, not by the shape or structure of the DNA itself.

Again, the human notes crucial details and names a negative control – loss of function mutants, defined by disabled HNH or RuvC domains – that Humata omitted. The key difference that this illustrates is how a human CERIC reviewer can trace the logic chain rather than gloss over and accept empirical correlation.

14.4.3.4 Implications

Humata predicted sweeping utility:

> The ability to engineer guide RNAs to target virtually any DNA sequence offers a powerful alternative to existing genome-editing tools.

The human reviewer accepted the alternative-to-ZFNs point but tempered the reach:

> Because Cas9 can be reprogrammed **using just a 100-nucleotide guide RNA**, there's **no need for labor-intensive protein redesign** for each new DNA target. This makes the system much more efficient and flexible, providing a foundation for creating **gene knockout models, testing gene therapy strategies, and building large-scale libraries for functional genomics research**.

Again, the devil is in the details, which Humata skips almost entirely. By specifying guide length and downstream applications already demonstrated in the literature, the human stays within verifiable bounds, whereas Humata's "virtually any DNA sequence" overgeneralizes the NGG constraint.

14.4.3.5 Context

Humata supplied a broad frame:

> The study is centered on harnessing the natural CRISPR-Cas9 immune system in bacteria for genome editing purposes.

The human version tied the work to unresolved miniaturization:

> Prior to 2012, Cas9's catalytic mechanism, RNA requirements, and DNA-targeting rules were unknown; by focusing on the **Streptococcus pyogenes system**, the authors asked **whether RNA molecules and Cas9 could be miniaturized into a single, user-programmable component**.

Only the human identified the animal model and the specific knowledge gap that the study addressed. The human also notes the study's main research question. Humata missed all of the Context for an easily readable, but overly generalized summary.

14.4.4 Synthesize the Results

Our head-to-head test reveals a clear pattern: Humata.ai delivers quick, broad-brush summaries that help readers get oriented, but it overgeneralizes and stumbles on the finer points of scholarship. Key shortcomings include:

- Missing figure or table references, so claims cannot be traced back to data;
- Vague, inflated language ("any dsDNA," "virtually any sequence") that overstates results;
- Poor attention to the step-by-step Reasoning that links Evidence to the Claim.

In contrast, the human CERIC review – although slower and more effortful – captured the article's argument and crucial details. It pinpointed the precise Claim, catalogued supporting Evidence, noted several lines of the Reasoning, located the study's gap and research question, and spelled out both immediate and longer-term Implications.

The takeaway here is consistent with the rest of this chapter. GenAI summaries are useful launchpads, never final destinations. When readers apply a deliberate CERIC pass – while also verifying each AI-generated point against the text and confirming that cited studies exist – we can combine machine speed with human rigor. Until LLM tools can routinely tag figures, generate counterfactuals, and accurately analyze argument structure, a human CERIC review remains a strong defense against seemingly fluent but superficial or overgeneralized output.

14.5 Conclusion and Future Directions

When it comes to how we adapt to the now genAI world we now live in, our task as researchers and educators is to walk an unknown path and decide what to carry forward: machine speed, human effort, or some combination of both. Here we adopt a pragmatist stance. We welcome genAI's knack for alphabetizing citations or generating a custom glossary at lightning speed. However, we should never surrender critical thinking, being aware that every genAI output removes crucial details about the research, both nuance and argument. The emerging literature on genAI in higher education is blunt: plagiarism, hallucination, and overreliance are the current tradeoffs for speed gains.

One remedy is structure. Prompt the model using a CERIC-based workflow, manually perform validation and cross-checks of all output, and keep the human thinker responsible for evaluating the argument. Treat the tool like a bright new intern: great at sorting multiple-choice sheets, yet utterly unqualified to write the exam. The open research question is whether long-term, scaffolded use of genAI increases our capacity to undertake challenging, time-consuming work or dulls us into a state of complacency.

To recap, the emerging research gives us some guidance:

- **When prompts are structured, genAI reasoning improves**. Well-scaffolded tasks with built-in verification allows for quality checks on output.
- **When structure is absent, genAI reasoning erodes**. Users who trust the first AI draft report fewer independent checks and weaker metacognitive reflection. Remember ABC: always be critical.
- **Instructional and workflow design, not the LLM, set the course**. CERIC-driven workflows can steer us toward skepticism; copy-and-paste tasks drift us toward plagiarism and overreliance.

Thus, genAI shifts the nature of critical thinking toward verification, response integration, and oversight of tasks (Lee et al., 2025). Because the models are increasingly fluent, and therefore, convincing at first glance, as researchers, we must be proactive to center and protect our critical reading and thinking practices.

14.5.1 Implications for Research and Learning

GenAI nudges research practice by generating methods digests, clarifying terminology, and suggesting experimental tweaks. Convenient, no doubt – but it can also flatten nuance, oversimplify narratives, and erase crucial experimental caveats (Cotton, Cotton, & Shipway, 2023). In our coursework, an effective countermeasure is deliberate assignment design: perform a CERIC analysis of both the source article and the genAI summary, checking all references, and manually writing notes and reflections that expose the line of thought. If instructors and advisors do not require this of you, require it of yourself. This is a pivotal moment when we can choose to level up our study strategies, recognizing that the deepest value of higher education is – and has always been – learning to read and think critically.

In research, today's students are tomorrow's mentors. Our habits as learners will write the next chapter of academic culture. The early data are encouraging: When we probe genAI output rather than simply copy-paste it, our Reasoning skills can strengthen, and our evaluation habits can mature from surviving to thriving (Kostka & Toncelli, 2023). When we treat genAI as scaffold, we perform better on transfer-of-learning tasks (Pallant et al., 2025). To ensure these gains are realized, we must integrate some of these genAI practices in our teaching and research:

1. **Build verification into every genAI task**. ABC: Validate and cross-check all output before acceptance.
2. **Trace every Claim to its Evidence and Reasoning**. Whether the assertion comes from *Nature*, a classmate, or a chatbot, the habit of critical thinking remains the same.
3. **Cultivate calibrated trust:** Use genAI as a bright intern, never an unquestioned source – and repeat that out loud in class and to colleagues and friends.

14.5.2 Looking Ahead from GenAI to AGI

LLMs, such as ChatGPT-4 and LLaMA-2, already permeate higher education, yet they still operate in what DeepMind terms the "emerging" tier of artificial general intelligence (AGI; Dickson, 2023). They display remarkable fluency inside well-defined prompts but lack the adaptive, autonomous reasoning that characterizes true AGI (Miao et al., 2021), which is well into development (Morris et al., 2023). As systems climb the capability ladder, higher education will confront fresh questions: Who owns an AI-assisted publication? How do we grade work partly written by a model? What counts as "original" insight?

Preparing ourselves for a future that likely includes true AGI begins with developing strong critical reading and thinking habits now. By keeping these skills at the center of our research and learning, we can help to ensure that faster machines lead to deeper minds – human minds – exactly the trajectory of critical thinking that higher education was built to develop.

14.6 Chapter Key Takeaways

1. **GenAI can be a support, not a substitute, for critical reading and thinking**. GenAI can accelerate the early stages of academic work by summarizing methods and jargon, identifying potential Claims, or explaining dense prose, but it cannot replace the deeper scrutiny that scholarly critical reading requires. Chapter 14's comparison between a Humata.ai summary and a human CERIC article review highlights this divide: the AI captured surface-level ideas, while the human reader traced precise methods, Claims, and Reasoning. Fluent text may appear convincing, but it often skips over figure references, overgeneralizes, or skips experimental limitations. CERIC helps keep us anchored in the internal logic of a study – tracking how Claims are justified, never assuming they are accurate.
2. **Structure is the safeguard against cognitive offloading**. Structured use of GenAI – through prompts, reflection checkpoints, and CERIC-based workflows – helps users stay intellectually engaged and reduces the risk of cognitive offloading. Without this scaffolding, we can easily default to accepting genAI output at face

value, skipping verification steps, or abandoning our own thinking. Studies reviewed in the chapter show that unstructured genAI use leads to poor source validation, factual errors, and weaker metacognitive reflection. In contrast, structured assignments require learners to verify summaries, support deeper processing, and always remain critical. The message is clear: structure does not just improve learning – it protects it. GenAI's value depends, not on the tool itself, but on the design choices surrounding its use.

3. **GenAI can scaffold comprehension but must be paired with critical habits**. GenAI tools can improve reading access by translating jargon, condensing complex methods or theories, and generating clarifying questions, especially helpful for readers new to a field and facing steep learning curves. When paired with verification tasks, genAI can improve comprehension and persistence, but these gains disappear when speed replaces skepticism. Auto-summaries, while convenient, risk eroding critical academic habits, such as tracing methods, comparing studies, or vetting assumptions. Critical reading habits – like verifying Claims, checking Evidence, and testing Reasoning – must remain central, especially when genAI makes the surface work feel easier.
4. **CERIC remains essential for verifying, evaluating, and deepening AI-assisted scholarship**. The chapter's head-to-head comparison of genAI and human CERIC reviews reveals, albeit anecdotally, that automated summaries miss key dimensions of scientific argument – specific experimental details, figure references, Reasoning chains, and more. While LLMs can quickly restate surface Claims, only manual review uncovers the architecture of an argument, allowing us to test its coherence. These distinctions matter, especially in scientific argumentation, where seemingly minor differences can carry major impacts. Until genAI tools can routinely and accurately displace field-specific insight and logic, CERIC remains an essential framework, ensuring that machine speed does not displace scholarly precision.

Knowledge Check Key

Chapter 1

Question 1.1 Correct Answer: A

Explanation: The text notes that many programs assume students "will figure it out on their own" without direct, structured training (Sverdlik, 2019). This creates an unspoken expectation, or hidden curriculum, wherein learners struggle with reading skills that are treated as automatic or self-evident.

Why are other options incorrect?:

B (Incorrect): The text does not mention publishers hiding guidelines behind paywalls as the chief reason for any hidden curriculum.

C (Incorrect): While some educational resources have expanded to include videos and webinars, Chapter 1 does not indicate a widespread abandonment of text-based assignments.

D (Incorrect): Organizations such as the National Association of Biology Teachers and the American Physical Society encourage critical reading of research articles, and they do not discourage active reading.

Question 1.2 Correct Answer: B

Explanation: A primary research article is an original investigation produced by the researchers, detailing new data, methodologies, and interpretations (Creswell & Plano-Clark, 2017).

Why are other options incorrect?:

A (Incorrect): Secondary sources, such as edited collections, often summarize others' work instead of offering original data.

C (Incorrect): Classroom handouts that compile excerpts from many sources are typically learning tools, not original studies.

D (Incorrect): While theoretical articles can be part of the primary literature, Chapter 1 emphasizes that standard primary research articles include evidence-based data and interpretations rather than purely editorial commentary.

Question 1.3 Correct Answer: B

Explanation: To screen and improve the quality of research articles before publication.

Why are other options incorrect?:

A (Incorrect): Answer A is incorrect because the purpose of peer review is to evaluate and improve the quality and credibility of research, not to influence journal competition.
C (Incorrect): Answer C is incorrect because the peer review process focuses on evaluating the quality of individual research articles, not on increasing the number of journals.
D (Incorrect): Answer D is incorrect because peer reviewers receive no standardized training, as highlighted in Section 1.5.2.

Chapter 2

Question 2.1 Correct Answer: B
Explanation: The Toulmin et al. three-part model (data, warrant, claim) can oversimplify the complexity of real-world scientific arguments involving multiple subclaims or lines of reasoning.

Why are other options incorrect?:

A (Incorrect): The model does address data and claim, but lacks explicit guidance on evaluating evidence quality.
C (Incorrect): The model was developed to analyze all kinds of scientific arguments, not just ethical or philosophical ones.
D (Incorrect): The model does not require specialized statistical knowledge by default; indeed, its shortcoming is oversimplification, implying deep statistical analysis would be unwarranted.

Question 2.2 Correct Answer: B
Explanation: Rationale is motivation based on past work, and thus belongs in the Context part of the analysis. Reasoning does indeed link evidence to the claim.

Why are other options incorrect?:

A (Incorrect): Funding does not provide a scientific motivation for a study, even if it may be of practical importance. Reasoning does "explain" results, but more specifically is used to justify the Claim.
C (Incorrect): Rationale is based on past work that motivates the study, whereas Implications encompasses future directions after the study. Data forms part of the Evidence, while Reasoning connects Evidence to the Claim.
D (Incorrect): These elements are distinguishable based on the arguments above.

Question 2.3 Correct Answer: D
Explanation: Chapter 2 notes that the Abstract is the "most CERIC-rich" part of an article, potentially containing all five elements, even in brief.

Why are other options incorrect?:

A (Incorrect): The Title might highlight the primary claim but rarely captures all of the CERIC elements.
B (Incorrect): The Methods section typically focuses on Evidence and Reasoning, but is unlikely to include the rest of the CERIC components.
C (Incorrect): The Discussion section usually has a Claim, Reasoning, and Implications but rarely addresses Evidence or Context.

Chapter 3

Question 3.1 Correct Answer: C
Explanation: A Claim, as described, must be a clear, declarative statement of new knowledge backed by evidence and reasoning.

Why are other options incorrect?:

A (Incorrect): Chapter 3 emphasizes clarity and plain language, not technical jargon.
B (Incorrect): A hypothesis is a potential answer to a research question, but a Claim is an established conclusion arising from evidence; they are not simply the same thing in different tenses.
D (Incorrect): A Claim must directly answer a research question and be supported by evidence, so it cannot be just an unrelated opinion.
(Incorrect)

Question 3.2 Correct Answer: C
Explanation: A cause-effect Claim states explicitly that one phenomenon produces or leads to another. Here, the exposure to UV light directly causing higher mutation rates is the clearest cause–effect statement.

Why are other options incorrect?:

A (Incorrect): This describes a correlation ("students who attend extra tutoring tend to have higher scores") without establishing causation.
B (Incorrect): This is a Discovery Claim, identifying a new reagent, not a cause–effect relationship.
D (Incorrect): It describes a positive association, indicating correlation rather than direct causation.

Question 3.3 Correct Answer: C
Explanation: Chapter 3 emphasizes that the main or primary Claim usually answers the study's principal research question and represents new knowledge significant enough to warrant publication. Subclaims or findings support that core Claim.

Why are other options incorrect?:

A (Incorrect): Placement of these elements varies, and subclaims can appear in multiple sections. The main Claim often appears in the title or abstract but not strictly always in the abstract alone.
B (Incorrect): Subclaims or supporting findings frequently appear in results or discussion sections to build the argument for the main Claim.
D (Incorrect): The main claims and subclaims emerge from the research process; neither is "proven" before the study begins.

Chapter 4

Question 4.1 Correct Answer: B
Explanation: Evidence is the subset of data or information used to prove or support the Claim; Context is data or information included to motivate, situate, or provide background for why the study was done.

Why are other options incorrect?:

A (Incorrect): Numerical data that does not help establish the Claim is part of Context or general background, not necessarily Evidence.
C (Incorrect): Context may appear in the Introduction, and Evidence often appears in the Methods or Results sections (or figures, tables, or appendices), so this strict division is inaccurate.
D (Incorrect): Archival data can be Evidence if it is recalibrated or used to help establish a new Claim, as Chapter 4 explains.

Question 4.2 Correct Answer: C
Explanation: Chapter 4 states archival data becomes new Evidence if reanalyzed or recalibrated so it directly establishes a new Claim.

Why are other options incorrect?:

A (Incorrect): Simply republishing data does not generate a new claim or reanalysis.
B (Incorrect): Referencing statistics to justify or motivate a study is Context, not Evidence.
D (Incorrect): Providing historical background contextualizes the research and does not generate new findings.

Question 4.3 Correct Answer: B
Explanation: Chapter 4 explains that Evidence is the input (data, measurements, models) used to establish the Claim, while Reasoning is the framework or logic used to connect that Evidence to the Claim.

Why are other options incorrect?:

A (Incorrect): Evidence can be numeric or qualitative data, and citations are typically references, not automatically Evidence.
C (Incorrect): Every claim requires both evidence and reasoning. Neither is described as optional.
D (Incorrect): Evidence often appears in Methods/Figures, and Reasoning may appear in Results/Discussion. They aren't restricted to just the Discussion section.

Chapter 5

Question 5.1 Correct Answer: C
Explanation: Evidence refers to the data or models needed for the claim, and Reasoning is the logical process linking that data to the conclusion.

Why are other options incorrect?:

A (Incorrect): This reverses the definitions; Evidence is the data, and Reasoning is the interpretation.
B (Incorrect): Reasoning and evidence can appear in multiple sections (e.g., methods, results, discussion), so it's not strictly divided by section.
D (Incorrect): Though often presented together in an article, Evidence and Reasoning have distinct roles.

Question 5.2 Correct Answer: C
Explanation: Mentioning statistical significance and rejecting the null hypothesis shows a logical step connecting evidence to a claim (reasoning).

Why are other options incorrect?:

A (Incorrect): Listing a measurement is evidence, not the logic behind it.
B (Incorrect): Describing how samples were chosen is part of method/evidence design – though borderline, it primarily addresses how data were collected rather than the interpretive logic behind it.
D (Incorrect): Referring to data figures is again describing evidence, not explaining its implications or bridging to a conclusion.

Question 5.3 Correct Answer: D
Explanation: Failing to address sample bias or alternate explanations leads to insufficient reasoning for the claim.

Why are other options incorrect?:

A (Incorrect): This scenario clearly describes strong reasoning – control group details and statistics help justify the claim.

B (Incorrect): The authors show thorough reasoning by explaining potential reasons and confounders.
C (Incorrect): Providing calibration steps and uncertainties is good practice for bridging evidence and conclusion.

Chapter 6

Question 6.1 Correct Answer: B
Explanation: Suggesting additional, larger-scale trials is a forward-looking element and directly builds on the new result, an example of implications that point to future verification or expansion of the present study.

Why are other options incorrect?:

A (Incorrect): Highlighting historical context is a backward-looking element (Context), not forward-looking.
C (Incorrect): Fundamental constants are background data, not next steps.
D (Incorrect): Summarizing the key finding is a conclusion or claim restatement, not a future direction.

Question 6.2 Correct Answer: C
Explanation: The phrases "we predict" and "might dramatically improve" indicate a forward-looking statement that goes beyond current results, suggesting speculation or a possible outcome.

Why are other options incorrect?:

A (Incorrect): This statement references past work, so it's more indicative of context than a future-based implication.
B (Incorrect): This is a declarative claim supported by current evidence ("prove that"), not a future speculation.
D (Incorrect): Describing a study's sample pertains to the existing methods/evidence, not the future-oriented speculation of an implication.

Question 6.3 Correct Answer: C
Explanation: Context distills the relevant results of past research, while implications go beyond immediate findings to suggest next steps and broader consequences of the work.

Why are other options incorrect?:

A (Incorrect): It's actually the opposite: context summarizes the past research; implications discuss future research or broader impact.
B (Incorrect): Context looks to past research conducted before the study; implications can address improvements to the study.
D (Incorrect): Implications and Context serve different functions rather than substituting for each other.

Chapter 7

Question 7.1 Correct Answer: B

Explanation: The passage repeatedly defines Context as the backward-looking narrative that establishes what is already known and identifies the gap the current study fills.

Why are other options incorrect?:

A (Incorrect): Calibration and measurement belong to Evidence or methods rather than Context.

C (Incorrect): Summarizing new data is part of the Evidence, while discussing future implications relates to Implications.

D (Incorrect): Forward-looking statements are typically Implications, not Context.

Question 7.2 Correct Answer: C

Explanation: A sudden leap into methods and results, without mentioning prior research or the motivation for this new study, indicates missing Context.

Why are other options incorrect?:

A (Incorrect): This is actually a good example of thorough Context.

B (Incorrect): Citing recent studies and a gap is precisely what Context should do.

D (Incorrect): Although it mainly refers to conclusions, referencing existing literature there also reflects some contextual understanding.

Question 7.3 Correct Answer: C

Explanation: Terms like "HPC," "QNN-embedding," and "preconditioning" can overwhelm novices and exemplify complex jargon found in Context sections.

Why are other options incorrect?:

A (Incorrect): This phrase is clear and to the point about a research gap – hardly a jargon maze.

B (Incorrect): Standard demographic language is relatively straightforward, not jargon-heavy.

D (Incorrect): While Cronbach's alpha is a technical term, it's a relatively common statistic in many fields and less likely to be a jargon maze compared to the specialized HPC example.

Chapter 8

Question 8.1 Correct Answer: A

Explanation: Rawlings (2019) specifically recommends juxtaposing the abstract's Claim with the discussion's expanded version and noting how the authors frame the "gap" their work fills; this comparison helps pinpoint the study's central Claim.

Why are other options incorrect?:

B (Incorrect): Spotting transitional phrases is a strategy credited to Pinninti (2024) for identifying reasoning patterns, not for isolating the primary Claim.

C (Incorrect): Validating measurement tools is part of evaluating Evidence (van der Sande et al., 2023), but it does not resolve which statement constitutes the article's main Claim.

D (Incorrect): Assessing implications reveals the broader significance of findings, yet implications can flow from either primary or secondary claims; the passage does not link this practice to Rawlings's guidance on claim identification.

Question 8.2 Correct Answer: A

Explanation: The passage labels verbatim reuse of an unchanged methods section as ethically acceptable when the original publication is properly cited, ensuring transparency and avoiding misrepresentation.

Why are other options incorrect?:

B (Incorrect): Copy-pasting literature-review text without citation is classic plagiarism: It misattributes another author's synthesis and violates the passage's emphasis on proper attribution.

C (Incorrect): Republishing the same results without disclosure constitutes undisclosed self-plagiarism, misleading readers and editors about novelty – explicitly identified as unethical.

D (Incorrect): Using text written by someone else without attribution breaches authorship ethics and fails the passage's requirement for transparency, making it unacceptable.

Chapter 9

Question 9.1 Correct answer: B

Explanation: When authors share detailed protocols, raw datasets, and analysis code, other researchers can replicate the study step-by-step, confirm (or challenge) the findings, and integrate the results confidently into meta-analyses. Without this transparency, even well-designed studies remain isolated anecdotes, limiting their comparative value.

Why are other options incorrect?:

A (Incorrect): Matching designs and controlling confounders certainly improve internal validity, but a study can still be impossible to verify if its raw data or procedures are unavailable. Comparability hinges on how variables are controlled and on whether another team can re-run the entire study under the same conditions. That reproducibility requirement is broader than variable control alone.

C (Incorrect): Spatial or temporal standardization reduces noise stemming from location-specific factors, which certainly aids cross-study comparison. Yet standard sampling alone cannot reveal hidden coding errors, analytical choices, or data anomalies – issues that only become evident when the study can be fully reproduced. Thus, it is helpful but not the foundational criterion emphasized.

D (Incorrect): Sound statistics are essential for judging whether observed differences are meaningful, but they operate on available data. If the underlying data and methods are opaque, no amount of statistical sophistication can rescue comparability. The passage treats statistical rigor as necessary but not sufficient; reproducibility underlies the entire enterprise.

E (Incorrect): This activity concerns publication-level metrics – how fields differ in citing patterns – not the verification of a study's empirical findings. Because it analyses citation outputs rather than experimental or observational inputs, it lies outside the study-level criterion that enables direct replication.

Question 9.2 Correct answer: B

Explanation: "Referential backtracking" is defined in the passage as the practice of examining an article's own reference list to locate earlier, foundational articles on the same topic. This backward look along the citation chain allows researchers to trace a subject's intellectual lineage.

Why are other options incorrect?:

A (Incorrect): PubMed's "Similar Articles" relies on keyword and MeSH-term matching, not a manual review of the cited works in the article itself.

C (Incorrect): Citation-network visualization is a broader mapping exercise that may use forward and backward links, but the term "referential backtracking" in the passage refers specifically to reading the reference list, not network analysis tools.

D (Incorrect): Restricting a Scopus search to non-English or review material is a database-filter strategy, unrelated to scanning the references of a single article.

E (Incorrect): Comparing Journal Impact Factors ranks journals, not individual cited studies, and therefore falls outside the definition of "referential backtracking."

Chapter 10

Question 10.1 Correct answer: A

Explanation: Citing retracted or discredited studies as valid evidence for setting up the study is a Context error because it shows the work is poorly situated within the existing literature.

Why are other options incorrect?:

B (Incorrect): Overextending findings from a narrow sample is an Implications flaw (overgeneralization), not a context issue.

C (Incorrect): Claiming to have "proven" something beyond doubt with weak data is a Claim flaw because it is an overstatement relative to evidence.

D (Incorrect): Treating correlation as causation is identified in Section 10.2.3 as a Reasoning error, reflecting a logical leap between evidence and claim.

E (Incorrect): Failing to detect a real effect because of an underpowered sample is described as an Evidence flaw (i.e., Type II error), not a context problem.

Question 10.2 Correct answer: E

Explanation: Preregistration, open data/code, and thorough methodological detail provide another competent researcher with everything needed to replicate the work, matching every hallmark of quality listed in the passage (Kelly et al., 2014; Korner, 2008).

Why other options are incorrect:

Brief explanations

A (Incorrect): Novelty alone is not enough; withholding raw data prevents independent teams from replicating the analyses, thereby violating the transparency principle emphasized by Resnik and Shamoo (2017).

B (Incorrect): Prestigious funding and enthusiastic language do not guarantee reproducibility; proprietary, unavailable reagents block-independent verification, undermining the study's reliability.

C (Incorrect): Extensive citations do not compensate for the absence of information on statistical power or procedures; without such details, the study cannot be readily reproduced.

D (Incorrect) Conflict-of-interest disclosure is valuable, but a skeletal methods description.

Chapter 11

Question 11.1 Correct answer: C

Explanation: Class reports require students (individually or in teams) to break down a research article and deliver their analysis to an audience, mirroring the one-way delivery – then-Q&A rhythm of conference presentations.

Why other options are incorrect:

A (Incorrect): Journal clubs are collaborative by design: One member presents, but the discussion quickly becomes a shared critique in which all participants probe methods, data, and conclusions.

B (Incorrect): Reading clubs emphasize low-pressure, exploratory dialogue across multiple disciplines; they lack the formal presentation structure that characterizes conference talks.

D (Incorrect): Lab meetings encourage dynamic, bidirectional exchanges that integrate the article's findings with ongoing research, rather than a single scripted presentation.
E (Incorrect): Casual conversations over coffee are informal, unscripted, and lack the formal structure and audience interrogation typical of conference settings.

Question 11.2 Correct answer: C
Explanation: Providing a structured set of prompts or an analytic framework keeps the group focused on core elements – claim, evidence, reasoning, implications – thereby eliminating aimless topic-hopping.

Why other options are incorrect:

A (Incorrect): A moderator who manages airtime tackles the dominant-voices issue, but it does not by itself, supply the thematic roadmap that stops discussion from meandering.
B (Incorrect): Premeeting summaries improve participant preparation, yet drifting can still occur if the live conversation lacks guiding questions.
D (Incorrect): An expert critique may enrich content, but without shared guiding questions, the discussion can still veer off course once the expert finishes speaking.
E (Incorrect): Starting with a statistics inventory dives into minutiae and may worsen drifting by anchoring debate on isolated details rather than the article's central argument.

Chapter 12

Question 12.1 Correct answer: E
Explanation: Core purpose is to contrast methods (manual vs. AI vs. automated), matching the Pathway arc that aligns sections by technique and critiques their underlying reasoning.

Why other options are incorrect:

A (Incorrect): Focus is a single gap (delivery), so the Gap-Focused Spiral arc fits better.
B (Incorrect): Forward-looking scenarios fits the Foresight Horizon arc.
C (Incorrect): Chronological milestones fit Historical-Development (Ladder) arc.
D (Incorrect): Interlacing two disciplines fit Cross-road Weave arc.

Question 12.2 Correct answer: B
Explanation: Rearticulating the logic in plain language disrupts sentence-level mimicry – exactly the mechanism that curbs patchwriting.

Why other options are incorrect:

A (Incorrect): Quoting verbatim does not reduce patchwriting; it merely shifts to direct quotation.

C (Incorrect): Raw numbers are rarely patch-written; the issue is textual phrasing.
D (Incorrect): Helpful for plagiarism detection, but not directly tied to patchwriting's sentence-structure problem.
E (Incorrect): Funding transparency combats bias, not patchwriting.

Chapter 13

Question 13.1 Correct answer: B
Explanation: The passage states that annotations "tend to focus on discrete parts of a text," leaving a "synthesis gap" that dampens critical-thinking gains.

Why other options are incorrect:

A (Incorrect): Ceiling effects are not mentioned; the text blames task design, not pre-existing skill levels.
C (Incorrect): The studies cited (Johnson et al., Chamberlain et al.) did involve peer interaction; missing features are not singled out.
D (Incorrect): Timing is discussed only for exposure length ("brief exposure and minimal training"), not multi-year longitudinal deficits.
E (Incorrect): Color-coding appears later (Section 13.1) as a helpful scaffold; no evidence it added extraneous load here.

Question 13.2 Correct answer: C
Explanation: Mirrors Section 13.3.1's synthesis: structured prompts/roles and low-pressure, exploratory space together drive durable engagement.

Why other options are incorrect:

A (Incorrect): Provides scaffolds but removes autonomy ("no free-form"), contrary to the text's call for a balance of "structured guidance and open-ended collaboration."
B (Incorrect): Maximizes autonomy but omits scaffolds Porter (2022) and Bjorn (2024) deem essential for deeper discourse.
D (Incorrect): Centers on instructor broadcast and assessment, not peer annotation; minimal resemblance to SCA principles.
E (Incorrect): Emphasizes quantity over quality, an approach cited as counter-productive (risks "annotation fatigue" and shallow engagement).

Chapter 14

Question 14.1 Correct answer: B
Explanation: The speed comparison highlights a possible shortcut mentality and reinforces the need for disciplined, slower analysis.

Why other options are incorrect:

A (Incorrect): Benchmarking studies are cited as evidence of capability, not dismissed as invalid.
C (Incorrect): AGI timelines are uncertain and speculative with the current best estimate as 2030.
D (Incorrect): Humans can miss errors if they rely uncritically on AI outputs.
E (Incorrect): Chapter 14 advocates retaining and even strengthening critical-thinking instruction to complement GenAI tools.

Question 14.2 Correct answer: E
Explanation: The juxtaposition serves to demonstrate that CERIC's systematic claim–evidence–reasoning checks balance GenAI's speed, warning readers not to accept AI summaries at face value.

Why other options are incorrect:

A (Incorrect): GenAI often misses logic chains, context, and caveats that CERIC surfaces; therefore, CERIC remains necessary.
B (Incorrect): The text advocates a partnership model: use GenAI for efficiency but keep CERIC-led human scrutiny at the core.
C (Incorrect): Chapter 14 calls future integration speculative; it does not predict imminent full CERIC automation.
D (Incorrect): Case studies show meaningful differences: GenAI captures headlines, while CERIC uncovers methodological and logical nuances.

Glossary

Argument completeness: The state in which all five CERIC elements – Context, Evidence, Reasoning, Implications, and Claim – are present and explicitly connected. Completeness is assessed at both the paragraph and whole-article levels. When an element is missing or underdeveloped, the book supports readers to locate it or identify the gap.

Argumentation core: A three-step analytical sequence in which readers (1) articulate the main Claim, (2) point to the specific Evidence that supports that Claim, and (3) explain the Reasoning that links the two. The loop prompts learners to move beyond vague reactions toward transparent argumentation. Because each turn in the loop is shared, the book encourages social learning (in Chapter 13) to improve the quality of subsequent contributions.

Bypass skimming: A surface strategy where readers skip dense passages and rely on abstracts, figures, or peer summaries. While sometimes efficient, habitual bypassing blunts critical reading growth. The book suggests "diagnostic browsing" as a structured alternative that keeps the whole article in view while hunting for individual CERIC elements.

CERIC method: An adaptation of Toulmin et al. (1984) reasoning that organizes scientific arguments into five interlocking elements: Claim, Evidence, Reasoning, Implications, and Context. CERIC serves as both an analytic lens for reading and a generative scaffold for writing. Its cyclical arrows remind readers that revisions in any element ripple through the others.

Chain of evidence: A sequence of experimental results that, taken together, support the article's Claim more powerfully than any single data point could. Readers learn to trace this chain by understanding the types of Evidence considered valid in their field. Weak or missing links become prime targets for discussion.

Citation overload: The gradual accumulation of references that do not meaningfully advance the argument. It often starts with well-intentioned thoroughness but ends in clutter that obscures the central chain of Evidence. The troubleshooting guide recommends a "one-sentence rationale" test for every citation to curb the creep.

Claim: The new knowledge that the authors want readers to accept. In empirical articles, it is usually anchored in the Results or Discussion section, though the title often hints at it. A strong Claim is both declarative and bounded. It tells what the study establishes and what it does not.

Comparative analysis: A sequence in which two articles on the same topic are read back-to-back and mapped with CERIC. Juxtaposing the maps makes disciplinary differences in Reasoning or Evidence visible.

Context: The background information that signals why the study matters – typically including prior findings, real-world stakes, and unanswered questions. Context appears early in the traditional publication order and may be revisited in the Discussion to frame implications. Readers are warned that jargon-heavy context sections can become traps that significantly stall comprehension.

Diagnostic browsing: A 5-minute skim in which readers preview headings, figures, and conclusion statements, predicting which CERIC elements will require the most effort. The predictions guide time allocation during follow-up close reading. The practice preserves strategic scanning for relevance without sliding into wholesale bypass skimming.

Evidence: The qualitative or quantitative observations that directly support the Claim. The book adopts a broad view – figures, statistical tables, ancillary experiments, models, and even null results count as Evidence if explicitly interpreted by the authors. The book recommends readers practice annotating each piece during notetaking.

Generative artificial intelligence (genAI): GenAI refers to large language models (LLMs) that can produce novel outputs – such as text, images, or code – by learning statistical patterns from large datasets. In the book, we recommend treating genAI tools as brainstorming partners whose suggestions must always be critically evaluated with the CERIC framework and human thinking.

Implications: The forward-looking consequences of accepting the Claim ranging from immediate practical applications to long-term theoretical shifts. Implications should be logically tethered to the Evidence and framed with appropriate hedging. Readers' critique overreach by matching each stated Implication to specific data.

Jargon maze: A passage – often in the Context section – so dense and with field-specific vocabulary that novice critical readers get lost and stop reading. The recommended fix is a two-column glossary note where readers paraphrase each term in everyday language. Eventually, readers can learn to anticipate and mentally bypass such traps without losing the thread of the argument.

Learning transfer: The ability to apply CERIC skills to unfamiliar articles or even popular-science pieces. Transfer is nurtured through "far-domain" assignments – using the framework on a literature review or genAI assignment, for instance. The book treats transfer as a capstone achievement of critical reading practice.

Reasoning: The logic that explains how and why the Evidence supports the Claim. Reasoning may be deductive, inductive, or abductive, but it must make the Evidentiary leap explicit. The book teaches that missing Reasoning is the silent killer of scientific arguments.

Scientific argument: A cohesive presentation of a Claim supported by Evidence and Reasoning, situated in past Context and pointing to future implications. The

CERIC model treats scientific arguments as rhetorical constructions, not mere data dumps. Recognizing this empowers readers to critique both structure and substance.

Social collaborative annotation (SCA): A low-stakes text marking method in which readers annotate a text together asynchronously. This method reinforces seeing other people's ideas in real time and the feeling of low-stakes learning together.

Bibliography

Aad, G., Abajyan, T., Abbott, B., et al. (2012). Observation of a new particle in the search for the Standard Model Higgs boson with the ATLAS detector at the LHC. *Physics Letters B*, *716*(1), 1–29. https://doi.org/10.1016/j.physletb.2012.08.020

Aarts, A. A., Anderson, J. E., Anderson, C. J., et al. (2015). Estimating the reproducibility of psychological science. *Science*, *349*(6251), aac4716. https://doi.org/10.1126/science.aac4716

Abachi, S., Abbott, B., Abolins, M., et al. (1995). Observation of the top quark. *Physical Review Letters*, *74*(14), 2632–2637. https://doi.org/10.1103/PhysRevLett.74.2632

Abbott, B. P., Abbott, R., Abbott, T., et al. (2016). Observation of gravitational waves from a binary black hole merger. *Physical Review Letters*, *116*(6), 061102. https://doi.org/10.1103/PhysRevLett.116.061102.

Abbott, B. P., Abbott, R., Abbott, T., et al. (2017). GW170817: Observation of gravitational waves from a binary neutron star inspiral. *Physical Review Letters*, *119*(16), 161101. https://doi.org/10.1103/PhysRevLett.119.161101

Abdullah, C., Parris, J., Lie, R., Guzdar, A., & Tour, E. (2015). Critical analysis of primary literature in a master's-level class: Effects on self-efficacy and science-process skills. *CBE – Life Sciences Education*, *14*(3), ar34. https://doi.org/10.1187/cbe.14-10-0180

Abe, F., Akimoto, H., Akopian, A., et al. (CDF Collaboration) (1995). Observation of top quark production in $\bar{p}p$ collisions with the collider detector at Fermilab. *Physical Review Letters*, *74*, 2626. https://doi.org/10.1103/PhysRevLett.74.2626

Abi, B., Albahri, T., Al-Kilani, S., et al. (2021). Measurement of the positive muon anomalous magnetic moment to 0.46 ppm. *Physical Review Letters*, *126*(14), 141801. https://doi.org/10.1103/PhysRevLett.126.141801

Abrami, P. C., Bernard, R. M., Borokhovski, E., Waddington, D. I., Wade, C. A., & Persson, T. (2015). Strategies for teaching students to think critically: A meta-analysis. *Review of Educational Research*, *85*(2), 275–314. https://doi.org/10.3102/0034654314551063

Abrams, P. A. (2000). The evolution of predator-prey interactions: Theory and evidence. *Annual Review of Ecology and Systematics*, *31*(1), 79–105. https://doi.org/10.1146/annurev.ecolsys.31.1.79

Abrams, P. A. (2005). The consequences of predator and prey adaptations for top-down and bottom-up effects. In P. Barbosa & I. Castellanos (Eds.), *Ecology of Predator-Prey Interactions* (pp. 279–297). United Kingdom: Oxford Academic. https://doi.org/10.1093/oso/9780195171204.003.0013

Adewumi, T., Alkhaled, L., Buck, C., et al. (2023). Procot: Stimulating critical thinking and writing of students through engagement with large language models (LLMS). *arXiv preprint*. https://doi.org/10.48550/arXiv.2312.09801

Ahmad, Q. R., Allen, R. C., Andersen, T. C., et al. (2002). Direct evidence for neutrino flavor transformation from neutral-current interactions in the Sudbury Neutrino Observatory. *Physical Review Letters*, *89*(1), 011301. https://doi.org/10.1103/physrevlett.89.011301

Aitchison, C., Catterall, J., Ross, P., & Burgin, S. (2012). "Tough love and tears": Learning doctoral writing in the sciences. *Higher Education Research & Development*, *31*(4), 435–447. https://doi.org/10.1080/07294360.2011.559195

Aitchison, C., & Guerin, C. (Eds.). (2014). *Writing Groups for Doctoral Education and Beyond: Innovations in Practice and Theory*. London, England: Routledge.

Akers, K. G. (2017). Being critical and constructive: A guide to peer reviewing for librarians. *Journal of the Medical Library Association*, *105*(1), 1–4. https://doi.org/10.5195/jmla.2017.100

Alexander, P. A. (2020). Methodological guidance paper: The art and science of quality systematic reviews. *Review of Educational Research*, *90*(1), 6–23. https://doi.org/10.3102/0034654319854352

Allard, F., Homeier, D., & Freytag, B. (2011). Stellar to Substellar Model Atmospheres. *Proceedings of the International Astronomical Union*, *7*(S282), 235–242. https://doi.org/10.1017/S1743921311027438

Allard, F., Homeier, D., & Freytag, B. (2012a). Models of very-low-mass stars, brown dwarfs and exoplanets. *Philosophical Transactions of the Royal Society A: Mathematical, Physical and Engineering Sciences*, *370*, 2765. https://doi.org/10.1098/rsta.2011.0269

Allard, F., Homeier, D., Freytag, B., & Sharp, C. M. (2012b). Atmospheres from very low-mass stars to extrasolar planets. In C. Reylé, C. Charbonnel, & M. Schultheis (Eds.), *Low-Mass Stars and the Transition Stars/Brown Dwarfs – EES2011, EAS Publications Series*, 57, 3–43. https://doi.org/10.1051/eas/1257001

Allen, A., Markou, S., Tebbutt, W., et al. (2025). End-to-end data-driven weather prediction. *Nature*, *641*, 1172–1179. https://doi.org/10.1038/s41586-025-08897-0

Allen, G. E., & Baker, J. J. W. (2016). The nature and logic of science. In G. E. Allen & J. J. W. Baker (Eds.), *Scientific Process and Social Issues in Biology Education* (pp. 29–82). Switzerland: Springer Cham. https://doi.org/10.1007/978-3-319-44380-5_2

Allers, K. N., & Liu, M. C. (2013). A near-infrared spectroscopic study of young field ultracool dwarfs. *The Astrophysical Journal*, *772*(2), 79. https://doi.org/10.1088/0004-637X/772/2/79

Allison, D. B., Brown, A. W., George, B. J., & Kaiser, K. A. (2016). Reproducibility: A tragedy of errors. *Nature*, *530*(7588), 27–29. https://doi.org/10.1038/530027a

Allon, S., Baggett, A., Hayes, B., et al. (2023). Association of a gamified journal club on internal medicine residents' engagement and critical appraisal skills. *Journal of Graduate Medical Education*, *15*(4), 475–480. https://doi.org/10.4300/JGME-D-22-00812.1

Almahmoud, M., Bawa, P., Zhu, H., & Goggins, S. (2025). Toward scalable social annotation: Managing cognitive load in large learning communities. *arXiv preprint*. https://arxiv.org/abs/2501.01545

Alonso, J. (2024, Sep 25). *How much do students really read?* Inside Higher Ed Report. www.insidehighered.com/news/students/academics/2024/09/25/students-turn-ai-do-their-assigned-readings-them. Accessed March 5, 2025.

Alvarez, L. W., Alvarez, W., Asaro, F., & Michel, H. V. (1980). Extraterrestrial cause for the Cretaceous–Tertiary extinction. *Science*, *208*(4448), 1095–1108. https://doi.org/10.1126/science.208.4448.1095

American Association for the Advancement of Science. (2025, July 30). *Science in the classroom*. www.scienceintheclassroom.org/

American Astronomical Society (2025, June 14). *AAS Journal Reference Instructions – AAS Journals. AAS Journals.* https://journals.aas.org/references/

American Geophysical Union (2025, Jul. 30). Grammar and Style Guide. *AGU Publications.* www.agu.org/publications/authors/journals/grammar-style-guide

American Institute of Physics Publishing LLC. (2025, Jul. 30). Author instructions. *AIP Publishing LLC.* Retrieved from https://publishing.aip.org/resources/researchers/author-instructions

American National Standards Institute (ANSI), & Council of National Library Associations. (1972). *American National Standard for the Preparation of Scientific Papers for Written or Oral Presentation.* (ANSI Z39 16-1972). American National Standards Institute.

American National Standards Institute (ANSI). Subcommittee 26 on the Preparation of Scientific Papers, & Council of National Library Associations. (1979). *American National Standard for the Preparation of Scientific Papers for Written Or Oral Presentation: Approved January 9, 1979.* (Revision of ANSI Z39 16-1972). American National Standards Institute.

American Psychological Association. (2020). *Publication manual of the American Psychological Association* (7th ed.). https://doi.org/10.1037/0000165-000

Anastasiou, D., Wirngo, C. N., & Bagos, P. (2024). The effectiveness of concept maps on students' achievement in science: A meta-analysis. *Educational Psychology Review, 36*(2), 39. https://doi.org/10.1007/s10648-024-09877-y

Andersen, E. (2012, Mar. 23). True Fact: The Lack of Pirates Is Causing Global Warming. Forbes. Retrieved July 31, 2025, from www.forbes.com/sites/erikaandersen/2012/03/23/true-fact-the-lack-of-pirates-is-causing-global-warming/

Angulo C., Arnould, M., Rayet, et al. (1999). A compilation of charged-particle induced thermonuclear reaction rates. *Nuclear Physics A, 656*, 3. https://doi.org/10.1016/S0375-9474(99)00030-5

Apetoh, L., Ghiringhelli, F., Tesniere, A., et al. (2007). The interaction between HMGB1 and TLR4 dictates the outcome of anticancer chemotherapy and radiotherapy. *Immunological Reviews, 220*, 47–59. https://doi.org/10.1111/j.1600-065X.2007.00573.x

Atkins, P. W., De Paula, J., Keeler, J. (2023). *Atkins' Physical Chemistry.* United Kingdom: Oxford University Press.

Atewell, S. (2025, May 22). Student perceptions of AI 2025 (Report). *Jisc.* www.jisc.ac.uk/reports/student-perceptions-of-ai-2025

Atkinson, R. C., & Shiffrin, R. M. (1968). Human memory: A proposed system and its control processes. In K. W. Spence & J. T. Spence (Eds.), *The Psychology of Learning and Motivation* (Vol. 2, pp. 89–195). Academic Press.

Atzema, C. (2004). Presenting at journal club: A guide. *Annals of Emergency Medicine, 44*(2), 169–174. https://doi.org/10.1016/j.annemergmed.2004.03.035

Austin, M. (2007). Species distribution models and ecological theory: A critical assessment and some possible new approaches. *Ecological Modelling, 200*(1–2), 1–19. https://doi.org/10.1016/j.ecolmodel.2006.07.005

Atwater, B. F., & Hemphill-Haley, E. (1997). *Recurrence intervals for great earthquakes of the past 3,500 years at northeastern Willapa Bay, Washington* (No. 1576). US Government Printing Office. https://doi.org/10.3133/pp1576

Avise, J. C., Arnold, J., Ball, R. M., et al. (1987). Intraspecific phylogeography: The mitochondrial DNA bridge between population genetics and systematics. *Annual Review of Ecology and Systematics, 18*, 489–522.

Ayambire, R. A., Rytwinski, T., Taylor, J. J., et al. (2024). Challenges in assessing the effects of environmental governance systems on conservation outcomes. *Conservation Biology, e14392*, 1–14. https://doi.org/10.1111/cobi.14392

Baddeley, A. D., & Hitch, G. (1974). Working memory. In G. H. Bower (Ed.), *The Psychology of Learning and Motivation* (Vol. 8, pp. 47–89). Academic Press.

Bae, S., Park, J., & Kim, J. S. (2014). Cas-OFFinder: A fast and versatile algorithm that searches for potential off-target sites of Cas9 RNA-guided endonucleases. *Bioinformatics*, *30*(10), 1473–1475. https://doi.org/10.1093/bioinformatics/btu048

Bai, Y., Yao, L., Wei, T., et al. (2020). Presumed asymptomatic carrier transmission of COVID-19. *JAMA*, *323*(14), 1406. https://doi.org/10.1001/jama.2020.2565

Bair, C. R., & Haworth, J. G. (2004). Doctoral student attrition and persistence: A meta-synthesis of research. In J. C. Smart (Ed.), *Higher Education: Handbook of Theory and Research* (Vol. 9, pp. 481–534). New York, NY: Agathon.

Baker, M. (2016). 1,500 scientists lift the lid on reproducibility. *Nature*, *533*(7604), 452–454. https://doi.org/10.1038/533452a

Bakkalbasi, N., Bauer, K., Glover, J., & Wang, L. (2006). Three options for citation tracking: Google Scholar, Scopus and Web of Science. *Biomedical Digital Libraries*, *3*(1), 1–8. https://doi.org/10.1186/1742-5581-3-7, 1–8.

Baldwin, M. (2018). Scientific autonomy, public accountability, and the rise of "peer review" in the Cold War United States. *Isis*, *109*(3), 538–558. https://doi.org/10.1086/699978

Baltimore, D. (1970). Viral RNA-dependent DNA polymerase: RNA-dependent DNA polymerase in virions of RNA tumour viruses. *Nature*, *226*(5252), 1209–1211. https://doi.org/10.1038/2261209a0

Ban, N., Nissen, P., Hansen, J., Moore, P. B., & Steitz, T. A. (2000). The complete atomic structure of the large ribosomal subunit at 2.4 Å resolution. *Science*, *289*(5481), 905–920. https://doi.org/10.1126/science.289.5481.90

Bandura, A. (1986). *Social Foundations of Thought and Action: A Social Cognitive Theory*. Prentice-Hall.

Banik, G. M., Baysinger, G., Kamat, P., & Pienta, N. (2020). *The ACS Guide to Scholarly Communication*. Washington, DC: ACS Publications. https://doi.org/10.1021/acsguide

Bara-Stolzenberg, E., Eagan, K., Zimmerman, H. B., et al. (2019). *Undergraduate Teaching Faculty: The HERI Faculty Survey 2016–2017*. Los Angeles, CA: Higher Education Research Institute at UCLA.

Barber, R. A., Yang, J., Yang, C., Barker, O., Janicke, T., & Tobias, J. A. (2024). Climate and ecology predict latitudinal trends in sexual selection inferred from avian mating systems. *PLoS Biology*, *22*(11), e3002856. https://doi.org/10.1371/journal.pbio.3002856

Barbour, M. A., & Gibert, J. P. (2021). Genetic and plastic rewiring of food webs under climate change. *Journal of Animal Ecology*, *90*(8), 1814–1830. https://doi.org/10.1111/1365-2656.13541

Barroga, E., & Matanguihan, G. J. (2021). Creating logical flow when writing scientific articles. *Journal of Korean Medical Science, 36*(40), e275. https://doi.org/10.3346/jkms.2021.36.e275

Bartlett, F. C. (1932). *Remembering: A Study in Experimental and Social Psychology*. Cambridge University Press.

Bauer, M. I., & Johnson-Laird, P. N. (1993). How diagrams can improve reasoning. *Psychological Science*, *4*(6), 372–378. https://doi.org/10.1111/j.1467-9280.1993.tb00584.x

Baxt, W. G., Waeckerle, J. F., Berlin, J. A., & Callaham, M. L. (1998). Who reviews the reviewers? Feasibility of using a fictitious manuscript to evaluate peer reviewer performance. *Annals of Emergency Medicine*, *32*(3), 310–317. https://doi.org/10.1016/s0196-0644(98)70006-x

Bayes, T., & Price, R. (1763). An Essay towards Solving a Problem in the Doctrine of Chances. By the Late Rev. *Mr. Bayes, F. R. S. Communicated by Mr. Price, in a Letter to John Canton, A. M. F. R. S. Philosophical Transactions of the Royal Society of London. 53*, 370–418. https://doi.org/10.1098%2Frstl.1763.0053

Beall, J. (2018). Scientific soundness and the problem of predatory journals. In A. Kaufman & J. C. Kaufman (Eds.), *Pseudoscience: The Conspiracy against Science* (pp. 283–300 Chapter 12). MIT Press.

Bean, J. L., Seifahrt, A., Hartman, H., et al. (2010). The proposed giant planet orbiting VB 10 does not exist. *The Astrophysical Journal Letters*, *711*(1), L19. https://doi.org/10.1088/2041-8205/711/1/L19

Becke, A. D. (1993). Density-functional thermochemistry. III. The role of exact exchange. *The Journal of Chemical Physics*, *98*(7), 5648–5652. https://doi.org/10.1063/1.464913.

Bell, J. S. (1964). On the Einstein Podolsky Rosen paradox. *Physics Physique Fizika*, *1*(3), 195. https://journals.aps.org/ppf/abstract/10.1103/PhysicsPhysiqueFizika.1.195

Benneke, B., Wong, I., Piaulet, C., et al. (2019). Water Vapor and Clouds on the Habitable-zone Sub-Neptune Exoplanet K2-18b. *The Astrophysical Journal Letters*, *887*(1), L14. https://doi.org/10.3847/2041-8213/ab59dc

Bennett, A. M., Benneke, B., Wong, I., et al. (2018). Nivolumab plus ipilimumab in lung cancer with a high tumor mutational burden. *New England Journal of Medicine*, *378*(22), 2093–2104. https://doi.org/10.1056/NEJMoa1801946

Bice, D. M., & Carr, R. H. (2017). Evaluating primary literature in an undergraduate environmental geology course. *Journal of Geoscience Education*, *65*(2), 136–148. https://doi.org/10.5408/16-175.1

Bjorn, G. (2023). The Power of Peer Engagement: Exploring the effects of social collaborative annotation on reading comprehension of primary literature. *AI Computer Science and Robotics Technology*, *2*. https://doi.org/10.5772/acrt.24

Bjorn, G. (2024). The CERIC method plus social collaborative annotation improves critical reading of the primary literature in an interdisciplinary graduate course. *Frontiers in Education*, *9*. https://doi.org/10.3389/feduc.2024.1257747

Bjorn, G. A., Quaynor, L., & Burgasser, A. J. (2022). Reading Research for Writing: Co-constructing core skills using Primary Literature. *Impacting Education Journal on Transforming Professional Practice*, *7*(1), 47–58. https://doi.org/10.5195/ie.2022.237

Bohannon, J. (2013). Who's afraid of peer review? *Science*, *342*(6154), 60–65. https://doi.org/10.1126/science.342.6154.60

Böhm-Vitense, E. (1956). Über die Wasserstoffkonvektionszone in Sternen verschiedener Effektivtemperaturen und Leuchtkräfte. Mit 5 Textabbildungen. *Zeitschrift für Astrophysik*, *46*, 108. https://ui.adsabs.harvard.edu/abs/1958ZA...46.108B/abstract

Böhm-Vitense, E. (1958). Über die Wasserstoffkonvektionszone in Sternen verschiedener Effektivtemperaturen und Leuchtkräfte. *Zeitschrift für Astrophysik*, *46*, 108–143.

Bond-Lamberty, B., Ballantyne, A., Berryman, E., et al. (2024). Twenty years of progress, challenges, and opportunities in measuring and understanding soil respiration. *Journal of Geophysical Research: Biogeosciences*, *129*(2), e2023JG007637. https://doi.org/10.1029/2023JG007637

Boote, D. N., & Beile, P. (2005). Scholars before researchers: On the centrality of the dissertation literature review in research preparation. *Educational Researcher*, *34*(6), 3–15.

Booth, A., Sutton, A. and Papaioannou, D. (2016) *Systematic Approaches to a Successful Literature Review*. London: Sage.

Borg, M., Kembro, J., Notander, J., Petersson, C., & Ohlsson, L. (2011). Conflict management in student groups–A teacher's perspective in higher education. *Högre utbildning*, *1*(2), 111–124. https://doi.org/10.23865/hu.v1.860

Borg, S. (2003). Teacher cognition in language teaching: A review of research on what language teachers think, know, believe, and do. *Language Teaching*, *36*(2), 81–109. https:/doi.org/10.1017/S0261444803001903

Bornmann, L. (2011). Scientific peer review. *Annual Review of Information Science and Technology*, *45*(1), 197–245. https://doi.org/10.1002/aris.2011.1440450112

Bornmann, L., & Leydesdorff, L. (2018). Count highly-cited papers instead of papers with h citations: Use normalized citation counts and compare "like with like"! *Scientometrics*, *115*(2), 1119–1123. 1Field-normalized impact factors: A comparison of using different methods for normalization. Journal of Informetrics, *12*(4), 1011–1016.

Bourdieu, P. (1986). The forms of capital. In J. Richardson (Ed.), *Handbook of Theory and Research for the Sociology of Education* (pp. 241–258). New York: Greenwood.

Bourrier, V., De Wit, J., Bolmont, E., et al. (2017). Temporal evolution of the high-energy irradiation and water content of TRAPPIST-1 exoplanets. *The Astronomical Journal*, *154*(3), 121. https://doi.org/10.3847/1538-3881/aa859c

Boyd, P. W., Jickells, T., Law, C. S., et al. (2007). Mesoscale iron enrichment experiments 1993–2005: Synthesis and future directions. *Science*, *315*(5812), 612–617. https://doi.org/10.1126/science.1131669

Brainard, J., & You, J. (2018). What a massive database of retracted papers reveals about science publishing's "death penalty." *Science*, *25*(1), 1–5. https://doi.org/10.1126/science.aav8384

Brenner, S., & Lerner, R. A. (1992). Encoded combinatorial chemistry. *Proceedings of the National Academy of Sciences*, *89*(12), 5381–5383. https://doi.org/10.1073/pnas.89.12.5381

Bresser, R., Melanese, K., Sphar, C. (2009). *Supporting English language Learners in Math Class, Grades K-2*. United States: Math Solutions Publications.

Brewin C. R. (2023). Inaccuracy in the scientific record and open postpublication critique. *Perspectives on Psychological Science: A journal of the Association for Psychological Science*, *18*(5), 1244–1253. https://doi.org/10.1177/17456916221141357

Brigham Young University Research & Writing Center (2025, Jul. 22). Scientific Writing: IMRAD format. https://rwc.byu.edu/00000188-e4bc-d222-a7ea-eefd769e0000/scientific-writing-imrad-format-pdf

Brokowski, C., & Adli, M. (2019). CRISPR ethics: Moral considerations for applications of a powerful tool. *Journal of Molecular Biology*, *431*(1), 88–101. https://doi.org/10.1016/j.jmb.2018.05.044

Brown, M., & Croft, B. (2020). Social annotation and an inclusive praxis for open pedagogy in the college classroom. *Journal of Interactive Media in Education*, *2020*(1), 8, 1–88. https://doi.org/10.5334/jime.561

Brown, P. C., Roediger III, H. L., & McDaniel, M. A. (2014). *Make It Stick: The Science of Successful Learning*. Harvard University Press.

Brown, R. A. J., & Renshaw, P. D. (2000). Collective argumentation: A sociocultural approach to reframing classroom teaching and learning. In H. Cowie & G. van der Aalsvoort (Eds.), *Social Interaction in Learning and Instruction: The Meaning of Discourse for the Construction of Knowledge* (pp. 52–66). Pergamon/Elsevier Science Inc.

Bubeck, S., Chandrasekaran, V., Eldan, R., et al. (2023). Sparks of artificial general intelligence: Early experiments with GPT-4. *arXiv preprint*. https://doi.org/10.48550/arXiv.2303.12712

Buckley, C., & Nerantzi, C. (2020). Effective use of visual representation in research and teaching within higher education. *International Journal of Management and Applied Research, 7*(3), 196–214. https://doi.org/10.18646/2056.54.cfp008

Bullock, O. M., Colon-Amill, D. C., Shulman, H. C., & Dixon, G. N. (2019). Jargon as a barrier to effective science communication: Evidence from metacognition. *Public Understanding of Science, 28*(7), 845–853. https://doi.org/10.1177/0963662519865687

Burchfield, C. M., & Sappington, J. (2000). Compliance with Required Reading Assignments. *Teaching of Psychology, 27*(1), 58–60. https://eric.ed.gov/?id=EJ613785

Burde, D., & Linden, L. L. (2013). Bringing education to Afghan girls: A randomized controlled trial of village-based schools. *American Economic Journal: Applied Economics, 5*(3), 27–40. https://doi.org/10.1257/app.5.3.27

Burde, D., Middleton, J. A., & Samii, C. (2016). The Assessment of Learning Outcomes and Social Effects of Community-Based Education: A Randomized Field Experiment in Afghanistan, Phase One Outcomes Report. Steinhardt School of Culture, Education, and Human Development, New York University.

Burgasser, A. J., Logsdon, S. E., Gagné, J., et al. (2015). The Brown Dwarf Kinematics Project (BDKP). IV. Radial velocities of 85 late-M and L dwarfs with MagE. *The Astrophysical Journal Supplement Series, 220*(1), 18. https://doi.org/10.1088/0067-0049/220/1/18

Burgasser, A. J., & Mamajek, E. E. (2017). On the Age of the TRAPPIST-1 System. *The Astrophysical Journal, 845*(2), 110. https://doi.org/10.3847/1538-4357/aa7fea

Burgess, M. L., Benge, C., Onwuegbuzie, A. J., & Mallette, M. H. (2012). Doctoral students' reasons for reading empirical research articles: A mixed analysis. *Journal of Effective Teaching, 12*(3), 5–33.

Buzan, T., & Buzan, B. (2006). *The Mind Map Book*. Pearson Education.

Caetano, C. M. P. F. (2012). G1/S Cell Cycle Regulated Transcription and Genome Stability [PhD Dissertation, University College London]. https://discovery.ucl.ac.uk/id/eprint/1390624/1/Catia%20Caetano%20PhD%20Thesis%20With%20Corrections.pdf

Callaham, M., Schriger, D., & Cooper, R. J. (2001). An instructional guide for peer reviewers of biomedical manuscripts. *Hospital Medicine, 62*, 172–175.

Capron, E., Rasmussen, S. O., Popp, T. J., et al. (2021). The anatomy of past abrupt warmings recorded in Greenland ice. *Nature Communications, 12*(1), 2106. https://doi.org/10.1038/s41467-021-22241-w

Carey, M. A., Steiner, K. L., & Petri Jr, W. A. (2020). Ten simple rules for reading a scientific paper. *PLoS Computational Biology, 16*(7), e1008032. https://doi.org/10.1371/journal.pcbi.1008032

Carter-Thomas, S., & Rowley-Jolivet, E. (2008). If-conditionals in medical discourse: From theory to disciplinary practice. *Journal of English for Academic Purposes, 7*(3), 191–205. https://doi.org/10.1016/j.jeap.2008.03.004

Carvalho-Filho, M. A., Santos, T. M., Ozahata, T. M., & Cecilio-Fernandes, D. (2018). Journal Club Challenge: Enhancing student participation through gamification. *Medical Education, 52*(5). https://doi.org/10.1111/medu.13552

Caruso, C., Hughes, K., & Drury, C. (2021). Selecting Heat-Tolerant corals for proactive reef restoration. *Frontiers in Marine Science, 8*. https://doi.org/10.3389/fmars.2021.632027

Casadevall, A., & Fang, F. C. (2012). Reforming science: Methodological and cultural reforms. *Infection and Immunity, 80*(3), 891–896. https://doi.org/10.1128/IAI.06183-11

Center of Open Science (2025, July 30). *Reproducibility Project: Cancer Biology*. www.cos.io/rpcb

Chakravarti, L. J., Beltran, V. H., & Oppen, M. J. H. (2017) Rapid thermal adaptation in photosymbionts of reef-building corals. *Global Change Biology*, *23*, 4675–4688. https://doi.org/10.1111/gcb.13702

Chamberlain, L. J., Bruce, J., De La Cruz, M., et al. (2021). A text-based intervention to promote literacy: An RCT. *Pediatrics*, *148*(4). https://doi.org/10.1542/peds.2020-048553

Chambers, C. D. (2017). *The seven deadly sins of psychology: A Manifesto for Reforming the Culture of Scientific Practice*. Princeton University Press.

Chang, S. H. (2011). Reading strategies of struggling readers. *Journal of Research in Education*, *21*(2), 2–13. https://files.eric.ed.gov/fulltext/EJ1098429.pdf

Chang, A., & Hu, H. C. M. (2018). Learning Vocabulary through Extensive Reading: Word Frequency Levels and L2 Learners' Vocabulary Knowledge Level. *TESL-EJ*, *22*(1), n1.

Channa, M. A., Nordin, Z. S., Siming, I. A., Chandio, A. A., & Koondher, M. A. (2015). Developing reading comprehension through metacognitive strategies: A review of previous studies. *English Language Teaching*, *8*(8), 181–186. https://doi.org/10.5539/elt.v8n8p181

Chatzimichail, T., & Hatjimihail, A. T. (2023). A Bayesian inference based computational tool for parametric and nonparametric medical diagnosis. *Diagnostics*, *13*(19), 3135. https://doi.org/10.3390/diagnostics13193135

Chatzimichail, T., & Hatjimihail, A. T. (2024). A software tool for estimating uncertainty of Bayesian posterior probability for disease. *Diagnostics*, *14*(4), 402. https://doi.org/10.3390/diagnostics14040402

Chekroud, S. R., Gueorguieva, R., Zheutlin, A. B., et al. (2018). Association between physical exercise and mental health in 1.2 million individuals in the USA between 2011 and 2015: A cross-sectional study. *Lancet Psychiatry*, *5*(9), 739–746. https://doi.org/10.1016/S2215-0366(18)30227-X

Chen, A., & Leitch, B. (2024). LLMs as academic reading companions: Extending HCI through synthetic personae. arXiv:2403.19506v2. https://doi.org/10.48550/arXiv.2403.19506

Chen, F., Li, S., Lin, L., & Huang, X. (2024). Identifying temporal changes in student engagement in social annotation during online collaborative reading. *Education and Information Technologies*, *29*(13), 16101–16124. https://doi.org/10.1007/s10639-024-12494-5

Chow, C. K., Atkins, E. R., Hillis, G. S., et al. (2021). Initial treatment with a single pill containing quadruple combination of quarter doses of blood pressure medicines versus standard dose monotherapy in patients with hypertension (QUARTET): A phase 3, randomised, double-blind, active-controlled trial. *The Lancet*, *398*(10305), 1043–1052. https://doi.org/10.1016/S0140-6736(21)01922-X

Church, G. M., Gao, Y., & Kosuri, S. (2012). Next-Generation Digital Information Storage in DNA. *Science*, *337*(6102), 1628. https://doi.org/10.1126/science.1226355

Church, K. (2024). Emerging Trends: When Can Users Trust GPT, and When Should They Intervene? *Natural Language Engineering*, *30*(2), 417–427. https://doi.org/10.1017/S1351324923000578

Clarivate. (2025a Analytics. (2025, July 25). Journal Citation Reports. https://clarivate.com/academia-government/scientific-and-academic-research/research-funding-analytics/journal-citation-reports/

Clarivate. (2025b, July 31). Web of Science platform. https://clarivate.com/academia-government/scientific-and-academic-research/research-discovery-and-referencing/web-of-science/

Clinton-Lisell, V. (2023). Social annotation: What are students' perceptions and how does social annotation relate to grades? *Research in Learning Technology*, *31*. https://doi.org/10.25304/rlt.v31.3050

Clowe, D., Bradač, M., Gonzalez, A. H., et al. (2006). A direct empirical proof of the existence of dark matter. *The Astrophysical Journal Letters*, *648*(2), L109–L113. https://doi.org/10.1086/508162

Clump, M. A., & Doll, J. (2007). Do the low levels of reading course material continue? An examination in a forensic psychology graduate program. *Journal of Instructional Psychology*, *34*(4), 242–247.

Cochon Drouet, O., Lentillon-Kaestner, V., & Margas, N. (2023). Effects of the Jigsaw method on student educational outcomes: Systematic review and meta-analyses. *Frontiers in Psychology*, *14*, 1216437. https://doi.org/10.3389/fpsyg.2023.1216437

Coffin, J. M. (2021). 50th anniversary of the discovery of reverse transcriptase. *Molecular Biology of the Cell*, *32*(2), 91–97. https://doi.org/10.1091/mbc.E20-09-0612

Cohn, J. (2018). Talking back to texts: An introduction to putting the "social" in "social annotation. In A. J. Reid (Ed.), *Marginalia in Modern Learning Contexts* (pp. 1–16). IGI Global.

Coleman, J. S. (1993). Rational reconstruction of society. *American Sociological Review*, *58*(1), 1–15.

Colquhoun, D. (2011, Sep. 5). Publish-or-perish: Peer review and the corruption of science. *The Guardian*. Retrieved July 25, 2025, from http://www.theguardian.com/science/2011/sep/05/publish-perish-peer-review-science

Committee on Professional Training. (2015). *Undergraduate Professional Education in Chemistry: ACS Guidelines and Evaluation Procedures for Bachelor's Degree Programs* (pp. 1–38). American Chemical Society. www.acs.org/content/dam/acsorg/about/governance/committees/training/2015-acs-guidelines-for-bachelors-degree-programs.pdf

Concordet, J. P., & Haeussler, M. (2018). CRISPOR: Intuitive guide selection for CRISPR/Cas9 genome editing experiments and screens. *Nucleic Acids Research*, *46*(W1), W242–W245. https://doi.org/10.1093/nar/gky354

Conlin, P. R., Chow, D., Miller, E. R., et al. (2000). The effect of dietary patterns on blood-pressure control in hypertensive patients: Results from the Dietary Approaches to Stop Hypertension (DASH) trial. *American Journal of Hypertension*, *13*(9), 949–955. https://doi.org/10.1016/S0895-7061(99)00284-8

Connor, C. M., Alberto, P. A., Compton, D. L., & O'Connor, R. E. (2014). Improving reading outcomes for students with or at risk for reading disabilities: A synthesis of the contributions from the Institute of Education Sciences Research Centers (Report NCSER 2014–3000). National Center for Special Education Research. https://files.eric.ed.gov/fulltext/ED544759.pdf

Cooke, S. J., Rice, J. C., Prior, K. A., et al. (2016). The Canadian context for evidence-based conservation and environmental management. *Environmental Evidence*, *5*, 1–9. https://doi.org/10.1186/s13750-016-0065-8

Corbyn, Z. (2008, Dec. 18). Call to scrap peer review in hunt for brilliant ideas. *Times Higher Education*. Retrieved July 25, 2025, from http://www.timeshighereducation.co.uk/404707.article

Cortez, M. H., & Ellner, S. P. (2010). Understanding rapid evolution in predator-prey interactions using the theory of fast-slow dynamical systems. *The American Naturalist*, *176*(5), E109–E127. https://doi.org/10.1086/656485

Cotton, D. R. E., Cotton, P. A., & Shipway, J. R. (2023). Chatting and cheating: Ensuring academic integrity in the era of ChatGPT. *Innovations in Education and Teaching International*, *61*(2), 228–239. https://doi.org/10.1080/14703297.2023.2190148

Council of Science Editors. (2006). CSE Citation Style – Quick Guide *7th Edition* [Guide]. www.tru.ca/library/pdf/csecitationstyle.pdf

Courtland, R. (2008). Planktos dead in the water. *Nature News*. https://doi.org/10.1038/news.2008.604

Cram, W. A., Templier, M., & Paré, G. (2020). (Re) considering the concept of literature review reproducibility. *Journal of the Association for Information Systems*, *21*(5), 10.

Creswell, J. W., & Plano-Clark, V. L. P. (2017). *Designing and Conducting Mixed Methods Research*. Sage Publications.

Crick, F. H. C. (1958) On protein synthesis. *Symposia of the Society for Experimental Biology*, *12*, 138–163.

Cronin, P., Ryan, F., & Coughlan, M. (2008). Undertaking a literature review: A step-by-step approach. *British Journal of Nursing*, *17*(1), 38–43. https://doi.org/10.12968/bjon.2008.17.1.28059

Crowther, P. A., Schnurr, O., Hirschi, R., et al. (2010). The R136 star cluster hosts several stars whose individual masses greatly exceed the accepted 150 M⊙ stellar mass limit. *Monthly Notices of the Royal Astronomical Society*, *408*(2), 731–751. https://doi.org/10.1111/j.1365-2966.2010.17167.x

Cui, H., Wang, C., Maan, H., et al. (2024). scGPT: Toward building a foundation model for single-cell multi-omics using generative AI. *Nature Methods*, *21*(8), 1470–1480. https://doi.org/10.1038/s41592-024-02201-0

Cui, T., & Wang, J. (2024). Empowering active learning: A social annotation tool for improving student engagement. In D. W. Johnson & R. T. Johnson (Eds.), (1987). *Learning Together and Alone: Cooperative, Competitive, and Individualistic Learning* (2nd ed.). Prentice-Hall, Inc. (2), 712–730. https://doi.org/10.1111/bjet.13403

Cutillas, A., Benolirao, E., Camasura, J., Golbin Jr, R., Yamagishi, K., & Ocampo, L. (2023). Does mentoring directly improve students' research skills? Examining the role of information literacy and competency development. *Education Sciences*, *13*(7), 694.

Daivadanam, M., Ingram, M., Annerstedt, K. S., et al. (2019). The role of context in implementation research for non-communicable diseases: Answering the 'how-to' dilemma. *PLoS ONE*, *14*(4), e0214454. https://doi.org/10.1371/journal.pone.0214454

Daly, H. E. (1996). Steady-state economics. In M. A. Cahn & R. O'Brien (Eds.), *Thinking about the Environment: Readings on Politics, Property, and the Physical World* (pp. 250–255). Routledge. https://doi.org/10.4324/9781315698724

Darling-Hammond, L., Hyler, M. E., & Gardner, M. (2017). *Effective Teacher Professional Development*. Learning Policy Institute.

Davis, M. B., Shaw, R. G., & Etterson, J. R. (2005). Evolutionary responses to changing climate. *Ecology*, *86*(7), 1704–1714. https://doi.org/10.1890/03-0788

Dawson, C., Julku, H., Pihlajamäki, M., et al. (2024). Evidence-based scientific thinking and decision-making in everyday life. *Cognitive Research Principles and Implications*, *9*(1). https://doi.org/10.1186/s41235-024-00578-2

Day, RA (1989). The Origins of the Scientific Paper: The IMRAD Format. *The American Medical Writers Association*, *4*(2), 16–18

De La Garza, H., & Vashi, N. A. (2022). The role of peer review in the scientific process. In J. Faintuch & S. Faintuch (Eds.), *Integrity of Scientific Research*. Cham: Springer. https://doi.org/10.1007/978-3-030-99680-2_41

De la Peña, C., & Luque-Rojas, M. J. (2021). Levels of reading comprehension in higher education: Systematic review and meta-analysis. *Frontiers in Psychology*, *12*, 712901. https://doi.org/10.3389/fpsyg.2021.712901

de la Puente, M., Torres, J., Troncoso, A. L. B., Meza, Y. Y. H., & Carrascal, J. X. M. (2024). Investigating the use of chatGPT as a tool for enhancing critical thinking and argumentation skills in international relations debates among undergraduate students. *Smart Learning Environments*, *11*(1), 55. https://doi.org/10.1186/s40561-024-00347-0

de la Puente Pacheco, M. A., Torres, J., Blanco Troncoso, A. L., Guzmán Murillo, H. J., & Carrascal, J. X. M. (2025). Enhancing critical thinking and argumentation skills in Colombian undergraduate diplomacy students: ChatGPT-assisted and traditional debate methods. *Journal of Political Science Education*, 1–11. https://doi.org/10.1080/15512169.2025.2449936

DeAngelo, L., Hurtado, S., Pryor, J. H., Kelly, K. R., Santos, J. L., & Korn, W. S. (2009). *The American College Teacher: National Norms for the 2007–2008 HERI Faculty Survey*. Los Angeles: Higher Education Research Institute, UCLA.

Deer, B. (2011, Jan 6). How the case against the MMR vaccine was fixed. *BMJ*, *342*. https://doi.org/10.1136/bmj.c5347

Dehairs, J., Talebi, A., Cherifi, Y., & Swinnen, J. V. (2016). CRISP-ID: Decoding CRISPR mediated indels by Sanger sequencing. *Scientific Reports*, *6*(1), 28973. https://doi.org/10.1038/srep28973

Demirci, T., & Oktay, M. (2021). The effectiveness of concept teaching using concept maps on academic achievement and elimination of misconceptions: Protein synthesis case. *Science Education International*, *32*(4), 390–399. https://doi.org/10.33828/sei.v32.i4.15

Denny, C. A. (2012). Impact of adult hippocampal neurogenesis on behavior [PhD Dissertation, Columbia University]. https://doi.org/10.7916/d8xk8nhw

Dias, B. G., & Ressler, K. J. (2014). Parental olfactory experience influences behavior and neural structure in subsequent generations. *Nature Neuroscience*, *17*, 89–96. https://doi.org/10.1038/nn.3594.

Díaz, C., Dorner, B., Hussmann, H., & Strijbos, J. (2021). Conceptual review on scientific reasoning and scientific thinking. *Current Psychology*, *42*(6), 4313–4325. https://doi.org/10.1007/s12144-021-01786-5

Díaz Quezada, V., & Sepúlveda Albornoz, M. (2023). Validation of an Instrument to Assess Deductive Reasoning in Solving Types of Problems. *International Journal of Engineering Pedagogy*, *13*(7). https://doi.org/10.3991/ijep.v13i7.40153.

Dickson, B. (2023, Nov 16). Here is how far we are to achieving AGI, according to DeepMind. VentureBeat. https://venturebeat.com/ai/here-is-how-far-we-are-to-achieving-agi-according-to-deepmind/. Accessed March 5, 2025.

Dockner, E. (2000). *Differential Games in Economics and Management Science*. Cambridge University Press.

Doudna, J. A., & Charpentier, E. (2014). The new frontier of genome engineering with CRISPR-Cas9. *Science*, *346*(6213), 1258096. https://doi.org/10.1126/science.1258096

Dowden, B. (2023). Fallacies. The Internet Encyclopedia of Philosophy. ISSN 2161-0002. Retrieved March 5, 2025, from https://iep.utm.edu/fallacy/#H6. Accessed on March 5, 2025.

Dragert, H., Wang, K., & James, T. S. (2001). A silent slip event on the deeper Cascadia subduction interface. *Science*, *292*(5521), 1525–1528. https://doi.org/10.1126/science.1060152

Drozdov, A. P., Eremets, M. I., Troyan, I. A., Ksenofontov, V., & Shylin, S. I. (2015). Conventional superconductivity at 203 Kelvin at high pressures in the sulfur hydride system. *Nature*, *525*(7567), 73–76. https://doi.org/10.1038/nature14964

Dunbar, K. N., & Klahr, D. (2012). Scientific thinking and reasoning. In K. J. Holyoak & R. G. Morrison (Eds.), *The Oxford Handbook of Thinking and Reasoning* (pp. 701–718). New York, NY: Oxford University Press. Faulkner, D. (1941). Logical Consistency as a Research Technique. Educational Research Bulletin, 42, 43.

Dzuali, F., Karikari, A., & Poa, T. (2024). ChatGPT may improve access to language-concordant care for patients with non-English language preferences. *JMIR Medical Education*, *10*, e51435. https://doi.org/10.2196/51435

Earman, J., & Glymour, C. (1980). Relativity and eclipses: The British eclipse expeditions of 1919 and their predecessors. *Historical Studies in the Physical Sciences*, *11*(1), 49–85.

Edao, H. G., Chang, C.-Y., Dilebo, W. B., et al. (2024). Nickel–Iron Layered Double Hydroxides/Nickel Sulfide Heterostructured Electrocatalysts on Surface-Modified Ti Foam for the Oxygen Evolution Reaction. *ACS Applied Materials & Interfaces 16*(38), 50602–50613. https://doi.org/10.1021/acsami.4c08215

Editors of The Lancet.P445.Einstein, A. (1905). Über einen die Erzeugung und Verwandlung des Lichtes betreffenden heuristischen Gesichtspunkt. *Annalen der Physik*, *17*(6): 132–148 https://doi.org/10.1002%2Fandp.19053220607

El Boghdady, M. (2025). Equality and diversity in research: Building an inclusive future. *BMC Research Notes*, *18*(14). https://doi.org/10.1186/s13104-025-07096-4

el-Guebaly, N., Foster, J., Bahji, A., & Hellman, M. (2022). The critical role of peer reviewers: Challenges and future steps. *Nordic Studies on Alcohol and Drugs*, *40*(1), 14–21. https://doi.org/10.1177/14550725221092862

Ellington, A. J. (2003). A meta-analysis of the effects of calculators on students' achievement and attitude levels in precollege mathematics classes. *Journal for Research in Mathematics Education*, *34*(5), 433–463. https://doi.org/10.2307/30034795

Elsevier. (2025, Jul. 31). *Scopus*. www.elsevier.com/products/scopus, Inc. (2025). *What is peer review?* www.elsevier.com/reviewer/what-is-peer-review

Emsley, R. (2023). ChatGPT: These are not hallucinations – they're fabrications and falsifications. *Schizophrenia*, *9*, 52. https://doi.org/10.1038/s41537-023-00379-4

Enfield, D. B., Mestas-Nuñez, A. M., & Trimble, P. J. (2001). The Atlantic multidecadal oscillation and its relation to rainfall and river flows in the continental US. *Geophysical Research Letters*, *28*(10), 2077–2080. https://doi.org/10.1029/2000GL012745

Enfield, D. B., Mestas-Nuñez, A. M. & Trimble, P. J. (2001). The Atlantic Multidecadal Oscillation and its relation to rainfall and river flows in the continental U.S. *Geophysical Research Letters 28*(10), 2077–2080. https://doi.org/10.1029/2000GL012745 https://agupubs.onlinelibrary.wiley.com/doi/10.1029/2000gl012745

Enyedy, N., & Stevens, R. (2006). Analyzing collaboration. In K. Sawyer (Ed.), *The Cambridge Handbook of the Learning Sciences* (pp. 191–212). Cambridge University Press. https://doi.org/10.1017/cbo9781139519526.013

Eppler, M. J. (2006). A comparison between concept maps, mind maps, conceptual diagrams, and visual metaphors as complementary tools for knowledge construction and sharing. *Information Visualization*, *5*(3), 202–210. https://doi.org/10.1057/palgrave.ivs.9500131

Eren-Zaffar, N. (2020, Dec). The Impact of Pre-reading Strategies on Reading Performance: An Action Research. In 2020 Sixth International Conference on e-learning (econf) (pp. 210–213). IEEE. https://doi.org/10.1109/econf51404.2020.9385456

Errington, T. M., Denis, A., Perfito, N., Iorns, E., & Nosek, B. A. (2021). Challenges for assessing replicability in preclinical cancer biology. *eLife*, *10*. https://doi.org/10.7554/elife.67995

Eryilmaz, E., Thoms, B., Mary, J., & Kim, R. (2014). Instructor-driven and peer-driven intelligent support for online collaborative learning. *Educational Technology Research and Development*, *62*(1), 79–101. https://doi.org/10.1007/s11423-013-9320-4

Euclid, & Heath, T. L. (1956). *The Thirteen Books of Euclid's Elements*. Courier Corporation.

Evagorou, M., Erduran, S., & Mäntylä, T. (2015). The role of visual representations in scientific practices: From conceptual understanding and knowledge generation to "seeing" how science works. *International Journal of Stem Education*, *2*, 1–13. https://doi.org/10.1186/s40594-015-0024-x

Falagas, M. E. (2007). Peer review in open access scientific journals. *PubMed*. https://pubmed.ncbi.nlm.nih.gov/20101291Open Medicine, *1*(1), 49–51.

Fass, L. (2008). Imaging and cancer: A review. *Molecular Oncology*, *2*(2), 115–152. https://doi.org/10.1016/j.molonc.2008.04.001

Faulkner, D. (1941). Logical consistency as a research technique. *Educational Research Bulletin*, *20*(2), 42–44 and 54. www.jstor.org/stable/1472900

Feiden, G. A., & Chaboyer, B. (2012). Self-consistent Magnetic Stellar Evolution Models of the Detached, Solar-type Eclipsing Binary EF Aquarii. *The Astrophysical Journal*, *761*(1), 30. https://doi.org/10.1088/0004-637X/761/1/30

Feil, R., & Fraga, M. F. (2012). Epigenetics and the environment: Emerging patterns and implications. *Nature Reviews Genetics*, *13*(2), 97–109. https://doi.org/10.1038/nrg3142

Feng, J. L. (2010). Dark matter candidates from particle physics and methods of detection. *Annual Review of Astronomy and Astrophysics*, *48*(1), 495–545. https://doi.org/10.1146/annurev-astro-082708-101659

Ferguson J. W., Alexander D. R., Allard F., et al. (2005). Low-Temperature Opacities. *The Astrophysical Journal*, *623*, 585. https://doi.org/10.1086/428642

Feynman, R. P. (1974, Jun. 14). *Cargo Cult Science* [Commencement Address] California Institute of Technology

Figueroa, J. S., Piro, N. A., Clough, C. R., & Cummins, C. C. (2006). A nitridoniobium (V) reagent that effects acid-chloride-to-organic-nitrile conversion: Synthesis via heterodinuclear (Nb/Mo) dinitrogen cleavage, mechanistic insights, and recycling. *Journal of the American Chemical Society*, *128*(3), 940–950. https://doi.org/10.1021/ja056408j

Finch, C. T., Zacharias, N., & Jao, W.-C. (2018). URAT South Parallax Results. *The Astronomical Journal*, *155*(4), 176. https://doi.org/10.3847/1538-3881/aab2b1

Fish, S. (1980). *Is There a Text in This Class?* Cambridge, MA: Harvard University Press.

Fister, I., & Perc, M. (2016). Toward the discovery of citation cartels in citation networks. *Frontiers in Physics*, *4*, 240569. https://doi.org/10.3389/fphy.2016.00049

Fitzpatrick, K. (2011). *Planned Obsolescence: Publishing, Technology, and the Future of the Academy*. New York, NY: New York University Press.

Filippazzo, J. C., Rice, E. L., Faherty, J., Cruz, K. L., Van Gordon, M. M., & Looper, D. L. (2015). Fundamental parameters and spectral energy distributions of young and field age objects with masses spanning the stellar to planetary regime. *The Astrophysical Journal*, *810*(2), 158. https://doi.org/10.1088/0004-637X/810/2/158

Fong, C. J., Adelugba, S. F., Garza, M., et al. (2024). A scoping review of the associations between sense of belonging and academic outcomes in postsecondary education. *Educational Psychology Review*, *36*(4), 138. https://doi.org/10.1007/s10648-024-09974-y

Ford, M. J. (2012). A dialogic account of Sense-Making in scientific argumentation and reasoning. *Cognition and Instruction*, *30*(3), 207–245. https://doi.org/10.1080/07370008.2012.689383

Forest, K., & Rayne, S. (2009). Incorporating primary-literature summary projects into a first-year chemistry curriculum. *Journal of Chemical Education*, *86*(5), 592–596. https://doi.org/10.1021/ed086p592

Fowler, J., & Aksnes, D. (2007). Does self-citation pay? *Scientometrics*, *72*(3), 427–437. https://doi.org/10.1007/s11192-007-1777-2

Fraine, J. D., Deming, D., Benneke, B., et al. (2014). Water vapour absorption in the clear atmosphere of a Neptune-sized exoplanet. *Nature*, *513*(7519), 526–529. https://doi.org/10.1038/nature13785

Franks, S. J., & Hoffmann, A. A. (2012). Genetics of climate change adaptation. *Annual Review of Genetics*, *46*(1), 185–208. https://doi.org/10.1146/annurev-genet-110711-155511

Fraunhofer, J. (1817). Bestimmung des Brechungs- und des Farbenzerstreuungs-Vermögens verschiedener Glasarten. *Annalen der Physik*, *56*(7), 264–313. https://doi.org/10.1002/andp.18170560706

Freedman R. S., Marley M. S., & Lodders K. (2008). Line and Mean Opacities for Ultracool Dwarfs and Extrasolar Planets, *The Astrophysical Journal Supplement Series*, *174*, 504. https://doi.org/10.1086/521793

Freedman R. S., Lustig-Yaeger J., Fortney J. J., Lupu R. E., Marley M. S., & Lodders K. (2014). Gaseous Mean Opacities for Giant Planet and Ultracool Dwarf Atmospheres over a Range of Metallicities and Temperatures. *The Astrophysical Journal Supplement Series*, *214*, 25. https://doi.org/10.1088/0067-0049/214/2/25

Freeman, R. B., & Huang, W. (2015). Collaborating with people like me: Ethnic coauthorship within the United States. *Journal of Labor Economics*, *33*(S1), S289–S318. https://doi.org/10.1086/678973

Franklin, R. E., & Gosling, R. G. (1953). Molecular configuration in sodium thymonucleate. *Nature*, *171*(4356), 740–741. https://doi.org/10.1038/171740a0

Frigg, R., & Hartmann, S. (2025). Models in science. In E. N. Zalta & U. Nodelman (Eds.), *The Stanford Encyclopedia of Philosophy* (Summer 2025 edition). Stanford University. https://plato.stanford.edu/archives/sum2025/entries/models-science/

Fukuda, Y., Hayakawa, T., Ichihara, E. et al. (1998). Evidence for oscillation of atmospheric neutrinos. *Physical Review Letters*, *81*(8), 1562–1567. https://doi.org/10.1103/physrevlett.81.1562

Gaia Collaboration, Babusiaux, C., van Leeuwen, F., et al. (2018). Gaia Data Release 2. Observational Hertzsprung-Russell diagrams. *Astronomy & Astrophysics*, *616*, A10. https://doi.org/10.1051/0004-6361/201832843

Gandhi, L., Rodríguez-Abreu, D., Gadgeel, S., et al. (2018). Pembrolizumab plus Chemotherapy in Metastatic Non–Small-Cell Lung Cancer. *New England Journal of Medicine*, *378*(22), 2078–2092. https://doi.org/10.1056/nejmoa1801005

Gao, F. (2013). A case study of using a social annotation tool to support collaboratively learning. *Internet and Higher Education*, *17*(2), 76–83. https://doi.org/10.1016/j.iheduc.2012.11.002

Garassino, M. C., Gadgeel, S., Esteban, E., et al. (2020). Patient-reported outcomes following pembrolizumab or placebo plus pemetrexed and platinum in patients with previously untreated, metastatic, non-squamous non-small-cell lung cancer (KEYNOTE-189): A multicentre, double-blind, randomised, placebo-controlled, phase 3 trial. *The Lancet Oncology*, *21*(3), 387–397. https://doi.org/10.1016/s1470-2045(19)30801-0

Garfield, E. (2006). The history and meaning of the Journal Impact Factor. *JAMA*, *295*(1), 90. https://doi.org/10.1001/jama.295.1.90-93

Garner, J., & Alley, M. (2013). How the design of presentation slides affects audience comprehension: A case for the assertion-evidence approach. International *Journal of Engineering Education*, *29*(6), 1564–1579. www.ijee.ie/articles/Vol29-6/23_ijee2791ns.pdf

Gautret, P., Lagier, J. C., Parola, P., et al. (2020). RETRACTED:[Retracted] Hydroxychloroquine and azithromycin as a treatment of COVID-19: Results of an open-label non-randomized clinical trial. *International Journal of Antimicrobial Agents*, *56*(1), 105949. https://doi.org/10.1016/j.ijantimicag.2020.105949

Gearing, N. (2024). What is the ideal methodological response for the learning and teaching of critical thinking and evaluative judgement in the age of generative artificial intelligence? *ATLAANZ Journal*, *7*(1). https://doi.org/10.26473/ATLAANZ.2024/006

General Medical Council. (2010, Jan. 28May 24). Fitness to Practise Panel hearing: Dr Andrew Jeremy Wakefield – Determination on serious professional misconduct and sanction. General Medical Council. Retrieved July 31, 2025, from https://cdn.factcheck.org/UploadedFiles/gmc-charge-sheet.pdf

Gerlich, N. W. (2025). AI tools in society: Impacts on cognitive offloading and the future of critical thinking. *Societies*, *15*(1), Article 6. https://doi.org/10.3390/soc15010006

Ghadirian, H., Salehi, K., & Ayub, A. F. M. (2018). Social annotation tools in higher education: A preliminary systematic review. *International Journal of Learning Technology*, *13*(2), 130–150. https://doi.org/10.1504/IJLT.2018.092096

Gibney, E. (2017). The scandal that rocked experimental physics. *Nature*, *551*(7680), 430–434. https://doi.org/10.1038/551430a

Gibson Jr., E. K., McKay, D. S., Thomas-Keprta, K. L., et al. (2001). Life on Mars: Evaluation of the evidence within Martian meteorites ALH84001, Nakhla, and Shergotty. *Precambrian Research*, *106*(1–2), 15–34. https://doi.org/10.1016/S0301-9268(00)00122-4

Gigerenzer, G. (2018). Statistical rituals: The replication delusion and how we got there. *Advances in Methods and Practices in Psychological Science*, *1*(2), 198–218. https://doi.org/10.1177/2515245918771329

Gihawi, A., Ge, Y., Lu, J., et al. (2023). Major data-analysis errors invalidate cancer-microbiome findings. *mBio*, *14*, e01607–23. https://doi.org/10.1128/mbio.01607-23

Gil-Flores, J., Torres-Gordillo, J. J., & Perera-Rodríguez, V. H. (2012). The role of online reader experience in explaining students' performance in digital reading. *Computers & Education*, *59*(2), 653–660. https://doi.org/10.1016/j.compedu.2012.03.014

Gilbert, J. K., Reiner, R., & Nakhleh, M. P. (Eds.). (2008). *Visualization: Theory and Practice in Science Education*. Springer. https://doi.org/10.1007/978-1-4020-5267-5

Gillon, M., Jehin, E., Lederer, S. M., et al. (2016). Temperate Earth-sized planets transiting a nearby ultracool dwarf star. *Nature*, *533*(7602), 221–224. https://doi.org/10.1038/nature17448

Gizis, J. E., Monet, D. G., Reid, I. N., Kirkpatrick, J. D., Liebert, J., & Williams, R. J. (2000). New Neighbors from 2MASS: Activity and Kinematics at the Bottom of the Main Sequence. *The Astronomical Journal*, *120*(2), 1085. https://doi.org/10.1086/301456

Glonti, K., Boutron, I., Moher, D., & Hren, D. (2019). Journal editors' perspectives on the roles and tasks of peer reviewers in biomedical journals: A qualitative study. *BMJ Open*, *9*(11), e033421.

Glonti, K., Cauchi, D., Cobo, E., Boutron, I., Moher, D., & Hren, D. (2019). A scoping review on the roles and tasks of peer reviewers in the manuscript review process in biomedical journals. *BMC Med*, *17*(1), 118. https://doi.org/10.1186/s12916-019-1347-0

Godlee, F., Smith, J., & Marcovitch, H. (2011). Wakefield's article linking MMR vaccine and autism was fraudulent. *BMJ*, *342*, c7452. https://doi.org/10.1136/bmj.c7452

Goldfarb, S., Tarver, W., & Sen, B. (2014). Family structure and risk behaviours: The role of the family meal in assessing likelihood of adolescent risk behaviours. *Psychology Research and Behavior Management*, *7*, 53–66. https://doi.org/10.2147/PRBM.S40461

Goldfinger, C., Nelson, C. H., Johnson, J. E., & Shipboard Scientific Party. (2003). Holocene earthquake records from the Cascadia subduction zone and northern San Andreas fault based on precise dating of offshore turbidites. *Annual Review of Earth and Planetary Sciences*, *31*(1), 555–577. https://doi.org/10.1146/annurev.earth.31.100901.141246

Gomes, G. (2023). Necessary and sufficient conditions, counterfactuals and causal explanations. *Erkenntnis*, *89*(8), 3085–3108. https://doi.org/10.1007/s10670-023-00668-5

Google (2025, Jul. 31). *Google Scholar*. https://scholar.google.com/

Gopalan, M., & Brady, S. T. (2019). College students' sense of belonging: A national perspective. *Educational Researcher*, *49*(2), 134–137. https://doi.org/10.3102/0013189X19897622

Gorzycki, M., Desa, G., Howard, P. J., & Allen, D. D. (2019). "Reading is important," but "I don't read": Undergraduates' experiences with academic reading. *Journal of Adolescent & Adult Literacy*, *63*(5), 499–508. https://doi.org/10.1002/jaal.1020.

Goudsouzian, L. K., & Hsu, J. L. (2023). Reading primary scientific literature: Approaches for teaching students in the undergraduate STEM classroom. *CBE – Life Sciences Education*, *22*(3), es3. https://doi.org/10.1187/cbe.22-10-0211

Gough, D. O., & Taylor, R. J. (1966). The influence of a magnetic field on Schwarzschild's criterion for convective instability in an ideally conducting fluid. *Monthly Notices of the Royal Astronomical Society*, *133*, 85. https://doi.org/10.1093/mnras/133.1.85

Green, B. N., Johnson, C. D., & Adams, A. (2006). Writing narrative literature reviews for peer-reviewed journals: Secrets of the trade. *Journal of Chiropractic Medicine*, *5*(3), 101–117.

Greenberg, S. A. (2009). How citation distortions create unfounded authority: analysis of a citation network. *BMJ*, 339. https://doi.org/10.1136/bmj.b2680

Gregory, A. T., & Denniss, A. R. (2019). Everything you need to know about peer review – The good, the bad and the ugly. *Heart, Lung and Circulation*, *28*(8), 1148–1153. https://doi.org/10.1016/j.hlc.2019.05.171

Griffiths, N., & Davila, Y. C. (2022). Embedding scaffolded reading practices into the first-year university science curriculum. In K. Manarin (Ed.), *Reading across the Disciplines* (pp. 143–165). Indiana University Press.

Grimmer, K. (2020). Using a checklist to improve the quality of research reporting. *Philippine Journal of Allied Health Sciences*, *3*(1), 5–8. https://doi.org/10.36413/pjahs.0302.002

Guerrier-Takada, C., Gardiner, K., Marsh, T., Pace, N., & Altman, S. (1983). The RNA moiety of ribonuclease P is the catalytic subunit of the enzyme. *Cell*, *35*, 849–857. https://doi.org/10.1016/0092-8674(83)90117-4

Guo, D., McTigue, E. M., Matthews, S. D., & Zimmer, W. (2020). The impact of visual displays on learning across the disciplines: A systematic review. *Educational Psychology Review*, *32*(3), 627–656. https://doi.org/10.1007/s10648-020-09523-3

Guo, Y., & Lee, D. (2023). Leveraging ChatGPT for enhancing critical-thinking skills. *Journal of Chemical Education*, *100*(12), 4876–4883. https://doi.org/10.1021/acs.jchemed.3c00505

Gwon, Y. N., Kim, J. H., Chung, H. S., et al. (2024). The use of generative AI for scientific literature searches for systematic reviews: ChatGPT and Microsoft Bing-AI performance evaluation. *JMIR Medical Informatics*, *12*, e51187. https://doi.org/10.2196/51187

Haenlein, M., & Kaplan, A. (2019). A brief history of artificial intelligence: On the past, present, and future of artificial intelligence. *California Management Review*, *61*(4), 5–14. https://doi.org/10.1177/0008125619864925

Hakiki, M., Fadli, R., Samala, A. D., et al. (2023). Exploring the impact of using Chat-GPT on student learning outcomes in technology learning: The comprehensive experiment. *Advances in Mobile Learning Educational Research*, *3*(2), 859–872. https://doi.org/10.25082/AMLER.2023.02.013

Hall, J. A., Tickle-Degnen, L., Rosenthal, R., & Mosteller, F. (1994). Hypotheses and problems in research synthesis. In H. Cooper & L. V. Hedges (Eds.), *The Handbook of Research Synthesis* (pp. 17–28). New York, NY: Russell Sage Foundation.

Halloun, I. A., & Hestenes, D. (1985). Common sense concepts about motion. *American Journal of Physics*, *53*, 1043–1055. https://doi.org/10.1119/1.14031

Hansen, J., Sato, M., Lacis, A., Ruedy, R., Tegen, I., & Matthews, E. (1998). Climate forcings in the Industrial era. *Proceedings of the National Academy of Sciences of the United States of America*, *95* (22), 12753–12758. https://doi.org/10.1073/pnas.95.22.12753

Hanson, M. A., Gómez Barreiro, P., Crosetto, P., & Brockington, D. (2024). The strain on scientific publishing. *Quantitative Science Studies*, *5*(4): 823–843. https://doi.org/10.1162/qss_a_00327

Hart, C. (1998). *Doing a Literature Review: Releasing the Social Science Research Imagination*. London: Sage.

Hartley, J. (1999). From structured abstracts to structured articles: A modest proposal. *Journal of Technical Writing and Communication*, *29*(3), 255–270. https://doi.org/10.2190/3rww-a579-hc8w-6866

Harvard College Writing Program. (2025, Jul. 31). Summarizing, Paraphrasing, and Quoting. Harvard Guide to Using Sources. https://usingsources.fas.harvard.edu/summarizing-paraphrasing-and-quoting

Harzing, A. W., & Alakangas, S. (2016). Google Scholar, Scopus and the Web of Science: A longitudinal and cross-disciplinary comparison. *Scientometrics*, *106*(1), 787–804. https://doi.org/10.1007/s11192-015-1798-9

Haynes, K. N. (2008). Reasons for doctoral attrition. *Health*, *8*(17), 1–5.

Heaven, W. D. (2023). ChatGPT is everywhere. Here's where it came from. MIT *Technology Review*, 10.

Hedges, L. V., & Olkin, I. (1985). *Statistical Methods for Meta-analysis*. Academic Press.

Hellmann, M. D., Ciuleanu, T.-. E., Pluzanski, A., et al. (2018). Nivolumab plus ipilimumab in lung cancer with a high tumor mutational burden. *New England Journal of Medicine*, *378*(22), 2093–2104. https://doi.org/10.1056/NEJMnejmoa1801946

Hembree, R., & Dessart, D. J. (1986). Effects of hand-held calculators in precollege mathematics education: A meta-analysis. *Journal for Research in Mathematics Education*, *17*(2), 83–99. https://doi.org/10.5951/jresematheduc.17.2.0083

Henry, T. J., Jao, W.-C., Winters, J. G., et al. (2018). The Solar Neighborhood XLIV: RECONS Discoveries within 10 parsecs. *The Astronomical Journal*, *155*, 265. https://doi.org/10.3847/1538-3881/aac262

Hérisson, J.-P., & Chauvin, Y. (1971). Catalyse de transformation des oléfines par les complexes du tungstène. II. Télomérisation des oléfines cycliques en présence d'oléfines acycliques. *Die Makromolekulare Chemie*, *141*(1), 161–176. doi:10.1002/macp.1971.021410112

Heron, S. F., Eakin, C. M., Douvere, F., et al. (2017). *Impacts of Climate Change on World Heritage Coral Reefs: A First Global Scientific Assessment*. UNESCO World Heritage Centre. Retrieved July 31, 2025, from https://unesdoc.unesco.org/ark:/48223/pf0000250596

Higgins, J. P. (2011). Cochrane handbook for systematic reviews of interventions. Version 5.1. The Cochrane Collaboration. www.cochrane-handbook.org.

Hiniker, S. M., Reddy, S. A., Maecker, H. T., et al. (2016). A prospective clinical trial combining radiation therapy with systemic immunotherapy in metastatic melanoma. *International Journal of Radiation Oncology • Biology • Physics*, *96*(3), 578–588. https://doi.org/10.1016/j.ijrobp.2016.07.005

Hirsch, J. E. (2005). An index to quantify an individual's scientific research output. *Proceedings of the National Academy of Sciences*, *102*(46), 16569–16572.

Hmelo-Silver, C. E., & Jeong, H. (2021). Benefits and challenges of interdisciplinarity in CSCL research: A view from the literature. *Frontiers in Psychology*, *11*, 579986. https://doi.org/10.3389/fpsyg.2020.579986

Hodgson, J., Kalir, J., & Andrews, C. D. (2023). Social annotation: Promising technologies and practices in writing. In O. Kruse, T. Binder, & M. Sennewald (Eds.), *Digital Writing Technologies in Higher Education* (pp. 141–160). Springer. https://doi.org/10.1007/978-3-031-36033-6_9

Hoegh-Guldberg, O., Mumby, P. J., Hooten, A. J., et al. (2007). Coral reefs under rapid climate change and ocean acidification. *Science*, *318*(5857), 1737–1742. https://doi.org/10.1126/science.1152509

Hollett, T., & Kalir, J. (2017). Mapping the emergence of a social-reading practice on Hypothesis and Slack: A case of graduate-student learning. In Proceedings of the 17th ACM conference on knowledge technologies and data-driven learning (pp. 170–179).

Holmes, W., & Tuomi, I. (2022). State of the art and practice in AI in education. *European Journal of Education*, *57*(4), 542–570. https://doi.org/10.1111/ejed.12533

Hoppin, F. G., Jr. (2002). How I review an original scientific article. *American Journal of Respiratory and Critical Care Medicine*, *166*(8), 1019–1023. https://doi.org/10.1164/rccm.200204-324oe374OE

Horan, S. M., & Booth-Butterfield, M. (2013). Understanding the routine expression of deceptive affection in romantic relationships. *Communication Quarterly*, *61*(2), 195–216. https://doi.org/10.1080/01463373.2012.751435

Hoskins, S. G., Lopatto, D., & Stevens, L. M. (2011). The C.R.E.A.T.E. approach to primary literature shifts undergraduates' self-assessed ability to read and analyze journal articles, attitudes about science, and epistemological beliefs. *CBE – Life Sciences Education*, *10*(4), 368–378. https://doi.org/10.1187/cbe.11-03-0027

Howard, R. M. (1995). Plagiarisms, authorships, and the academic death penalty. *College English*, *57*(7), 788–806. https://doi.org/10.58680/ce19959094

Howard, K. N., Stapleton, E. K., Nelms, A. A., Ryan, K. C., & Segura-Totten, M. (2021). Insights on biology-student motivations and challenges when reading and analyzing primary literature. *PLoS ONE*, *16*(5), e0251275. https://doi.org/10.1371/journal.pone.0251275

Hu, Z., Song, C., Xu, C., et al. (2020). Clinical characteristics of 24 asymptomatic infections with COVID-19 screened among close contacts in Nanjing, China. *Science China Life Sciences*, *63*(5), 706–711. https://doi.org/10.1007/s11427-020-1661-4

Huang, Y. T., Shih, S. M., & Tseng, S. S. (2019). Enhancing student critical literacy through social annotations. In R. Branch, H. Lee, & S. Tseng (Eds.), *Educational Media and Technology Yearbook* (Vol. 42, pp. 25–35). Springer. https://doi.org/10.1007/978-3-030-27986-8_3

Hubbard, K., & Dunbar, S. D. (2017). Perceptions of scientific research literature and strategies for reading papers depend on academic career stage. *PLoS ONE*, *12*(12), e0189753. https://doi.org/10.1371/journal.pone.0189753

Hughes, T. P., Anderson, K. D., Connolly, S. R., et al. (2018). Spatial and temporal patterns of mass bleaching of corals in the Anthropocene. *Science*, *359*(6371), 80–83. https://doi.org/10.1126/science.aan8048

Humanes, A., Beauchamp, E. A., Bythell, J. C., et al. (2021). An experimental framework for selectively breeding corals for assisted evolution. *Frontiers in Marine Science*, *8*, 669995. https://doi.org/10.3389/fmars.2021.669995

Humanes, A., Lachs, L., Beauchamp, et al. (2024). Selective breeding enhances coral heat tolerance to marine heatwaves. *Nature Communications*, *15*(1). https://doi.org/10.1038/s41467-024-52895-1

Hwang, G. J., Wu, T. T., & Chen, Y. J. (2007). Ubiquitous computing technologies in education. *International Journal of Distance Education Technologies*, *5*(4), 1–4. https://doi.org/10.4018/jdet.2007100101

Iben Jr., I., Fujimoto, M. Y., & MacDonald, J. (1992). Diffusion and Mixing in Accreting White Dwarfs. *The Astrophysical Journal*, *388*, 521. https://doi.org/10.1086/171171

IEEE Author Center Journals. (2025, Jul. 25). About the peer review process. https://journals.ieeeauthorcenter.ieee.org/submit-your-article-for-peer-review/about-the-peer-review-process/

Iglesias, C. A., & Rogers, F. J. (1996). Updated Opal Opacities. *The Astrophysical Journal*, *464*, 943. https://doi.org/10.1086/177381

Ingram, L., Hussey, J., Tigani, M. & Hemmelgarn, M. (2006). *Writing a Literature Review and Using a Synthesis Matrix* [Handout]. Raleigh, NC: North Carolina State University Writing and Speaking Tutorial Service Tutors. https://case.fiu.edu/writingcenter/online-resources/_assets/synthesis-matrix-2.pdf

Intersalt Cooperative Research Group. (1988). Intersalt: An international study of electrolyte excretion and blood pressure. Results for 24 hour urinary sodium and potassium excretion. BMJ: British Medical Journal, 319–328. www.jstor.org/stable/29700360

Institute for the Dissemination of Arts and Science. (2011). *Reviewers Information Pack*. Retrieved July 25, 2025, from http://www.icmsquare.net/FileStore/reviewerGuides.pdf

Institute of Educational Sciences. (2025, Jul. 31). Education Resources Information Center (ERIC). https://eric.ed.gov/

Ioannidis, J. P. A. (2005). Why most published research findings are false. *PLoS Medicine*, *2*(8), e124. https://doi.org/10.1371/journal.pmed.0020124

Ioannidis, J. P. A. (2017). The reproducibility wars: Successful, unsuccessful, uninterpretable, exact, conceptual, triangulated, contested replication. *Clinical Chemistry*, *63*(5), 943–945. https://doi.org/10.1373/clinchem.2017.271965

Ioannidis, J. P. A. (2019). Why most published research findings are false. *CHANCE*, *32*(1), 4–13. https://doi.org/10.1080/09332480.2019.1579573

IPBES. (2019). Global assessment report on biodiversity and ecosystem services. Intergovernmental Science-Policy Platform on Biodiversity and Ecosystem Services. https://doi.org/10.5281/zenodo.3831673

IPCC. (2021). Climate Change 2021: The Physical Science Basis. Contribution of Working Group I to the Sixth Assessment Report of the Intergovernmental Panel on Climate Change. Cambridge University Press. www.ipcc.ch/report/ar6/wg1/

Iqbal, S. A., Wallach, J. D., Khoury, M. J., Schully, S. D., & Ioannidis, J. P. A. (2016). Reproducible research practices and transparency across the biomedical literature. *PLoS Biology*, *14*(1), e1002333. https://doi.org/10.1371/journal.pbio.1002333

Ishikawa, T. (2018). Satellite navigation and geospatial awareness: Long-term effects of using navigation tools on wayfinding and spatial orientation. *The Professional Geographer*, *71*(2), 197–209. https://doi.org/10.1080/00330124.2018.1479970

Iwai, Y. (2016). The effect of explicit instruction on strategic reading in a literacy methods course. *International Journal of Teaching and Learning in Higher Education*, *28*(1), 110–118. www.isetl.org/ijtlhe/pdf/IJTLHE2140.pdf

Jaenisch, R., & Bird, A. (2003). Epigenetic regulation of gene expression: How the genome integrates intrinsic and environmental signals. *Nature Genetics*, *33*, 245–254. https://doi.org/10.1038/ng1089

Jairam, D., Kiewra, K. A., Rogers-Kasson, S., Patterson-Hazley, M., & Marxhausen, K. (2014). SOAR versus SQ3R: A test of two study systems. *Instructional Science*, *42*(3), 409–420.

Janick-Buckner, D. (1997). Getting undergraduates to critically read and discuss primary literature. *Journal of College Science Teaching*, *27*, 340–347.

Jarrett, C. (2024, Dec. 16). Best practices for facilitating group discussions with Hypothesis social annotation. *Hypothesis*. Retrieved July 25, 2025, from https://web.hypothes.is/blog/best-practices-for-facilitating-group-discussions-with-hypothesis-social-annotation/?utm_source=chatgpt.com

Jao, W. C., Henry, T. J., Gies, D. R., & Hambly, N. C. (2018). A gap in the lower main sequence revealed by Gaia DR 2. *The Astrophysical Journal Letters*, *861*(1), L11. https://doi.org/10.3847/2041-8213/aacdf6

Jasanoff, S., Hurlbut, J. B., & Saha, K. (2019). Democratic governance of human germline genome editing. *The CRISPR journal*, *2*(5), 266–271. https://doi.org/10.1089/crispr.2019.004

Jasny, B. R., Chin, G., Chong, L., & Vignieri, S. (2011). Again, and again, and again…. *Science*, *334*(6060), 1225.

Jefferson, T., Alderson, P., Wager, E., & Davidoff, F. (2002). Effects of editorial peer review: A systematic review. *JAMA*, *287*(21), 2784–2786. https://doi.org/10.1001/jama.287.21.2784

Jennings, C. G. (2006, Jun. 29). Quality and value: The true purpose of peer review. Nature Blogs. Retrieved July 25, 2025, from http://blogs.nature.com/peer-to-peer/2006/06/quality_and_value_the_true_pur.html

Jensen, M. E., Pease, E. A., Lambert, K., et al. (2013). Championing person-first language: A call to psychiatric mental health nurses. *Journal of the American Psychiatric Nurses Association*, *19*(3), 146–151. https://doi.org/10.1177/107839031348972

Ji, Z., Lee, N., Frieske, R., et al. (2023). Survey of hallucination in natural language generation. *ACM Computing Surveys*, *55*(12), 1–38. https://doi.org/10.1145/3571730

Jinek, M., Chylinski, K., Fonfara, I., Hauer, M., Doudna, J. A., & Charpentier, E. (2012). A programmable dual-RNA–guided DNA endonuclease in adaptive bacterial immunity. *Science*, *337*(6096), 816–821. https://doi.org/10.1126/science.1225829

Johnson, D. W., & Johnson, R. T. (1987). *Learning Together and Alone: Cooperative, Competitive, and Individualistic Learning* (2nd ed.). Prentice-Hall, Inc.

Johnsen, S. J., Clausen, H. B., Dansgaard, W., et al. (1997). The δ^{18}O record along the Greenland Ice Core Project deep ice core and the problem of possible Eemian climatic instability. *Journal of Geophysical Research: Oceans*, *102*(C12), 26397–26410. https://doi.org/10.1029/97JC00167

Johnson, T. E., Archibald, T. N., & Tenenbaum, G. (2010). Individual and team annotation effects on students' reading comprehension, critical thinking, and metacognitive skills. *Computers in Human Behavior*, *26*(6), 1496–1507.

Joint Task Force on Undergraduate Physics Programs. (2015). *Phys 21: Preparing Physics Students for 21st-century Careers*. American Physical Society.

Jones, R. (2010). Finding the good argument OR why bother with logic? In C. Lowe & P. Zemliansky (Eds.), *Writing Spaces: Readings on Writing Volume 1* (Vol. 1, pp. 156–179). Parlor Press.

Joseph, S. (2023). Large language model-based tools in language teaching to develop critical thinking and sustainable cognitive structures. *Rupkatha Journal on Interdisciplinary Studies in Humanities*, *15*(4). https://doi.org/10.21659/rupkatha.v15n4.13

Justice, A. C., Cho, M. K., Winker, M. A., Berlin, J. A., & Rennie, D. (1998). Does masking author identity improve peer review quality? *JAMA*, *280*(3), 240–242. https://doi.org/10.1001/jama.280.3.240

Kaendler, C., Wiedmann, M., Rummel, N., & Spada, H. (2015). Teacher competencies for the implementation of collaborative learning in the classroom: A framework and research review. *Educational Psychology Review*, *27*, 505–536. https://doi.org/10.1007/s10648-014-9288-9

Kalichman, M. (2014). Rescuing responsible peer review. *Accountability in Research*, *21*(2), 92–101. https://doi.org/10.1080/08989621.2013.848800

Kalir, J. H. (2020a). Social annotation enabling collaboration for open learning. *Distance Education*, *41*(2), 245–260. https://doi.org/10.1080/01587919.2020.1757413

Kalir, J. H. (2020b). "Annotation is first draft thinking": Educators' marginal notes as brave writing. *English Journal*, *110*(2), 62–68. https://doi.org/10.58680/ej202030968

Kalir, J. H., & Garcia, A. (2019). Civic writing on digital walls. *Journal of Literacy Research*, *51*(4), 420–443. https://doi.org/10.1177/1086296X19877208

Kalir, J. H., Morales, E., Fleerackers, A., & Alperin, J. P. (2020). "When I saw my peers annotating" Student perceptions of social annotation for learning in multiple courses. *Information and Learning Sciences*, *121*(3–4), 207–230. https://doi.org/10.1108/ILS-12-2019-0128

Kalir, R. H., & Garcia, A. (2021). *Annotation*. Cambridge, MA: MIT Press. https://mitpress.mit.edu/9780262539920/annotationKim

Kamler, B., & Thomson, P. (2014). *Helping Doctoral Students Write* (2nd ed.). Routledge.

Kararo, M., & McCartney, M. (2019). Annotated primary scientific literature: A pedagogical tool for undergraduate courses. *PLoS Biology*, *17*(1), e3000103. https://doi.org/10.1371/journal.pbio.3000103

Karikó, K., Buckstein, M., Ni, H., & Weissman, D. (2005). Suppression of RNA recognition by Toll-like receptors: The impact of nucleoside modification and the evolutionary origin of RNA. *Immunity*, *23*(2), 165–175. https://doi.org/10.1016/j.immuni.2005.06.008

Kashiyama, Y., Ishizuka, Y., Terauchi, I., et al. (2021). Engineered chlorophyll catabolism conferring predator resistance for microalgal biomass production. *Metabolic Engineering*, *66*, 79–86. https://doi.org/10.1016/j.ymben.2021.03.018

Kasneci, E., Seßler, K., Küchemann, S., et al. (2023). ChatGPT for good? On opportunities and challenges of large language models for education. *Learning and Individual Differences*, *103*, 102274. https://doi.org/10.1016/j.lindif.2023.102274

Keeling, C. D. (1960). The concentration and isotopic abundances of carbon dioxide in the atmosphere. *Tellus*, *12*(2), 200–203. https://doi.org/10.3402/tellusa.v12i2.9366

Kelly, J., Sadeghieh, T., & Adeli, K. (2014). Peer review in scientific publications: Benefits, critiques, & a survival Guide. *PubMed*, *25*(3), 227–243. https://pubmed.ncbi.nlm.nih.gov/27683470

Kepler, J., & Baumgardt, C. (1951). *Johannes Kepler: Life and Letters*. Philosophical Library.

Keranen, K. M., Weingarten, M., Abers, G. A., Bekins, B. A., & Ge, S. (2014). Sharp increase in central Oklahoma seismicity since 2008 induced by massive wastewater injection. *Science*, *345*(6195), 448–451. https://doi.org/10.1126/science.125580

Keshav, S. (2007). How to read a paper. *ACM SIGCOMM Computer Communication Review*, *37*(3), 83–84. https://doi.org/10.1145/1273445.127345

Kim, E., & Chan, M. H. W. (2004). Observation of superflow in solid helium. *Science*, *305*(5692), 1941–1944. https://doi.org/10.1126/science.1101501

Kim, K. J., Li, B., Winer, J., et al. (1993). Inhibition of vascular endothelial growth factor-induced angiogenesis suppresses tumour growth in vivo. *Nature*, *362*(6423), 841–844. https://doi.org/10.1038/362841a0

Kincaid, J. P., Fishburne, R. P., Rogers, R. L., & Chissom, B. S. (1975). *Derivation of new readability formulas (automated readability index, fog count, and Flesch reading ease formula) for Navy enlisted personnel* (Report). Naval Air Station Memphis: Chief of Naval Technical Training. Research Branch Report 8–75. https://doi.org/10.21236/ADA006655

Kmietowicz, Z. (2010, May 24). Wakefield is struck off for the "serious and wide-ranging findings against him." *BMJ*, *340*, c2803. https://doi.org/10.1136/bmj.c2803

Kobos, P. H., Erickson, J. D., & Drennen, T. E. (2006). Technological learning and renewable energy costs: Implications for US renewable energy policy. *Energy Policy*, *34*(13), 1645–1658. https://doi.org/10.1016/j.enpol.2004.12.008

Koeneman, M., Goedhart, M., & Ossevoort, M. (2013). Introducing pre-university students to primary scientific literature through argumentation analysis. *Research in Science Education*, *43*(5), 2009–2034. https://doi.org/10.1007/s11165-012-9341-y

Kohnke, L., & Moorhouse, B. L. (2025). How generative-AI-assistance impacts cognitive load during academic writing tasks. In K. Smith & J. Doe (Eds.), *Advances in Artificial Intelligence in Education* (pp. 345–360). Springer. https://doi.org/10.1007/978-3-031-71385-9_31

Kohnke, L., Moorhouse, B. L., & Zou, D. (2023). ChatGPT for language teaching and learning. *RELC Journal*, *54*(2), 537–550. https://doi.org/10.1177/00336882231162868

Körner, A. M. (2008). *Guide to Publishing a Scientific Paper* (1st ed.). Routledge. https://doi.org/10.4324/9780203938751

Kostka, I., & Toncelli, R. (2023). Exploring applications of ChatGPT to English language teaching: Opportunities, challenges, and recommendations. *Tesl-Ej*, *27*(3), n3. https://doi.org/10.55593/ej.27107int

Kozeracki, C. A., Carey, M. F., Colicelli, J., & Levis-Fitzgerald, M. (2006). An intensive Primary-Literature–based teaching program directly benefits undergraduate science majors and facilitates their transition to doctoral programs. *CBE – Life Sciences Education*, *5*(4), 340–347. https://doi.org/10.1187/cbe.06-02-0144

Krajcik, J., & McNeill, K. L. (2015). Designing and assessing scientific explanation tasks. In R. Gunstone (Eds.), *Encyclopedia of Science Education* (pp. 285–291). Springer eBooks. https://doi.org/10.1007/978-94-007-2150-0_48

Krieger, J. L., & Gallois, C. (2017). Translating science: Using the science of language to inform the language of science. *Journal of Language and Social Psychology*, *36*(1), 3–13. https://doi.org/10.1177/0261927X16663256

Krishnan, A. (2009). What are academic disciplines? Some observations on the disciplinarity vs. interdisciplinarity debate. NCRM Working Paper Series, ESRC National Centre for Research Methods.

Krontiris-Litowitz, J. (2013). Using primary literature to teach science literacy to introductory biology students. *Journal of Microbiology & Biology Education*, *14*(1), 66–77. https://doi.org/10.1128/jmbe.v14i1.538

Kroupa, P., Tout, C. A., & Gilmore, G. (1990). The low-luminosity stellar mass function. *Monthly Notices of the Royal Astronomical Society*, *244*, 76–85. https://ui.adsabs.harvard.edu/abs/1990MNRAS.244...76K/abstract

Kruger, K., Grabowski, P. J., Zaug, A. J., Sands, J., Gottschling, D. E., & Cech, T. R. (1982). Self-splicing RNA: Autoexcision and autocyclization of the ribosomal RNA intervening sequence of Tetrahymena. *Cell*, *31*(1), 147–157. https://doi.org/10.1016/0092-8674%2882%2990414-7

Kulkarni, A. V., Aziz, B., Shams, I., & Busse, J. W. (2009). Comparisons of citations in Web of Science, Scopus, and Google Scholar for articles published in general medical journals. *JAMA*, *302*(10), 1092–1096. https://doi.org/10.1001/jama.2009.1307

Kumar, M. (2009). A review of the review process: Manuscript peer-review in biomedical research. *Biology and Medicine*, *1*(4), 1–16.

Kwan, B. S. C. (2008). The nexus of reading, writing and researching in the doctoral undertaking of humanities and social sciences: Implications for literature reviewing. *English for Specific Purposes*, *27*(1), 42–56.

Kwan, B. S. C. (2009). Reading in preparation for writing a Ph.D. thesis: Case studies of experiences. *Journal of English for Academic Purposes*, *8*(3), 180–191. https://doi.org/10.1016/j.jeap.2009.02.001

Lamman, C. (2024). Translating scientific papers for the public. *Physics Today*, *77*(2), 52–55. https://doi.org/10.1063/PT.6.3.20231218a

Landy, N. I., Sajuyigbe, S., Mock, J. J., Smith, D. R., & Padilla, W. J. (2008). Perfect metamaterial absorber. *Physical Review Letters*, *100*(20), 207402. https://doi.org/10.1103/PhysRevLett.100.207402

Langer, C. J., Gadgeel, S. M., Borghaei, H., et al. (2016). Carboplatin and pemetrexed with or without pembrolizumab for advanced, non-squamous non-small-cell lung cancer: A randomised, phase 2 cohort of the open-label KEYNOTE-021 study. *The Lancet Oncology*, *17*(11), 1497–1508. https://doi.org/10.1016/s1470-2045(16)30498-3

Larson, B. Z., Moser, C., Caza, A., Muehlfeld, K., & Colombo, L. A. (2024). Critical thinking in the age of generative AI. *Academy of Management Learning & Education*, *23*(3), 373-378. https://doi.org/10.5465/amle.2024.0338

Lauer, P. A., Christopher, D. E., Firpo-Triplett, R., & Buchting, F. (2013). The impact of short-term professional development on participant outcomes: A review of the literature. *Professional Development in Education*, *40*(2), 207–227. https://doi.org/10.1080/19415257.2013.776619

Lee, C. C., & Low, M. Y. H. (2024). Using genAI in education: The case for critical thinking. *Frontiers in Artificial Intelligence*, *7*, 1452131. https://doi.org/10.3389/frai.2024.1452131

Lee, H. P., Sarkar, A., Tankelevitch, L., et al. (2025, April). The impact of generative AI on critical thinking: Self-reported reductions in cognitive effort and confidence effects from a survey of knowledge workers. In *Proceedings of the 2025 CHI Conference on Human Factors in Computing Systems* (pp. 1–22). https://doi.org/10.1145/3706598.3713778

Lee, H. P. H., Sarkar, A., Tankelevitch, L., et al. (2025, April). The impact of generative AI on critical thinking: Self-reported reductions in cognitive effort and confidence. In CHI '25 Proceedings. https://doi.org/10.1145/3706598.3713778

Lee, J. J., Hitchcock, C., & Casal, J. E. (2018). Citation practices of L2 university students in first-year writing: Form, function, and stance. *Journal of English for Academic Purposes*, *33*, 1–11. https://doi.org/10.1016/j.jeap.2018.01.001

Lee, S., Foster, C., Zhong, M., et al. (2023). Annotations serve as an on ramp for introductory biology students learning to read primary scientific literature. *Journal of Microbiology & Biology Education*, *24*(1), Article e00214-22. https://doi.org/10.1128/jmbe.00214-22

Leggett, S. K. (1992). Infrared colors of low-mass stars. *Astrophysical Journal Supplement Series*, *82*(1), 351–394.

Letchford, J., Corradi, H., & Day, T. (2017). A flexible e-learning resource promoting the critical reading of scientific papers for science undergraduates. *Biochemistry and Molecular Biology Education*, *45*(6), 483–490. https://doi.org/10.1002/bmb.21072

Liang, P., Bommasani, R., Lee, T., et al. (2022). Holistic evaluation of language models. *arXiv preprint*. arXiv:2211.09110.

Lie, R., Abdullah, C., He, W., & Tour, E. (2016). Perceived challenges in primary literature in a Master's Class: Effects of experience and instruction. *CBE – Life Sciences Education*, *15*(4), ar77. https://doi.org/10.1187/cbe.15-09-0198

Lim, L., & Renshaw, P. (2001). The relevance of sociocultural theory to culturally diverse partnerships and communities. *Journal of Child and Family Studies*, *10*(1), 9–21. https://doi.org/10.1023/A:1016625432567

Lin, L., Li, S., Huang, X., & Chen, F. (2024). Longitudinal changes of student engagement in social annotation. *Distance Education*, *45*(1), 103–121. https://doi.org/10.1080/01587919.2024.2303488

Lindegren, L., Hernández, J., Bombrun, A., et al. (2018). Gaia data release 2-the astrometric solution. *Astronomy & Astrophysics*, *616*, A2. https://doi.org/10.1051/0004-6361/201832727

Liotta, L. J., & Almeida, C. A. (2005). Organic chemistry of the cell: An interdisciplinary approach to learning with a focus on reading, analyzing, and critiquing primary literature. *Journal of Chemical Education*, *82*(12), 1794. https://doi.org/10.1021/ed082p1794

Liss, T. M., & Tipton, P. L. (1997). The discovery of the top quark. *Scientific American*, *277*(3), 54–59.

List, A., & Lin, C. J. (2023). Content and quantity of highlights and annotations predict learning from multiple digital texts. *Computers & Education*, *199*, 104791. https://doi.org/10.1016/j.compedu.2023.104791

Liu, Y., Guo, Z., Liang, T., Vulić, I., & Collier, N. (2024). *Aligning with Logic: Measuring, Evaluating, and Improving Logical Consistency in Large Language Models*. University of Cambridge.

Liumbruno, G. M., Velati, C., Pasaualetti, P., & Franchini, M. (2012). How to write a scientific manuscript for publication. *Blood Transfusion*, *11*(2), 217–226. https://doi.org/10.2450/2012.0247-12

Livingston, G., Sommerlad, A., Orgeta, V., et al. (2017). Dementia prevention, intervention, and care. *The Lancet*, *390*(10113), 2673–2734. https://doi.org/10.1016/S0140-6736(17)31363-6

Livingston, G., Sommerlad, A., Ortega, V., et al. (2020). Dementia prevention, intervention, and care. *The Lancet*, *390*(10113), P2673–2734. https://doi.org/10.1016/S0140-6736(17)31363-6

Livingston, G., Huntley, J., Sommerlad, A., et al. (2020). Dementia prevention, intervention, and care: 2020 report of the Lancet Commission. *The Lancet*, *396*(10248), 413-446. https://doi.org/10.1016/S0140-6736(20)30367-6

Livingstone, M. B. E., & Black, A. E. (2003). Markers of the validity of reported energy intake. *The Journal of Nutrition*, *133*(3), 895S–920S. https://doi.org/10.1093/jn/133.3.895S

Lloyd, Z. T., Kim, D., Cox, J. T., Doepker, G. M., & Downey, S. E. (2022). Using the annotating strategy to improve students' academic achievement in social studies. *Journal of Research in Innovative Teaching & Learning*, *15*(2), 218–231. https://doi.org/10.1108/JRIT-09-2021-0065

London Protocol Contracting Parties (2013). *Resolution LP.4(8) On the amendment to the London Protocol to regulate the placement of matter for ocean fertilization and other marine geoengineering activities*. Retrieved July 31, 2025, from the International Maritime Organization, www cdn.imo.org/localresources/en/KnowledgeCentre/IndexofIMOResolutions/LCLPDocuments/LP.4(8).pdf

Long, Q. X., Tang, X. J., Shi, Q. L., et al. (2020). Clinical and immunological assessment of asymptomatic SARS-CoV-2 infections. *Nature Medicine*, *26*(8), 1200–1204. https://doi.org/10.1038/s41591-020-0965-6

Louis, Y. D., Bhagooli, R., Kenkel, C., Baker, A. C., & Dyall, S. D. (2017). Gene expression biomarkers of heat stress in scleractinian corals: Promises and limitations. *Comparative Biochemistry and Physiology Part C: Toxicology & Pharmacology*, *191*, 63–77. https://doi.org/10.1016/j.cbpc.2016.08.007

Lu, E. P., Fischer, B. G., Plesac, M. A., & Olson, A. P. (2022). Research methods: How to perform an effective peer review. *Hospital Pediatrics*, *12*(11), e409–e413.

Lu, S., Zhou, Q., Zuo, Y., et al. (2022). Inverse design with deep generative models: Next step in materials discovery. *National Science Review*, *9*(8), nwac111. https://doi.org/10.1093/nsr/nwac111

Lucas, P. J., Baird, J., Arai, L., Law, C., & Roberts, H. M. (2007). Worked examples of alternative methods for the synthesis of qualitative and quantitative research in systematic reviews. *BMC Medical Research Methodology*, *7*, 4. https://doi.org/10.1186/1471-2288-7-4

Lucey, B. (2013, Sep. 27). Peer review: How to get it right – 10 tips. *The Guardian*. Retrieved July 25, 2025, from http://www.theguardian.com/higher-education-network/blog/2013/sep/27/peer-review-10-tips-research-paper

Luger, R., Sestovic, M., Kruse, E., et al. (2017). A seven-planet resonant chain in TRAPPIST-1. *Nature Astronomy*, *1*(6), 0129. https://doi.org/10.1038/s41550-017-0129

Luo, L., & Kiewra, K. A. (2019). Soaring to successful synthesis writing. *Journal of Writing Research*, *11*(1), 163–209. https://doi.org/10.17239/jowr-2019.11.01.06

Macadam, A., Morgans, C., Cheok, J., et al. (2025) Assessing the potential for "assisted gene flow" to enhance heat tolerance of multiple coral genera over three key phenotypic traits. *Biological Conservation*, *306*, 111155. https://doi.org/10.1016/j.biocon.2025.111155

Macagno, F., & Rapanta, C. (2019). *The Logic of Academic Writing*. New York, NY: Wessex Press, Inc. https://philpapers.org/archive/MACTLO-54.pdf

MacDonald, J. (2023). Dude, where's my citations? ChatGPT's hallucination of citations. *Mind Pad*, *11*(1), 21–26.

MacDonald, J., & Gizis, J. (2018). An explanation for the gap in the Gaia HRD for M dwarfs. *Monthly Notices of the Royal Astronomical Society*, *480*(2), 1711–1714. https://doi.org/10.1093/mnras/sty1888

Machi, L. A., & McEvoy, B. T. (2016). *The Literature Review: Six Steps to Success* (3rd ed.). Thousand Oaks, CA: Corwin, a SAGE Company.

Madsen, K. M., Hviid, A., Vestergaard, M., et al. (2002). A population-based study of measles, mumps, and rubella vaccination and autism. *New England Journal of Medicine*, *347*(19), 1477–1482. https://doi.org/10.1056/NEJMoa021134

Maeng, J., & Bell, R. (2013). Theories, laws, and hypotheses: Teaching overarching science concepts through gas laws and kinetic molecular theory. *The Science Teacher*, *80*(7), 38–43. https://doi.org/10.2505/4/tst13_080_07_38

Madhusudhan, N., Nixon, M. C., Welbanks, L., Piette, A. A. & Booth, R. A. (2020). The Interior and Atmosphere of the Habitable-zone Exoplanet K2-18b. *The Astrophysical Journal Letters*, *891*(1), L7. https://doi.org/10.3847/2041-8213/ab7229

Mahaux, O., Bauchau, V., & Van Holle, L. (2016). Pharmacoepidemiological considerations in observed-to-expected analyses for vaccines. *Pharmacoepidemiology and Drug Safety*, *25*(2), 215–222. https://doi.org/10.1002/pds.3918

Mahmood, S. S., Levy, D., Vasan, R. S., & Wang, T. J. (2014). The Framingham Heart Study and the epidemiology of cardiovascular disease: A historical perspective. *The Lancet*, *383*(9921), 999–1008. https://doi.org/10.1016/S0140-6736(13)61752-3

Maldacena, J. (1999). The large-N limit of superconformal field theories and supergravity. *International Journal of Theoretical Physics*, *38*(4), 1113–1133. https://doi.org/10.1023/A:1026654312961

Maldonado, V., Contreras, M. & Melnick, D. (2021). A comprehensive database of active and potentially-active continental faults in Chile at 1:25,000 scale. *Sci Data*, *8*, 20. https://doi.org/10.1038/s41597-021-00802-4

Malhi, G. S., Kaur, M., & Kaushik, P. (2021). Impact of climate change on agriculture and its mitigation strategies: A review. *Sustainability*, *13*(3), 1318. https://doi.org/10.3390/su13031318

Manabe, S., & Wetherald, R. T. (1967). Thermal equilibrium of the atmosphere with a given distribution of relative humidity. *Journal of the Atmospheric Sciences*, *24*(3), 241–259. https://doi.org/10.1175/1520-0469(1967)024<0241:TEOTAW>2.0.CO;2

Manchikanti, L., Kaye, A. M., Boswell, M. V., & Hirsch, J. A. (2015). Medical journal peer review: Process and bias. *Pain physician*, *18*(1), E1.

Margolis, E. (Ed.). (2001). *The Hidden Curriculum in Higher Education*. New York, NY: Routledge.

Martin, J. H. (1990). Glacial–interglacial CO_2 change: The iron hypothesis. *Paleoceanography*, *5*(1), 1–13. https://doi.org/10.1029/PA005i001p00001

Martín-Martín, A., Orduna-Malea, E., Thelwall, M., & López-Cózar, E. D. (2018). Google Scholar, Web of Science and Scopus: A systematic comparison of citations in 252 subject categories. *Journal of Informetrics*, *12*(4), 1160–1177. https://doi.org/10.1016/j.joi.2018.09.002

Matarese, V. (2013). 5. Using strategic, critical reading of research papers to teach scientific writing: The reading–research–writing continuum. In V. Matarese (Ed.), *Supporting Research Writing* (pp. 73–89). Oxford: Chandos Publishing. https://doi.org/10.1016/B978-1-84334-666-1.50005-9

Mayr, E. (1982). *The Growth of Biological Thought: Diversity, Evolution, and Inheritance*. Cambridge, MA: Harvard University Press.

McAlpine, L. (2012). Shining a light on doctoral reading: Implications for doctoral identities and pedagogies. *Innovations in Education & Teaching International*, *49*(4), 351–361.

McCann, T. M. (2010). Gateways to writing logical arguments. *English Journal*, *99*(6), 33–39.

McCormack, M. P., & Keating, J. G. (2024). Integrating collaborative annotation into higher education courses for social engagement. In M. E. Auer, U. R. Cukierman, E. V. Vidal, & E. T. Caro (Eds.), *Towards a Hybrid, Flexible and Socially Engaged Higher Education. ICL 2023. Lecture Notes in Networks and Systems*, 899. Cham: Springer. https://doi.org/10.1007/978-3-031-51979-6_9

McCormick, K. (2003). Closer than close reading: Historical analysis, cultural analysis, and symptomatic reading in the undergraduate classroom. Intertexts: Reading pedagogy in college writing classrooms, 27–49.

McGill University. (2017, Sep.). Bayes Diagnostic Tests (Version 3.11.1). Retrieved January 25, 2025, from www.medicine.mcgill.ca/epidemiology/joseph/Bayesian-Software-Diagnostic-Testing.html

The McGraw Center. (2025). Cultivating active learning. Princeton University. Accessed on March 1, 2025, from https://mcgraw.princeton.edu/faculty/teaching-princeton/designing-your-course/cultivating-active-learning.

McGraw Center for Teaching and Learning. (2025, Jul. 25). *Active reading strategies: Techniques for engaging deeply with texts*. https://mcgraw.princeton.edu/undergraduates/resources/resource-library/active-reading-strategies.

McMinn, M. R., Tabor, A., Trihub, B. L., Taylor, L., & Dominguez, A. W. (2009). Reading in graduate school: A survey of doctoral students in clinical psychology. *Training and Education in Professional Psychology*, *3*(4), 233–239. https://doi.org/10.1037/a0016405

McNutt, R. A., Evans, A. T., Fletcher, R. H., & Fletcher, S. W. (1990). The effects of blinding on the quality of peer review. *JAMA*, *263*(10), 1371–1376. https://doi.org/10.1001/jama.1990.034401000789012

Meadows, A. (2013, Sep. 17). A new approach to peer review – an interview with Keith Collier, co-founder of Rubriq. *Wiley Exchanges*. Retrieved July 25, 2025, from http://exchanges.wiley.com/blog/2013/09/17/a-new-approach-to-peer-review-an-interview-with-keith-collier-co-founder-of-rubria

Medvecky, F., & Leach, J. (2019). *An Ethics of Science Communication*. Germany: Springer International Publishing.

Miao, F., Holmes, W., Huang, R., & Zhang, H. (2021). *AI and Education: Guidance for Policymakers*. Paris, France: UNESCO Publishing.

Miller, G. A. (1956). The magical number seven, plus or minus two: Some limits on our capacity for processing information. *Psychological Review*, *63*(2), 81–97. https://doi.org/10.1037/h0043158

Miller, S., & Murillo, M. (2012). 4. Why don't students ask librarians for help? Undergraduate help-seeking behavior in three undergraduate libraries. In L. M. Duke, & A. D. Asher, (Eds.), *College Libraries and Student Culture: What We Now Know* (pp. 49–70). Chicago, IL: American Library Association.

Milosevic, N., Gregson, C., Hernandez, R., & Nenadic, G. (2019). A framework for information extraction from tables in biomedical literature. *International Journal on Document Analysis and Recognition (IJDAR)*, *22*, 55–78. https://doi.org/10.1007/s10032-019-00317-0

Milosevic, N., Gregson, C., Hernandez, R., & Nenadic, G. (2016). Disentangling the structure of tables in scientific literature. In E. Métais, F. Meziane, M. Saraee, V. Sugumaran, & S. Vadera (Eds.), *Natural Language Processing and Information Systems. NLDB 2016. Lecture Notes in Computer Science*, 9612 (pp. 162–174). Cham: Springer. https://doi.org/10.1007/978-3-319-41754-7_14

Moher, D., Liberati, A., Tetzlaff, J., Altman, D. G., & PRISMA Group. (2010). Preferred reporting items for systematic reviews and meta-analyses: The PRISMA statement. *International Journal of Surgery*, *8*(5), 336–341.

Moher, D., Shamseer, L., Cobey, K. D., et al. (2017). Stop this waste of people, animals, and money. *Nature*, *549*(7670), 23–25211–213. https://doi.org/10.1038/54902311a

Mohty, M., & Melo, J. V. (2025). How to perform a high-quality peer review. *Clinical Hematology International*, *7*(1) 10–13. https://doi.org/10.46989/001c.128601

Morgans, C. A., Hung, J. Y., Bourne, D. G., & Quigley,M. (2020). Symbiodiniaceae probiotics for use in bleaching recovery. *Restoration Ecology*, *28*(2), 282–288. https://doi.org/10.1111/rec.13069

Morilak, D. A. (2015, Jun. 24). Journal club for graduate students: Does format matter? Neuronline. Retrieved March 19, 2025, from https://neuronline.sfn.org/training/journal-club-for-graduate-students-does-format-matter

Morris, M. R., Sohl-Dickstein, J., Fiedel, N., et al. (2023). Levels of AGI for operationalizing progress on the path to AGI. *arXiv preprint*. https://doi.org/10.48550/arXiv.2311.02462

Morse, J. M. (2009). "Cherry Picking": Writing from Thin Data. *Qualitative Health Research*, *20*(1), 3. https://doi.org/10.1177/1049732309354285

Moulton-Tetlock, E., & Shah, G. (2025). Students' perceptions of AI in the global south and north. *AXIS: Journal of Lasallian Higher Education*, *16*(1). https://axisjournal.org/wp-content/uploads/2025/07/16.1e-AI-2025.pdf

Mryglod, O., Kenna, R., Holovatch, Y., & Berche, B. (2013). Comparison of a citation-based indicator and peer review for absolute and specific measures of research-group excellence. *Scientometrics*, *97*, 767–777. https://doi.org/10.1007/s11192-013-1058-9

Mulligan, A. (2005). Is peer review in crisis? *Oral Oncology*, *41*(2), 135–141. https://doi.org/10.1016/j.oraloncology.2004.11.001

Munafò, M. R., Nosek, B. A., Bishop, D. V., et al. (2017). A manifesto for reproducible science. *Nature Human Behaviour*, *1*(1), 1–9. https://doi.org/10.1038/s41562-016-0021

Murguia-Berthier, A., Ramirez-Ruiz, E., Kilpatrick, C. D., et al. (2017). A neutron star binary merger model for GW170817/GRB 170817A/SSS17a. *The Astrophysical Journal Letters*, *848*(2), L34, https://doi.org/10.3847/2041-8213/aa91b3

Nair, P. K. R., & Nair, V. D. (2014). Organization of a Research Paper: The IMRAD Format. In: *Scientific Writing and Communication in Agriculture and Natural Resources* (pp. 13–25). Springer eBooks. https://doi.org/10.1007/978-3-319-03101-9_2

Nandakumar, S., Eggl, S., Tregloan-Reed, J., et al. (2023). The high optical brightness of the BlueWalker 3 satellite. *Nature*, *623*(7989), 938–941. https://doi.org/10.1038/s41586-023-06672-7

NASA (2025, Jul. 30). Space Cloud Watch. https://science.nasa.gov/citizen-science/space-cloud-watch/

National Academies of Sciences, Engineering, and Medicine. 2023. *Pathways to Discovery in Astronomy and Astrophysics for the 2020s*. Washington, DC: The National Academies Press. https://doi.org/10.17226/26141.

National Association of Biology Teachers. (2008). Guidelines for the evaluation of four-year undergraduate biology (pp. 1–6). https://nabt.org/files/galleries/Four_Year_Guidelines.pdf

National Human Genome Research Institute. (2025, July 31). *The Human Genome Project*. National Institutes of Health. www.genome.gov/human-genome-project

National Institutes of Health (NIH). (2025, Jul. 31). PubMed. https://pubmed.ncbi.nlm.nih.gov/

Nature. (2018). *Science Benefits from Diversity*. Nature. www.nature.com/articles/d41586-018-05326-3

Nelms, A. A., & Segura-Totten, M. (2019). Expert–novice comparison reveals pedagogical implications for students' analysis of primary literature. *CBE – Life Sciences Education*, *18*(4), ar56.

Ness, M. (2016). When readers ask questions: Inquiry-Based Reading Instruction. *The Reading Teacher*, *70*(2), 189–196. https://doi.org/10.1002/trtr.1492

Newman, I., & Covrig, D. M. (2013). Building consistency between title, problem statement, purpose, & research questions to improve the quality of research plans and reports. *New Horizons in Adult Education & Human Resource Development*, *25*(1), 70–79.

Newman, A., Donohue, R., & Eva, N. (2017). Psychological safety: A systematic review of the literature. *Human Resource Management Review*, *27*(3), 521–535. https://doi.org/10.1016/j.hrmr.2017.01.001

Nguyen, K. A., Borrego, M., Finelli, C. J., et al. (2021). Instructor strategies to aid implementation of active learning: A systematic literature review. *International Journal of STEM Education*, *8*(1), 1–18.

Nichols, N. L., & Sasser, J. M. (2014). The other side of the submit button: How to become a reviewer for scientific journals. *The Physiologist*, *57*(2), 88–91.

Nissen, P., Hansen, J., Ban, N., Moore, P. B., & Steitz, T. A. (2000). The structural basis of ribosome activity in peptide bond synthesis. *Science*, *289*(5481), 920–930. https://doi.org/10.1126/science.289.5481.920

Nobel Prize Outreach (2025, Dec. 21). Ribonucleic acid (RNA) – The biomolecule which can do it all. NobelPrize.org. Accessed July 1, 1025 at www.nobelprize.org/prizes/chemistry/1989/8986-ribonucleic-acid-rna-the-biomolecule-which-can-do-it-all

Nokelainen, P., Miettinen, M., Kurhila, J., Floréen, P., & Tirri, H. (2005). A shared document-based annotation tool to support learner-centred collaborative learning. *British Journal of Educational Technology*, *36*(5), 757–770.

Nordhaus, W. D. (2017). Revisiting the social cost of carbon. *Proceedings of the National Academy of Sciences*, *114*(7), 1518–1523. https://doi.org/10.1073/pnas.1609244114

Norris, S. P., & Phillips, L. M. (2003). How literacy in its fundamental sense is central to scientific literacy. *Science Education*, *87*(2), 224–240. https://doi.org/10.1002/sce.10066

Nosek, B. A., Alter, G., Banks, G. C., et al. (2015). Promoting an open research culture. *Science*, *348*(6242), 1422–1425. https://doi.org10.1126/science.aab2374

Nosek, B. A., & Errington, T. M. (2017). Making sense of replications. *eLife*, *6*. https://doi.org/10.7554/elife.23383

Nosek, B. A., Ebersole, C. R., DeHaven, A. C., & Mellor, D. T. (2018). The preregistration revolution. *Proceedings of the National Academy of Sciences*, *115*(11), 2600–2606. https://doi.org/10.1073/pnas.1708274114

Novak, E., Razzouk, R., & Johnson, T. E. (2012). The educational use of social annotation tools in higher education: A literature review. *The Internet and Higher Education*, *15*(1), 39–49. https://doi.org/10.1016/j.iheduc.2011.09.002

Novak, J. D. & Cañas, A. J. (2008). The theory underlying concept maps and how to construct and use them. Technical Report IHMC CmapTools 2006-01 Rev 01-2008, Florida Institute for Human and Machine Cognition. Available at: http://cmap.ihmc.us/Publications/ResearchPapers/TheoryUnderlyingConceptMaps.pdf

Novoselov, K. S., Geim, A. K., Morozov, S. V., et al. (2004). Electric field effect in atomically thin carbon films. *Science*, *306*(5696), 666–669. https://doi.org/10.1126/science.1102896

NSF. (2012, Aug. 17). NSF 12–600: Ocean Acidification (OA). *National Science Foundation (NSF)*. Retrieved March 28, 2025, from www.nsf.gov/funding/opportunities/oa-ocean-acidification/503477/nsf12-600/solicitation

NSF. (2022, Dec. 12). Biological Oceanography (BioOCE). *National Science Foundation (NSF)*. Retrieved March 28, 2025, from www.nsf.gov/funding/opportunities/biooce-biological-oceanography

NSF. (2023a, Jan. 3). 23-542: Biodiversity on a Changing Planet (BOCP). *National Science Foundation (NSF)*. Retrieved March 28, 2025, from www.nsf.gov/funding/opportunities/bocp-biodiversity-changing-planet/505929/nsf23-542/solicitation

NSF. (2023b, Dec. 21). NSF 24-520: Long-Term Ecological Research (LTER). *National Science Foundation (NSF)*. Retrieved March 28, 2025, from www.nsf.gov/funding/opportunities/lter-long-term-ecological-research/7671/nsf24-520/solicitation

Nst, A. R., & Dewi, U. (2025). Undergraduate students' perception of Humata AI as a writing tool for critical journal review (CJR). *ELLITE: Journal of English Language, Literature, and Teaching*, *10*(1). https://doi.org/10.32528/ellite.v10i1.3049

Nuffield Trust (2025, Feb. 27). *Vaccination Coverage for Children and Mothers: This Indicator Looks at Vaccination Coverage for Children and Mothers in the UK and Internationally [Report]*. London: Nuffield Trust. Retrieved July 31, 2025, from www.nuffieldtrust.org.uk/resource/vaccination-coverage-for-children-and-mothers-1

Nurjannah, N., Hardiah, M., & Fadhli, M. (2023). The effect of Jigsaw method on reading comprehension in non-major English basic science students. *Journal of English Education and Teaching*, *7*(2), 412–428. https://doi.org/10.33369/jeet.7.2.412-428

OCC. (2016, Jan. 4). *Media Advisory, Edmond area earthquakes*. Oklahoma Corporation Commission. Retrieved July 31, 2025, from https://oklahoma.gov/content/dam/ok/en/occ/documents/ajls/news/2016/01-04-16eq-advisory.pdf

O'Grady, C. (2024, Dec. 17). Infamous paper that popularized unproven COVID-19 treatment finally retracted. *Science*. https://doi.org/10.1126/science.z8aky7n

O'Malley-James, J. T., & Kaltenegger, L. (2017). UV surface habitability of the TRAPPIST-1 system. *Monthly Notices of the Royal Astronomical Society: Letters*, *469*(1), L26–L30. https://doi.org/10.1093/mnrasl/slx047

OpenAI (2024, Sep. 12). Learning to reason with LLMs.

Orsi, A. J., Cornuelle, B. D., & Severinghaus, J. P. (2014). Magnitude and temporal evolution of Dansgaard–Oeschger event 8 abrupt temperature change inferred from nitrogen and argon isotopes in GISP2 ice using a new least-squares inversion. *Earth and Planetary Science Letters*, *395*, 81–90. https://doi.org/10.1016/j.epsl.2014.03.030

Osborne, J., Erduran, S., & Simon, S. (2004). Enhancing the quality of argumentation in school science. *Journal of Research in Science Teaching*, *41*(10), 994–1020. https://doi.org/10.1002/tea.20035

Pallant, J. L., Blijlevens, J., Campbell, A., & Jopp, R. (2025). Mastering knowledge: The impact of generative AI on student learning outcomes. *Studies in Higher Education*, 1–22. https://doi.org/10.1080/03075079.2025.2487570

Pan, R., Hogdal, L. J., Benito, J. M., et al. (2014). Selective BCL-2 inhibition by ABT-199 causes on-target cell death in acute myeloid leukemia. *Cancer Discovery*, *4*(3), 362–375. https://doi.org/10.1158/2159-8290.CD-13-0609

Panitz, T., & Panitz, P. (2018). Encouraging the use of collaborative learning in higher education. In *University Teaching* (pp. 161–202). Routledge.

Paré, G., & Kitsiou, S. (2017, Feb. 27). Chapter 9. Methods for literature reviews. In F. Lau & C. Kuziemsky (Eds.), *Handbook of eHealth Evaluation: An Evidence-Based Approach [Internet]*. Victoria: University of Victoria. Available from: www.ncbi.nlm.nih.gov/books/NBK481583/

Paré, G., Trudel, M. C., Jaana, M., & Kitsiou, S. (2015). Synthesizing information systems knowledge: A typology of literature reviews. *Information & Management*, *52*(2), 183–199. https://doi.org/10.1016/j.im.2014.08.008

Parkman, A. (2016). The imposter phenomenon in higher education: Incidence and impact. *Journal of Higher Education Theory and Practice*, *16*(1), 51–60.

Parravano, A., McKee, C. F., & Hollenbach, D. J. (2011). An Initial Mass Function for Individual Stars in Galactic Disks. I. Constraining the Shape of the Initial Mass Function. *The Astrophysical Journal*, *726*, 27. https://doi.org/10.1088/0004-637X/726/1/27

Pascal, B. (1670). *Pensées*. Guillaume Desprez.

Passi, S., & Vorvoreanu, M. (2022). *Overreliance on AI: An Assessment of Risk and Recommendations for Governance [Report]*. Microsoft Research. www.microsoft.com/en-us/research/wp-content/uploads/2022/06/Aether-Overreliance-on-AI-Review-Final-6.21.22.pdf

Patten, M. L. (2013). *Understanding Research Methods: An Overview of the Essentials* (9th ed.). Routledge. https://doi.org/10.4324/9781315266312

Pecorari, D., & Shaw, P. (Eds.). (2019). *Student Plagiarism in Higher Education*. New York: Routledge.

Penzias, A. A., & Wilson, R. W. (1965). A Measurement of Excess Antenna Temperature at 4080 Mc/s. *The Astrophysical Journal*, *142*, 419. https://doi.org/10.1086/148307

Petticrew, M., & Roberts, H. (2008). *Systematic Reviews in the Social Sciences: A Practical Guide*. Germany: Wiley.

Petty, R. E., Cacioppo, J. T., Petty, R. E., & Cacioppo, J. T. (1986). The elaboration likelihood model of persuasion. *Advances in Experimental Social Psychology*, *19*, 123–205. https://doi.org/10.1016/s0065-2601(08)60214-2 (pp. 1–24).

Piaget, J. (1952). *The Origins of Intelligence in Children*. (M. Cook, Trans.). W. W. Norton & Company. https://doi.org/10.1037/11494-000

Pijeira-Díaz, H. J., van de Pol, J., Channa, F., & de Bruin, A. (2023). Scaffolding self-regulated learning from causal-relations texts: Diagramming and self-assessment to improve metacomprehension accuracy? *Metacognition and Learning*, *18*(3), 631–658.

Pilcher, N., & Cortazzi, M. (2024). "Qualitative" and "quantitative" methods and approaches across subject fields: Implications for research values, assumptions, and practices. *Quality & Quantity*, *58*(3), 2357–2387. https://doi.org/10.1007/s11135-023-01734-4

Pinninti, L. R. (2024). The Impact of peer-collaborative strategic reading and reflective journal writing on orchestrated reading strategy use and comprehension. *Teaching English as a Second Language Electronic Journal (TESL-EJ)*, *27*(4). https://doi.org/10.55593/ej.27108a3

Polack, F. P., Thomas, S. J., Kitchin, N., et al. (2020). Safety and efficacy of the BNT162b2 mRNA Covid-19 vaccine. *New England Journal of Medicine*, *383*(27), 2603–2615. https://doi.org/10.1056/NEJMoa2034577

Poore, G. D., Kopylova, E., Zhu, Q., et al. (2020). Microbiome analyses of blood and tissues suggest cancer diagnostic approach. *Nature*, *579*(7800), 567.

Popper, K. (1945). *The Open Society and Its Enemies*. Routledge. https://doi.org/10.4324/9780203439913

Porter, G. W. (2022). Collaborative online annotation: Pedagogy, assessment and platform comparisons. *Frontiers in Education*, *7*, 852849. https://doi.org/10.3389/feduc.2022.852849

Potekhin, A. Y., & Chabrier, G. (2012). Thermonuclear fusion in dense stars. Electron screening, conductive cooling, and magnetic field effects. *Astronomy & Astrophysics*, *538*, A115. https://doi.org/10.1051/0004-6361/201117938

Publishing Research Consortium. (2016). Peer review survey *2015*. [Report]. Bristol: Mark Ware Consulting. Accessed March 1, 2025, from https://assets.ctfassets.net/o78em1y1w4i4/5aqlvrhd07Kcl4X4FjpL5w/49877edd7058156421af83dc30ce0ca7/PRC-peer-review-survey-report-Final-2016-05-19.pdf

Precise characterization of a corridor-shaped structure in Khufu's Pyramid by observation of cosmic-ray muons. (2023). Nature Communications 14, 1144. https://doi.org/10.1038/s41467-023-36351-0

Procureur, S., Morishima, K., Kuno, M., et al. (2023). Precise characterization of a corridor-shaped structure in Khufu's Pyramid by observation of cosmic-ray muons. *Nature Communications*, *14*(1), 1144. https://doi.org/10.1038/s41467-023-36351-0

Purdue University Writing Lab. (2025, Mar. 28). Writing a literature review. Purdue Online Writing Lab (OWL). https://owl.purdue.edu/owl/research_and_citation/conducting_research/writing_a_literature_review.html

Quigley, K. M., Bay, L. K., & van Oppen, M. J. H. (2020). Genome-wide SNP analysis reveals an increase in adaptiveAssisted gene-flow using cryopreserved gametes increases genetic variation through selective breedingdiversity of coral. *Molecular Ecology*, *29*(12), 2176–2188. https://doi.org/10.1111/mec.15482

Qin, H., Zhang, W., Zhang, S., et al. (2023). Vision rescue via unconstrained in vivo prime editing in degenerating neural retinas. *Journal of Experimental Medicine*, *220*(5), e20220776. https://doi.org/10.1084/jem.20220776

Rahmat, N. H. (2020). Conflict resolution strategies in class discussions. *International Journal of Education*, *12*(3), 49–66. https://doi.org/10.5296/ije.v12i3.16914

Rajpurohit, A. S., Reyle, C., Allard, F., et al. (2013). The effective temperature scale of M dwarfs. *Astronomy & Astrophysics*, *556*, A15.

Rakedzon, T., Segev, E., Chapnik, N., Yosef, R., & Baram-Tsabari, A. (2017). Automatic jargon identifier for scientists engaging with the public and science communication educators. *PLoS ONE*, *12*(8), e0181742. https://doi.org/10.1371/journal.pone.0181742

Ram, A. (2025, Mar. 19). *How to present a paper.* www.ee.columbia.edu/~sfchang/course/svia-F03/papers/how-to-present-a-paper.html

Randolph, J. (2009). A guide to writing the dissertation literature review. Practical Assessment, *Research, and Evaluation*, *14*(1). https://doi.org/10.7275/b0az-8t74

Rao, C., Zhong, L., Guo, Y., Li, M., Zhang, L., & Wei, K. (2024). Astronomical adaptive optics: A review. *PhotoniX*, *5*(16). https://doi.org/10.1186/s43074-024-00118-7

Rao, T. S. S., & Andrade, C. (2011). The MMR vaccine and autism: Sensation, refutation, retraction, and fraud. *Indian Journal of Psychiatry*, *53*(2), 95–96. https://doi.org/10.4103/0019-5545.82529

Ravšelj, D., Keržič, D., Tomaževič, N., Umek, L., & Aristovnik, A. (2024). Higher-education students' perceptions of ChatGPT: A global study of early reactions. *PLoS ONE*, *19*(4), e0315011. https://doi.org/10.1371/journal.pone.0315011

Rawlings, J. S. (2019). Primary literature in the undergraduate immunology curriculum: Strategies, challenges, and opportunities. *Frontiers in Immunology*, *10*, 1857. https://doi.org/10.3389/fimmu.2019.01857

Redish, E. F. (2003). Teaching physics with the physics suite. *American Journal of Physics*, *71*(6), 1065–1073. https://doi.org/10.1119/1.1596170

Reid, A. J. (2014). A case study in social annotation of digital text. *Journal of Applied Learning Technology*, *4*(2).

Resnik, D. B., & Shamoo, A. E. (2017). Reproducibility and research integrity. *Accountability in Research*, *24*(2), 116–123. https://doi.org/10.1080/08989621.2016.1257387

Retraction Watch. (2015, Sep. 30). *Predatory journals published 400,000 papers in 2014: Report.* Retrieved July 31, 2025, from https://retractionwatch.com/2015/09/30/most-predatory-publishing-occurs-in-asia-africa-report/

Ridley, D. (2012). *The Literature Review: A Step-by-Step Guide for Students.* United Kingdom: SAGE Publications.

Riess, A. G., Macri, L. M., Hoffmann, S. L., et al. (2016). A 2.4% determination of the local value of the Hubble constant. *The Astrophysical Journal*, *826*(1), 56. https://doi.org/10.3847/0004-637X/826/1/56

Risko, E. F., & Gilbert, S. J. (2016). Cognitive offloading. *Trends in Cognitive Sciences*, *20*(9), 676–688.

Rizvi, N. A., Hellmann, M. D., Snyder, A., et al. (2015). Mutational landscape determines sensitivity to PD-1 blockade in non-small-cell lung cancer. *Science*, *348*(6230), 124–128. https://doi.org/10.1126/science.aaa1348

Robertson, J., Jepson, R., Macvean, A., & Gray, S. (2016). Understanding the importance of context: A qualitative study of a location-based exergame to enhance school childrens physical activity. *PloS One*, *11*(8), e0160927. https://doi.org/10.1371/journal.pone.0160927

Rocco, T. S., & Plakhotnik, M. S. (2009). Literature reviews, conceptual frameworks, and theoretical frameworks: Terms, functions, and distinctions. *Human Resource Development Review*, *8*(1), 120–130. https://doi.org/10.1177/1534484309332617

Rochon, P. A., Bero, L. A., Bay, A. M., et al. (2002). Comparison of review articles published in peer-reviewed and throwaway journals. *JAMA*, *287*(21), 2853–2856.

Rosa, J., & Flores, N. (2017). Unsettling race and language: Toward a raciolinguistic perspective. *Language in Society*, *46*(5), 621–647. https://doi.org/10.1017/S0047404517000562

Rost, M., & Knuuttila, T. (2022). Models as epistemic artifacts for scientific reasoning in science Education research. *Education Sciences*, *12*(4), 276. https://doi.org/10.3390/educsci12040276

Rowley, J., & Slack, F. (2004). Conducting a literature review. *Management Research News*, *27*(6), 31–39. https://doi.org/10.1108/01409170410784185

Rudolph, J., Tan, S., & Tan, S. (2023). ChatGPT: Bullshit spewer or the end of traditional assessments in higher education? *Journal of Applied Learning and Teaching*, *6*(1), 342–363. https://doi.org/10.37074/jalt.2023.6.1.9

Rumelhart, D. E., Hinton, G. E., & Williams, R. J. (1986). Learning representations by back-propagating errors. *Nature*, *323*(6088), 533–536. https://doi.org/10.1038/323533a0

Rüter, C. E., Makris, K. G., El-Ganainy, R., Christodoulides, D. N., Segev, M., & Kip, D. (2010). Observation of parity–time symmetry in optics. *Nature Physics*, *6*(3), 192–195. https://doi.org/10.1038/nphys1515

Ryan, R. M., & Deci, E. L. (2000). Self-determination theory and the facilitation of intrinsic motivation, social development, and well-being. *American Psychologist*, *55*(1), 68. https://doi.org/10.1037/0003-066X.55.1.68

Sagan, C. (1994). *Pale Blue Dot: A Vision of the Human Future in Space*. Random House.

Sagan, C. (1995). *The Demon-Haunted World: Science as a Candle in the Dark*. Ballantine Books.

Salpeter, E. E. (1954). Electrons screening and thermonuclear reactions. *Australian Journal of Physics*, *7*, 373. https://doi.org/10.1071/PH540373

Sample, I. (2014, Feb. 26). How computer-generated fake papers are flooding academia. *The Guardian*. Retrieved July 30, 2025, from www.theguardian.com/technology/shortcuts/2014/feb/26/how-computer-generated-fake-papers-flooding-academia

Sandefur, C. I., & Gordy, C. (2016). Undergraduate journal club as an intervention to improve student development in applying the scientific process. *Journal of College Science Teaching*, *45*(4), 52–58. https://doi.org/10.2505/4/jcst16_045_04_)

Sato, B. K., Kadandale, P., He, W., Murata, P. M. N., Latif, Y., & Warschauer, M. (2014). Practice makes pretty good: Assessment of primary literature reading abilities across multiple large-enrollment biology laboratory courses. *CBE – Life Sciences Education*, *13*(4), 677–686. https://doi.org/10.1187/cbe.14-02-0025

Sattary, L. (2009, Sep. 9). Peer review under the microscope. *Royal Society of Chemistry*. Retrieved July 30, 2025, from www.chemistryworld.com/news/peer-review-under-the-microscope/3003127.article

Saumon, D., Chabrier, G., & van Horn, H. M. (1995). An Equation of State for Low-Mass Stars and Giant Planets. *The Astrophysical Journal Supplement*, *99*, 713. https://doi.org/10.1086/192204

Saunders, M. N. K., & Rojon, C. (2011). On the attributes of a critical literature review. *Coaching: An International Journal of Theory, Research and Practice*, *4*(2), 156–162. https://doi.org/10.1080/17521882.2011.596485

Saxby, C., & Richardson, M. (2006). *Assessing the Impact of Open Access*. Oxford Journals Preliminary Report.

Saxby, C., & Richardson, M. (2006, June). Assessing the impact of open access: Preliminary findings from the Oxford Journals. *Oxford University Press*. Retrieved July 30, 2025, from www.homepages.ucl.ac.uk/~uczciro/nar.pdf

Schanche, N., Cameron, A. C., Hébrard, G., et al. (2018). Machine-learning approaches to exoplanet transit detection and candidate validation in wide-field ground-based surveys. *Monthly*

Notices of the Royal Astronomical Society, *483*(4), 5534–5547. https://doi.org/10.1093/mnras/sty3146

Schechter, A. N., Wyngaarden, J. B., Edsall, J. T., et al. (1989). Colloquium on scientific authorship: Rights and responsibilities. *The FASEB Journal: Official Publication of the Federation of American Societies for Experimental Biology*, *3*(2), 209–217. https://doi.org/10.1096/fasebj.3.2.2914630

Schellens, T., Van Keer, H., & Valcke, M. (2005). The impact of role assignment on knowledge construction in asynchronous discussion groups: A multilevel analysis. *Small Group Research*, *36*(6), 704–745. https://doi.org/10.1177/1046496405281771

Schley, D. (2009, Sep. 9). Peer reviewers satisfied with system. *Times Higher Education*. Retrieved July 25, 2025, from www.timeshighereducation.co.uk/408108.article

Schrock, R. R., Murdzek, J. S., Bazan, G. C., Robbins, J., Dimare, M., & O'Regan, M. (1990). Synthesis of molybdenum imido alkylidene complexes and some reactions involving acyclic olefins. *Journal of the American Chemical Society*, *112*(10), 3875–3886. https://doi.org/10.1021/ja00166a023

Schulman, H. C., Dixon, G. N., Bullock, O. M., & Colon-Amill, D. (2020). The effects of jargon on processing fluency, self-perceptions, and scientific engagement. *Journal of Language and Social Psychology*, *39*(5–6), 579–601. https://doi.org/10.1177/0261927x20902177

Schulz, T., Knierim, M. T., & Weinhardt, C. (2024). How generative-AI-assistance impacts cognitive load during knowledge work: A study proposal. In *NeuroIS Retreat* (pp. 357–365). Cham: Springer Nature Switzerland. https://doi.org/10.1007/978-3-031-71385-9_31

Schunk, D. H. (2012). *Learning Theories: An Educational Perspective*. Addison Wesley Longman.

Schwab, P., France, M. B., Ziller, J. W., & Grubbs, R. H. (1995). A series of well-defined metathesis catalysts – synthesis of $[RuCl_2(=CHR')(PR_3)_2]$ and its reactions. *Angewandte Chemie International Edition*, *34*(18), 2039–2041. https://doi.org/10.1002/anie.199520391

Schwarz, N. (2011) Feelings-as-information theory. In: P. Van Lange, A. Kruglanski, & E. T. Higgins (Eds.), *Handbook of Theories in Social Psychology* (pp. 289–308). Thousand Oaks, CA: SAGE Publications. https://doi.org/10.4135/9781446249215.n15289–308.

Schweitzer, M. D., & Schweitzer, L. (2020). Critical analysis of climate models: Strengths, weaknesses, and common errors. *Earth System Science Data*, *12*(3), 2253–2267. https://doi.org/10.5194/essd-12-2253-2020

Sentmanat, M. F., Peters, S. T., Florian, C. P., Connelly, J. P., & Pruett-Miller, S. M. (2018). A survey of validation strategies for CRISPR-Cas9 editing. *Scientific Reports*, *8*(1), 888. https://doi.org/10.1038/s41598-018-19441-8

Seralini, G. E., Clair, E., Mesnage, R., et al. (2012). RETRACTED: Long-term toxicity of a Roundup herbicide and a Roundup-tolerant genetically modified maize. *Food and Chemical Toxicology*, *50*(11), 4221–4231. https://doi.org/10.1016/j.fct.2012.08.005

Séralini, G.-. E., Mesnage, R., Defarge, N., & Spiroux de Vendômois, J. (2014). Conflicts of interests, confidentiality and censorship in health risk assessment: The example of an herbicide and a GMO. *Environmental Sciences Europe*, *26*, 13. https://doi.org/10.1186/s12302-014-0013-6

Sestito, P., & Randich, S. (2005). Time scales of Li evolution: a homogeneous analysis of open clusters from ZAMS to late-MS. *Astronomy & Astrophysics*, *442*(2), 615–627. https://doi.org/10.1051/0004-6361:20053482

Shadish, W. R., Cook, T. D., & Campbell, D. T. (2002). *Experimental and Quasi-experimental Designs for Generalized Causal Inference*. Houghton: Mifflin and Company.

Shah, R., Irpan, A., Turner, A. M., et al. (2025). An approach to technical AGI safety and security. *arXiv preprint*. https://doi.org/10.48550/arXiv.2504.01849

Shallue, C. J., & Vanderburg, A. (2018). Identifying exoplanets with deep learning: A five-planet resonant chain around Kepler-80 and an Eighth Planet around Kepler-90. *The Astronomical Journal*, *155*(2), 94. https://doi.org/10.3847/1538-3881/aa9e09

Shanahan, C., Shanahan, T., & Misischia, C. (2011). Analysis of expert readers in three disciplines: History, mathematics, and chemistry. *Journal of Literacy Research*, *43*(4), 393–429.

Shaw University Writing Center. (2024). How to Avoid Plaigarism. Retrieved July 31, 2025, from www.shawu.edu/wp-content/uploads/2024/07/How-to-Avoid-Plagiarism.pdf

Shean, D. E., Bhushan, S., Montesano, P., Rounce, D. R., Arendt, A., & Osmanoglu, B. (2020). A systematic, regional assessment of high mountain Asia glacier mass balance. *Frontiers in Earth Science*, *7*, 363.

Sherboboevna, K. A. (2020). The concept of discourse and its definition. *International Journal of Progressive Sciences and Technologies*, *20*(2), 126–128. https://doi.org/10.52155/ijpsat.v20.2.1824

Shin, E. S., Jung, H. J., Kim, J., Cho, M., & Kim, T. S. (2025). Skin autofluorescence is associated with glycemic variability in type 2 diabetes patients. *Scientific Reports*, *15*(1), 24105. https://doi.org/10.1038/s41598-025-09517-7

Shreffler, J., & Huecker, M. R. (2023). Type I and Type II errors and statistical power. In: *StatPearls. Treasure*. Island, FL: StatPearls Publishing. https://europepmc.org/article/MED/32491462/NBK430685#free-full-text

Shulman, H. C., Dixon, G. N., Bullock, O. M., & Colon-Amill, D. C. (2020) The effects of jargon on processing fluency, self-perceptions, and scientific engagement. *Journal of Language and Social Psychology*, *39*(5–6), 579–597. https://doi.org/10.1177/0261927x20902177J Lang Soc Psych. Epub Jan 29.

Shulman, H. C., & Sweitzer, M. D. (2018, Apr. 2) Advancing Framing Theory framing theory: Designing an equivalency frame to improve political information processing. *Human Communication Research*, *44*(2), 155–175. https://doi.org/10.1093/hcr/hqx006

Sigmon, A. J., & Bodek, M. J. (2022). Use of an online social annotation platform to enhance a flipped organic chemistry course. *Journal of Chemical Education*, *99*(2), 538–545. https://doi.org/10.1021/acs.jchemed.1c00889

Simkin, M. V., & Roychowdhury, V. P. (2005). Stochastic modeling of citation slips. *Scientometrics*, *62*(3), 367–384. https://doi.org/10.1007/s11192-005-0028-2

Simmons, J. P., Nelson, L. D., & Simonsohn, U. (2011). False-positive psychology: Undisclosed flexibility in data collection and analysis allows presenting anything as significant. *Psychological Science*, *22*(11), 1359–1366. https://doi.org/10.1177/0956797611417632

Sloane, J. D. (2021). Primary literature in undergraduate science courses: What are the outcomes? *Journal of College Science Teaching*, *50*(3), 51–60. https://doi.org/10.1080/0047231x.2021.12290508

Smith, R. (2006). Peer review: A flawed process at the heart of science and journals. *Journal of the Royal Society of Medicine*, *99*(4), 178–182. https://doi.org/10.1177/014107680609900414

Smithsonian Astrophysical Observatory (2025, July 31). NASA Astrophysics Data System. https://ui.adsabs.harvard.edu/

Snow, K. (2007). People first language. *Disability is Natural*. Retrieved July 25, 2025, from www.disabilityisnatural.com/images/PDF/pfl09.pdf

Sobol, R. W., Horton, J. K., Kühn, R., et al. (1996). Requirement of mammalian DNA polymerase-β in base-excision repair. *Nature*, *379*(6561), 183–186. https://doi.org/10.1038/379183a0

Soderblom, D. R. (2010). The ages of stars. *Annual Review of Astronomy and Astrophysics, 48*(1), 581–629. https://doi.org/10.1146/annurev-astro-081309-130806

Sollaci, L. B., & Pereira, M. G. (2004). The introduction, methods, results, and discussion (IMRAD) structure: A fifty-year survey. *PubMed, 92*(3), 364–367. https://pubmed.ncbi.nlm.nih.gov/15243643

Solomon, S., Plattner, G., Knutti, R., & Friedlingstein, P. (2009). Irreversible climate change due to carbon dioxide emissions, *Proceedings of the National Academy of Sciences of the United States of America, 106*(6), 1704–1709. https://doi.org/10.1073/pnas.0812721106

Soltis, J. F. (1984). On the nature of educational research. *Educational Researcher, 13*(10), 5–10. https://doi.org/10.3102/0013189x013010005

Somayazulu, M., Ahart, M., Mishra, A. K., et al. (2019). Evidence for superconductivity above 260 K in Lanthanum Superhydride at Megabar pressures. *Physical Review Letters, 122*(2). https://doi.org/10.1103/physrevlett.122.027001

Sorokin, C., & Krauss, R. W. (1958). The effects of light intensity on the growth rates of green algae. *Plant Physiology, 33*(2), 109–113. https://doi.org/10.1104/pp.33.2.109

Sowell, R. S. (2009). *Ph.D. Completion and Attrition: Findings from Exit Surveys for Ph.D. Completers*. Washington, DC: Council of Graduate Schools.

Sparrow, B., Liu, J., & Wegner, D. M. (2011). Google effects on memory: Cognitive consequences of having information at our fingertips. *Science, 333*(6043), 776–778. https://doi.org/10.1126/science.1207745

Spaulding, L. S., & Rockinson-Szapkiw, A. (2012). Hearing their voices: Factors doctoral candidates attribute to their persistence. *International Journal of Doctoral Studies, 7*, 199–219. https://doi.org/10.28945/1589

Speake, J., & Axon, S. (2012). "I never use 'maps' anymore": Engaging with Sat Nav technologies and the implications for cartographic literacy and spatial awareness. *The Cartographic Journal, 49*(4), 326–336. https://doi.org/10.1179/1743277412Y.0000000021

Spiro, R. J. (1980). Schema Theory and Reading Comprehension: New Directions. *Technical Report No. 191*. Center for the Study of Reading.

Stadler, M., Bannert, M., & Sailer, M. (2024). Cognitive ease at a cost: LLMs reduce mental effort but compromise depth in student scientific inquiry. *Computers in Human Behavior, 160*, 108386. https://doi.org/10.1016/j.chb.2024.108386

Stark, B. C., Kole, R., Bowman, E. J., & Altman, S. (1978). Ribonuclease P: An enzyme with an essential RNA component. *Proceedings of the National Academy of Sciences, 75*(8), 3717–3721. https://doi.org/10.1073/pnas.75.8.3717

Stauffer, J. R., Schultz, G., & Kirkpatrick, J. D. (1998). Keck spectra of Pleiades brown dwarf candidates and a precise determination of the lithium depletion edge in the Pleiades. *The Astrophysical Journal, 499*(2), L199. https://doi.org/10.1086/311379

Steer, P. J., & Ernst, S. (2021). Peer review – Why, when and how. *International Journal of Cardiology Congenital Heart Disease, 2*, 100083. https://doi.org/10.1016/j.ijcchd.2021.100083

STM. (2021, Oct. 19). STM global brief 2021 – Economics and market size, an STM report supplement (Report). International Association of Scientific, Technical and Medical Publishers. Retrieved July 25, 2025, from https://coilink.org/20.500.12592/4jnnjk on 30 Nov 2024. COI: 20.500.12592/4jnnjk.

Su, A. Y., Yang, S. J., Hwang, W. Y., & Zhang, J. J. (2010). A Web 2.0-based collaborative annotation system for enhancing knowledge sharing in collaborative learning environments. *Computers & Education, 55*(2), 752–766. https://doi.org/10.1016/j.compedu.2010.03.008

Subar, A. F., Freedman, L. S., Tooze, J. A., et al. (2015). Addressing current criticism regarding the value of self-report dietary data. *The Journal of Nutrition, 145*(12), 2639–2645. https://doi.org/10.3945/jn.115.219634

Sun, C., Hwang, G. J., Yin, Z., Wang, Z., & Wang, Z. (2023). Trends and issues of social annotation in education: A systematic review from 2000 to 2020. *Journal of Computer Assisted Learning*, *39*(2), 329–350. https://doi.org/10.1111/jcal.12764

Sun, T.-T. (2020). Active versus passive reading: How to read scientific papers? *National Science Review*, *7*(9), 1422–1427. https://doi.org/10.1093/nsr/nwaa130

Sun, S. (2020). Predictive reading using the Q-P/C method for scientific research. *National Science Review*, *7*(9), 1422–1430. https://doi.org/10.1093/nsr/nwaa053

Sun, X., Zhong, X., Xu, X., et al. (2025). A data-to-forecast machine learning system for global weather. *Nature Communications*, *16*(1), 6658. https://doi.org/10.1038/s41467-025-62024-1

Surmiak, Adrianna (2018). Confidentiality in qualitative research involving vulnerable participants: Researchers' perspectives. *Forum Qualitative Sozialforschung/Forum: Qualitative Social Research*, *19*(3), Art. 12, http://doi.org/10.17169/fqs-19.3.3099.

Sverdlik, A. (2019). *A Comprehensive Evaluation of the Doctoral Experience: Exploring Ph.D. Students' Socialization, Motivation, and Well-Being*. Canada: McGill University.

Sweller, J. (1988). Cognitive load during problem solving: Effects on learning. *Cognitive Science*, *12*(2), 257–285. https://doi.org/10.1207/s15516709cog1202_4

Szyf, M., Weaver, I. C., Champagne, F. A., Diorio, J., & Meaney, M. J. (2005). Maternal programming of steroid-receptor expression and phenotype through DNA methylation in the rat. *Frontiers in Neuroendocrinology*, *26*(3–4), 139–162. https://doi.org/10.1016/j.yfrne.2005.10.002

Tay, A. C.-H. (2025, May 20). Deep dive into three AI academic search tools. *Katina Magazine*. https://doi.org/10.1146/katina-052025-2

Taslim, M., Putra, R. P., Daulay, N., & Bulut, S. (2025). Academic Cheating with Generative AI in Higher Education: An Extended Model of the Theory of Planned Behavior with Motivational Antecedents. *Indonesian Journal of Psychology/Jurnal Psikologi*, 52(3). https://doi.org/10.22146/jpsi.107932

Taylor & Francis. (2025, July 31). Understanding peer review. Author Services: *What is peer review? A guide for authors*. https://authorservices.taylorandfrancis.com/publishing-your-research/peer-review/

Taylor, B., Miller, E., Farrington, C. P., et al. (1999). Autism and measles, mumps, and rubella vaccine: No epidemiological evidence for a causal association. *The Lancet*, *353*(9169), 2026–2029. https://doi.org/10.1016/S0140-6736(99)01239-8

Temin, H. M., & Mizutani, S. (1970). Viral RNA-dependent DNA polymerase: RNA-dependent DNA polymerase in virions of Rous sarcoma virus. *Nature*, *226*(5252), 1211–1213. https://doi.org/10.1038/2261211a0

Teng, M. F. (2023). Scientific writing, reviewing, and editing for open-access TESOL journals: The role of ChatGPT. *International Journal of TESOL Studies*, *5*(1), 87–91. https://doi.org/10.58304/ijts.20230107

Teong, X. T., Liu, K., Vincent, A. D., et al. (2023). Intermittent fasting plus early time-restricted eating versus calorie restriction and standard care in adults at risk of type 2 diabetes: A randomized controlled trial. *Nature Medicine*, *29*(4), 963–972. https://doi.org/10.1038/s41591-023-02287-7

Tesniere, A., Schlemmer, F., Boige, V., et al. (2010). Immunogenic death of colon cancer cells treated with oxaliplatin. *Oncogene*, *29*(4), 482–491. https://doi.org/10.1038/onc.2009.356

Thaler, R. (1980). Toward a positive theory of consumer choice. *Journal of Economic Behavior & Organization*, *1*(1), 39–60. https://doi.org/10.1016/0167-2681(80)90051-7

The American Physiological Society. (2025). *Peer Review Process*. https://journals.physiology .org/peer-review-process

The Editors of the Lancet. (2010). Retraction – Ileal-lymphoid-nodular hyperplasia, non-specific colitis, and pervasive developmental disorder in children. *The Lancet*, *375*(9713), 445. https://doi.org/10.1016/S0140-6736(10)60175-4

Tollefson, J. (2012). Ocean-fertilization project off Canada sparks furore. *Nature*, *490*(7421), 458. https://doi.org/10.1038/490458a

Toosi, A., Bottino, A. G., Saboury, B., Siegel, E., & Rahmim, A. (2021). A brief history of AI: How to prevent another winter (a critical review). *PET Clinics*, *16*(4), 449–469. https://doi .og/10.1016/j.cpet.2021.07.001

Topalian, S. L., Hodi, F. S., Brahmer, J. R., et al. (2012). Safety, activity, and immune correlates of anti-PD-1 antibody in cancer. *New England Journal of Medicine*, *366*(26), 2443–2454. https://doi.org/10.1056/NEJMoa1200690

Torraco, R. J. (2016). Writing integrative literature reviews: Using the past and present to explore the future. *Human Resource Development Review*, *15*(4), 404–428. https://doi .org/10.1177/1534484316671606

Toulmin, S., Rieke, R. D., & Janik, A. (1984). *An Introduction to Reasoning* (2nd ed.). United Kingdom: Macmillan.

Traag, V. A. (2021). Inferring the causal effect of journals on citations. *Quantitative Science Studies*, *2*(2), 496–504. https://doi.org/10.1162/qss_a_00128

Tranfield, D., Denyer, D., & Smart, P. (2003). Towards a methodology for developing evidence-informed management knowledge by means of systematic review. *British Journal of Management*, *14*(3), 207–222. https://doi.org/10.1111/1467-8551.00375

Trifonov, T., Reffert, S., Zechmeister, M., Reiners, A., & Quirrenbach, A. (2015). Precise radial velocities of giant stars – VIII. Testing for the presence of planets with CRIRES infrared radial velocities. *Astronomy and Astrophysics*, *582*, A54. https://doi .org/10.1051/0004-6361/201526196

Tsiaras, A., Waldmann, I. P., Tinetti, G., Tennyson, J., & Yurchenko, S. N. (2019). Water vapour in the atmosphere of the habitable-zone eight-Earth-mass planet K2-18 b. *Nature Astronomy*, *3*, 1086–1091. https://doi.org/10.1038/s41550-019-0878-9

Tukey, J. W. (1977). *Exploratory Data Analysis*. Addison-Wesley Publishing Company,

Tyton Partners. (2023). *Generative AI in Higher Education: Fall 2023 Update of the Time for Class Study* (Report). https://tytonpartners.com/app/uploads/2023/10/GenAI-IN-HIGHER-EDUCATION-FALL-2023-UPDATE-TIME-FOR-CLASS-STUDY.pdf

UCLA Undergraduate Writing Center. (2016). *Writing the literature review*. Retrieved July 30, 2025, from https://drive.google.com/file/d/186SkA2DNGprLvTXFz2xzxomMdBdf2kYr/view

U.S. Department of Education. (2021). *Condition of Education 2021* (NCES 202144). National Center for Education Statistics.

van der Sande, L., van Steensel, R., Fikrat-Wevers, S., & Arends, L. (2023). Effectiveness of interventions that foster reading motivation: A meta-analysis. *Educational Psychology Review*, *35*(1), 21. https://doi.org/10.1007/s10648-023-09719-3

Van Dis, E. A., Bollen, J., Zuidema, W., Van Rooij, R., & Bockting, C. L. (2023). ChatGPT: Five priorities for research. *Nature*, *614*(7947), 224–226. https://doi.org/10.1038/d41586-023-00288-7

Van Dusen, B., & Nissen, J. (2022). How statistical model development can obscure inequities in STEM student outcomes. *Journal of Women and Minorities in Science and Engineering*, *28*(3), 27–58.

van Lacum, E., Koeneman, M., Ossevoort, M., & Goedhart, M. (2016). Scientific argumentation model (SAM): A heuristic for reading research articles by science students. In N. Papadouris, A. Hadjigeorgiou, & C. P. Constantinou (Eds.), *Insights from Research in Science Teaching and Learning. Contributions from Science Education Research* (pp. 169–183). Cham: Springer International Publishing. https://doi.org/10.1007/978-3-319-20074-3_12

van Oppen, M. J. H., Oliver, J. K., Putnam, H. M., & Gates, R. D. (2015). Building coral reef resilience through assisted evolution. *Proceedings of the National Academy of Sciences*, *112*(8), 2307–2313. https://doi.org/10.1073/pnas.1422301112

van Pletzen, E. (2006). 5. A body of reading: Making "visible" the reading experiences of first-year medical students. In L. Thesen, E. van Pletzen, & N. S. Ndebele (Eds.), *Academic Literacy and the Languages of Change* (pp. 104–129). New York, NY: Continuum International Publishing Group.

van Saders, J. L., & Pinsonneault, M. H. (2012). An ^{3}He-driven Instability near the Fully Convective Boundary. *The Astrophysical Journal*, *751*(2), 98. https://doi.org/10.1088/0004-637X/751/2/98

Varney, J. J. (2010). The role of dissertation self-efficacy in increasing dissertation completion: Sources, effects, and viability of a new self-efficacy construct. *College Student Journal*, *44*(4), 932–948.

Varpio, L. (2018). Using rhetorical appeals to credibility, logic, and emotions to increase your persuasiveness. *Perspectives on Medical Education*, *7*, 207–210. https://doi.org/10.1007/s40037-018-0420-2

Vygotsky, L., Cole, M., & John-Steiner, V. (1978). *Mind in Society: Development of Higher Psychological Processes*. Cambridge, MA: Harvard University Press.

Waerden, B. L. V. D., & Taisbak, C. M. (2025, Jul. 10). Euclid. Encyclopedia Britannica. www.britannica.com/biography/Euclid-Greek-mathematician

Wager, E., & Jefferson, T. (2001). *How to Survive Peer Review*. BMJ Books.

Wakefield, A. J., Murch, S. H., Anthony, A., et al. (1998). RETRACTED: Ileal-lymphoid-nodular hyperplasia, non-specific colitis, and pervasive developmental disorder in children. *The Lancet*, *351*(9103), 637–641. [Retracted]. https://doi.org/10.1016/S0140-6736(97)11096-0

Wakeford, H. R., Sing, D. K., Deming, D., et al. (2013). HST hot Jupiter transmission spectral survey: etection of water in HAT-P-1b from WFC3 near-IR spatial scan observations. *Monthly Notices of the Royal Astronomical Society*, *435*(4), 3481–3493. https://doi.org/10.1093/mnras/stt1536

Wallach, J. D., Boyack, K. W., & Ioannidis, J. P. A. (2018). Reproducible research practices, transparency, and open access data in the biomedical literature, 2015–2017. *PLoS Biology*, *16*(11), e2006930. https://doi.org/10.1371/journal.pbio.2006930

Walsh, F. R., & Zoback, M. D. (2015). Oklahoma's recent earthquakes and saltwater disposal. *Science Advances*, *1*(5). https://doi.org/10.1126/sciadv.1500195

Waltman, L., Vvan Eck, N. J., & Vvan Raan, A. F. J. (2012). Universality of citation distributions revisited. *Journal of the American Society for Information Science and Technology*, *63*(1), 72–77. https://doi.org/10.1002/asi.21671

Walton, G. M., & Cohen, G. L. (2011). A brief social-belonging intervention improves academic and health outcomes of minority students. *Science*, *331*(6023), 1447–1451. https://doi.org/10.1126/science.1198364

Wang, N., Wang, X., & Su, Y. S. (2024). Critical analysis of the technological affordances, challenges and future directions of Generative AI in education: a systematic review. *Asia Pacific Journal of Education*, 44(1), 139-155. https://doi.org/10.1080/02188791.2024.2305156

Wang, Y., Zhang, Y., Wang, X., Yang, S., Li, S., & Zhang, H. (2023). A double-stranded RNA binding protein enhances drought resistance by modulating root system architecture in rice. *Nature Communications*, *14*, 46754. https://doi.org/10.1038/s41467-024-46754-2

Watson, J. D., & Crick, F. H. C. (1953a). Molecular structure of nucleic acids: A structure for deoxyribose nucleic acid. *Nature, 171*(4356), 737–738. https://doi.org/10.1038/171737a0

Watson, J. D., & Crick, F. H. C. (1953b). Genetical implications of the structure of deoxyribonucleic acid. *Nature, 171*(4361), 964–967. https://doi.org/10.1038/171964b0

Watson, N. T., Watson, K. L., & Stanley, C. A. (2017). Conflict management and dialogue in higher education: A global perspective. IAP.

Webster, J., & Watson, R. T. (2002). Analyzing the past to prepare for the future: Writing a literature review. *MIS Quarterly, 26*(2), xiii–xxiii. https://web.njit.edu/~egan/Writing_A_Literature_Review.pdf

Weier, J. (2001, Jul. 10). *John Martin (1935–1993)*, On the Shoulders of Giants. NASA Earth Observatory. Retrieved July 31, 2025, from https://earthobservatory.nasa.gov/features/Martin

Weissgerber, T. L., Milic, N. M., Winham, S. J., & Garovic, V. D. (2015). Beyond bar and line graphs: Time for a new data-presentation paradigm. *PLoS Biology, 13*(4), e1002128. https://doi.org/10.1371/journal.pbio.1002128

West, A. A., Hawley, S. L., Bochanski, J. J., et al. (2008). Constraining the age–activity relation for cool stars: the sloan digital sky survey data release 5 low-mass star spectroscopic sample. *The Astronomical Journal, 135*(3), 785.

Wheeldon, J., & Faubert, J. (2009). Framing experience: Concept maps, mind maps, and data collection in qualitative research. *International Journal of Qualitative Methods, 8*(3), 68–83. https://doi.org/10.1177/160940690900800307

White, E. B., & Strunk, W. (2023). *The Elements of Style* (4th ed.). Open Road Media.

Wiens, J. A. (1989). Spatial scaling in ecology. *Functional Ecology, 3*(4), 385–397.

Wiggins, G. P., & McTighe, J. (2005). *Understanding by Design.* United States: Association for Supervision and Curriculum Development.

Wikipedia. (2023). Earth Optimism. Wikipedia. https://en.wikipedia.org/wiki/Earth_Optimism

Wiley, K. (n.d.). *1998 Wakefield MMR Controversy: The Australian Experience* [Report]. Sydney, Australia: The University of Sydney. www.fondation-merieux.org/wp-content/uploads/2019/02/7th-vaccine-acceptance-meeting-2019-workshop-2-2.pdf?utm_source=chatgpt.com

Wilhite, A. W., & Fong, E. A. (2012). Coercive citation in academic publishing. *Science, 335*(6068), 542–543. https://doi.org/10.1126/science.1212540

Wilkins, M. H. F., Stokes, A. R., & Wilson, H. R. (1953). Molecular structure of nucleic acids: Molecular structure of deoxypentose nucleic acids. *Nature, 171*(4356), 738–740. https://doi.org/10.1038/171738a0

Wilkinson, M. D., Dumontier, M., Aalbersberg, I. J., et al. (2016). The FAIR Guiding Principles for scientific data management and stewardship. *Scientific Data, 3*, 160018. https://doi.org/10.1038/sdata.2016.18

Wilkinson, C., & Wilkinson, S. (2024). Principles of participatory research. In: I. Coyne & B. Carter (Eds.), *Being Participatory: Researching with Children and Young People.* Cham: Springer. https://doi.org/10.1007/978-3-031-47787-4_2

Wilkinson, C., & Wilkinson, S. (2017). Doing it write: Representation and responsibility in writing up participatory research involving young people. *Social Inclusion, 5*(3), 219–227. https://doi.org/10.17645/si.v5i3.957

Wisker, G. (2015). Developing doctoral authors: Engaging with theoretical perspectives through the literature review. *Innovations in Education and Teaching International, 52*(1), 64–74. https://doi.org/10.1080/14703297.2014.981841

Wölfle, T. (2025). Local Citation Network. https://localcitationnetwork.github.io/

Wolszczan, A., & Frail, D. A. (1992). A planetary system around the millisecond pulsar PSR 1257+12. *Nature*, *355*(6356), 145–147. https://doi.org/10.1038/355145a0

Wu, J. (2011). Improving the writing of research papers: IMRAD and beyond. *Landscape Ecology*, *26*(10), 1345–1349. https://doi.org/10.1007/s10980-011-9674-3

X-Chem (2025, Jul. 31). DNA-Encoded Library (DEL) Technology Overview. www.x-chemrx.com/del-overview/

Xing, Y.-F., Xu, Y.-H., Shi, M.-H., & Lian, Y.-X. (2016). The impact of PM2.5 on the human respiratory system. *Journal of Thoracic Diseases*, *8*(1), E69–74. https://doi.org/10.3978/j.issn.2072-1439.2016.01.19

Xu, W., & Dai, W. A. (2024). "You are good annotators": Investigating how social reading based on social annotations and role-assignment strategies facilitate learners' social interaction and knowledge construction. *Education and Information Technologies*. Advance online publication. https://doi.org/10.1007/s10639-024-13268-9

Yan, B., Li, Y., Cao, W., et al. (2024). Efficient and rapid hydrogen extraction from ammonia–water via laser under ambient conditions without catalyst. *Journal of the American Chemical Society*, *146*(7), 4864–4871. https://doi.org/10.1021/jacs.3c13459

Yan, J., Qian, J., Li, Y., Li, L., Wu, F., & Chen R. (2024). Toward sustainable lithium iron phosphate in lithium-ion batteries: Regeneration strategies and their challenges. *Advanced Functional Materials*, *34*, 2405055. https://doi.org/10.1002/adfm.202405055

Yang, S. J., Zhang, J., Su, A. Y., & Tsai, J. J. (2011). A collaborative multimedia annotation tool for enhancing knowledge sharing in CSCL. *Interactive Learning Environments*, *19*(1), 45–62. https://doi.org/10.1080/10494820.2011.528881

Yeh, H. C., Hung, H. T., & Chiang, Y. H. (2017). The use of online annotations in reading instruction and its impact on students' reading progress and processes. *ReCALL*, *29*(1), 22–38. https://doi.org/10.1017/S0958344016000021

Younas, A., & Ali, P. (2021). Five tips for developing useful literature-summary tables for writing review articles. *Evidence-Based Nursing*, *24*(2), 32–34. https://doi.org/10.1136/ebnurs-2021-103417

Younas, A., Shahzad, S., & Inayat, S. (2022). Data analysis and presentation in integrative reviews: A narrative review. *Western Journal of Nursing Research*, *44*(12), 1124–1133. https://doi.org/10.1177/01939459211030344

Zaporozhetz, L. E. (1987). The dissertation literature review: How faculty advisors prepare their doctoral candidates (Doctoral dissertation). University of Oregon.

Zarzour, H., & Sellami, M. (2017). A linked data-based collaborative annotation system for increasing learning achievements. *Educational Technology Research and Development*, *65*(2), 381–397. https://doi.org/10.1007/s11423-016-9473-7

Zhu, Q., Wang, X., Clowes, R., et al. (2020a). 3D Cage COFs: A dynamic three-dimensional covalent organic framework with high-connectivity organic cage nodes. *Journal of the American Chemical Society*, *142*(39), 16842–16848. https://doi.org/10.1021/jacs.0c07732

Zhu, X., Chen, B., Avadhanam, R. M., Konomi, S., & Chan, M. (2020b). Reading and connecting: Using social annotation in online classes. *Information and Learning Sciences*, *121*(5/6), 261–271. https://doi.org/10.1108/ILS-04-2020-0117

Zhu, X., Shui, H., & Chen, B. (2023). Beyond reading together: Facilitating knowledge construction through participation roles and social annotation in college classrooms. *The Internet and Higher Education*, *59*, 100919. https://doi.org/10.1016/j.iheduc.2023.100919

Index

For EU product safety concerns, contact us at Calle de José Abascal, 56–1°,
28003 Madrid, Spain or eugpsr@cambridge.org.

www.ingramcontent.com/pod-product-compliance
Ingram Content Group UK Ltd.
Pitfield, Milton Keynes, MK11 3LW, UK
UKHW050727090726
473066UK00013B/1049
* 9 7 8 1 0 0 9 6 3 1 2 7 3 *